ENCYCLOPEDIA OF GENOMICS

ENCYCLOPEDIA OF GENOMICS

Vol. 2

MICROBIAL GENOMICS

By

Dr. Shiv Kant Prasad

School of Studies of Zoology & Biotechnology

Vikram University

Ujjain

(India)

D P H

DISCOVERY PUBLISHING HOUSE PVT. LTD.

NEW DELHI-110 002

Published by:
Tilak Wasan
DISCOVERY PUBLISHING HOUSE PVT. LTD.
4383/4A, Ansari Road, Darya Ganj
New Delhi-110 002 (India)
Phone : +91-11-23279245, 43596064-65
Fax : +91-11-23253475
E-mail : parul.wasan@gmail.com
discoverypublishinghouse@gmail.com
web : www.discoverypublishinggroup.com

***First Edition:* 2012**

ISBN: 978-93-5056-032-7 (Set)
978-93-5056-033-4 (Vol. 1)
978-93-5056-034-1 (Vol. 2)
978-93-5056-035-8 (Vol. 3)
978-93-5056-036-5 (Vol. 4)
978-93-5056-037-2 (Vol. 5)

Encyclopedia of Genomics

Printed at:
Shree Balaji Art Press
Delhi

Preface

Progress in any scientific discipline is dependent on the availability of techniques and methods that extend the range and sophistication of experiments which may be performed. Over the last 30 years or so this has been demonstrated in a spectacular way by the emergence of genetic engineering. This field has grown rapidly to the point where, in many laboratories around the world, it is now routine practice to isolate a specific DNA fragment from the genome of an organism, determine its base sequence, and assess its function. The technology is also now used in many other applications, including pharmaceutical analysis of samples, paternity disputes, medical diagnosis, genomic and the biotechnology industry.

The present title **"Encyclopedia of Genomics"** is a fast growing area of research for undergraduate and post-graduate students in life sciences, chemistry, engineering and medicine. The overall objective of this publication is to provide a professional level reference work with comprehensive coverage of the molecular basis of genome and the application of that knowledge in genetics, evolution, medicine, and agriculture. It deals with the life processes at a molecular level genetic disease diagnosis and genetic therapy; the theory and techniques for understanding manipulating, and synthesizing biological molecules and their aggregates; and the application of biological processes to make or modify products to improve plants or animals or to develop microorganisms for specific uses.

The development of techniques for genomic science is based on an understanding of the molecular aspect of genomic processes occurring in nature. These techniques have, in turn, enabled expansion of our knowledge and understanding of how genes are organized and expressed in living cells and has provided major impetus to the development of biotechnological enterprises.

The present title has been written to provide readers with a thorough understanding of the techniques which underpin the developments in the field of genomic science. By gaining this understanding, readers will be able to apply the techniques of genomic manipulation to tackle specific problems or to achieve desired objectives in a wide variety of fields.

There can be no claim to originality except in the manner or treatment and much of the information has been obtained from the books and scientific journals available in the different libraries.

The author express his thanks to his friends and colleagues whose continue inspirations have initiated him to bring out this book.

The author expresses her gratitude to Mr. Wasan and staff of M/s Discovery Publishing House Pvt. Ltd. for their whole hearted co-operation in the publication of this book.

S.K. Prasad

Contents

Chapter 1

Human Retrovirus Pseudogenes

Repetitive elements or repeats are sequences present in multiple copies in. the genome but, unlike *multigene* families, they do not have any clear function in the host. Low-copy repeats *generated* by large-scale genomic duplications represent a class by themselves.

In general, repetitive elements are divided into tandem repeats and interspersed repeats. Tandem repeats are head-to-tail repetitions of the same sequence *motif*. *Interspersed* repeats are active or inactive copies of TEs dispersed throughout the genome. *Repetitive* elements can be grouped into sets of similar copies, called families or subfamilies. Families of TEs that encode enzymes necessary for their replication are termed *autonomous*.

Nonautonomous elements do not encode all necessary proteins, and their replication (amplification) depends on proteins provided by the autonomous elements. *Nonautonomous* elements can, thus, be viewed as *parasitic* elements competing for replication machinery with the *autonomous* copies.

Amplification of both autonomous and nonautonomous elements depends also on additional factors provided by the host cell. Based on their *replication* strategy, interspersed repeats are broadly divided into DNA *transposons* and *re-trotransposons*.

The DNA transposons amplify using the host DNA replication machinery, and their transcripts serve solely as mRNAs participating in translation of transposon-encoded proteins involved in the transposon *insertions* and *excisions*.

The retrotransposons replicate via an RNA intermediate and, thus, their transcripts serve both as mRNAs for protein translation and as templates for DNA synthesis. Before integration into the genome, retrotransposon RNA must be copied into cDNA using RNA-dependent DNA polymerase, also known as reverse transcriptase (RT). *Recognizable* copies of all repetitive elements constitute approx 50% of the human genome.

During the course of the human genome sequencing, more than 600 repeat families and subfamilies have been discovered. All are *systematically* organized in Repbase Update (RU).

Table 1.1 : Repetitive Elements in the Human Genome.

Type/class		*Superfamily*	*Further division*	*Family*	*Genome (%)*
Tandem repeats	Satellites				Approx 20?
	Telomeric and subtelomeric repeats				<0.01
	Microsatellites and minisatellites				1.4
Interspersed repeats	DNA transposons	Mariner/Tc 1		11 families	2.8
		hAT		14 families	2.4
		PiggyBac		1 families	0.02
		MuDr		2 families	0.03
		Harbinger		*HARBI1* gene	—
		P		Single-copy genes	—
	Non-LTR retrotransposons	L1	LINE	L1	17
			SINE	*Alu*	10.5
				SVA (SINE-R)	0.13
			Retropseudogenes	—	0.1-0_3'
		CRI	LINE	L2	3.1
				L3	0.31
			SINE	MIR	1
				MIR3	0.29
	LTR retro-transposons	Copia (HERVs)	Class I (gamma retroviruses)	>90 families	2.5
			Class II (beta retroviruses)	11 families	0.5
			Class III	>30 families	6

NON-LONG TERMINAL REPEAT

Non-long terminal repeat (LTR) retrotransposons are the most abundant *repetitive* elements in human genomic DNA and represent approx one-third of the genome. Their DNA copies are *co-linear* with the RNA transcripts and they lack LTRs present in retroviruses. Autonomous non-LTR *retrotransposons* are often referred to as LINEs and their nonautonomous *counterparts* as SINEs.

The human genome contains two superfamilies of LINEs: active L1 (LINE 1) elements, and extinct families of L2 and L3 elements. The latter belong to the CR1 superfamily. Human SINEs are represented by the active *Alu* and SVA repeats retrotransposed by L1 elements, and by the extinct *mammalian-wide interspersed* repeat (MIR) and MIR3 SINEs that coamplified with L2 and L3 families, respectively. Here, we *concentrate* on ancient LINE *(see* Ancient LINEs) and SINE *(see* Ancient SINEs) elements.

We also review L1-retroposed copies of cellular transcripts known as processed *pseudogenes* (see Processed Pseudogenes).

Line Elements

A typical structure of LINE elements is shown in Figure elsewhere in this chapter. LINEs usually contain two open reading frames referred to *as ORF1* and *ORF2*. The L3 ORF1 protein shares similarity with esterase domains. L2 elements apparently lack *ORF1,* although it is possible that the very old age of L2 copies and frequent 5' truncation typical for LINEs *prevented reconstruction* of a full-length L2 element.

The ORF2 protein contains the RT and *apurinic-apyrimidinic* endonuclease enzymatic domains. The transcription of LINE elements starts from a *poorly* characterized internal promoter for RNA polymerase II. After translation, the complex of LINE RNA and protein(s) enters the nucleus, where an *endonucleolytic* nick at a DNA target serves to prime reverse *transcription.*

Target-primed reverse transcription is another feature distinguishing non-LTR from LTR retrotransposons, which normally use cellular tRNA as primers for reverse transcription *(see* Human *Endogenous Retroviruses*). Unlike in L1 and L1-dependent elements, no target site duplications are created during integration of L2 and L3 elements.

Furthermore, L2 and L3 carry microsatellite-like 3' tails instead of the polyA tail found in L1 insertions Based on high sequence diversity, the age of L2 and L3 elements in the human genome is estimated to be approx 200-300 million years, corresponding to the early radiation of reptiles, birds, and mammals.

Recognizable L2 and L3 copies, together with their nonautonomous counterparts MIR and MIR3, represent approx 5% of the human genome. The potential of L2 and L3 copies to stimulate genomic *rearrangements* is very limited owing to their ancient origin, and no such case has been identified in humans so far.

The proliferation of L2 and L3 elements stopped long before the mammalian radiation, and recombination between highly diverged sequences is very unlikely. No protein-coding gene derived from L2 or L3 elements has been detected.

Ancient SINEs

Successful amplification of SINEs by the LINE machinery requires their transcription to he initiated by internal promoters. Typical SINEs are derived from *cellular* genes transcribed by RNA polymerase III, containing internal *promoters.*

For example, L 1-dependent Alu elements are derived from *7SL RNA* genes encoding the RNA scaffold of the signal recognition particle, whereas MIRs are derived from tRNA genes.

The 5' parts of MIRs are homologous to tRNA genes, and the 3' terminal portions are *homologous* to the respective LINE elements on which their *proliferation* depends. MIR shares its 3' terminus with L2 elements, MIR3 with L3 elements.

MIR and MIR3 SINEs, like their LINE2 and LINES counterparts, are very old and their contribution to genomic rearrangements in the human genome seems to be minimal because no homologous recombination between ancient SINEs associated with human disease has been reported. The figure shows elsewhere in this chapter typical structures of integrated interspersed repeals. (A)

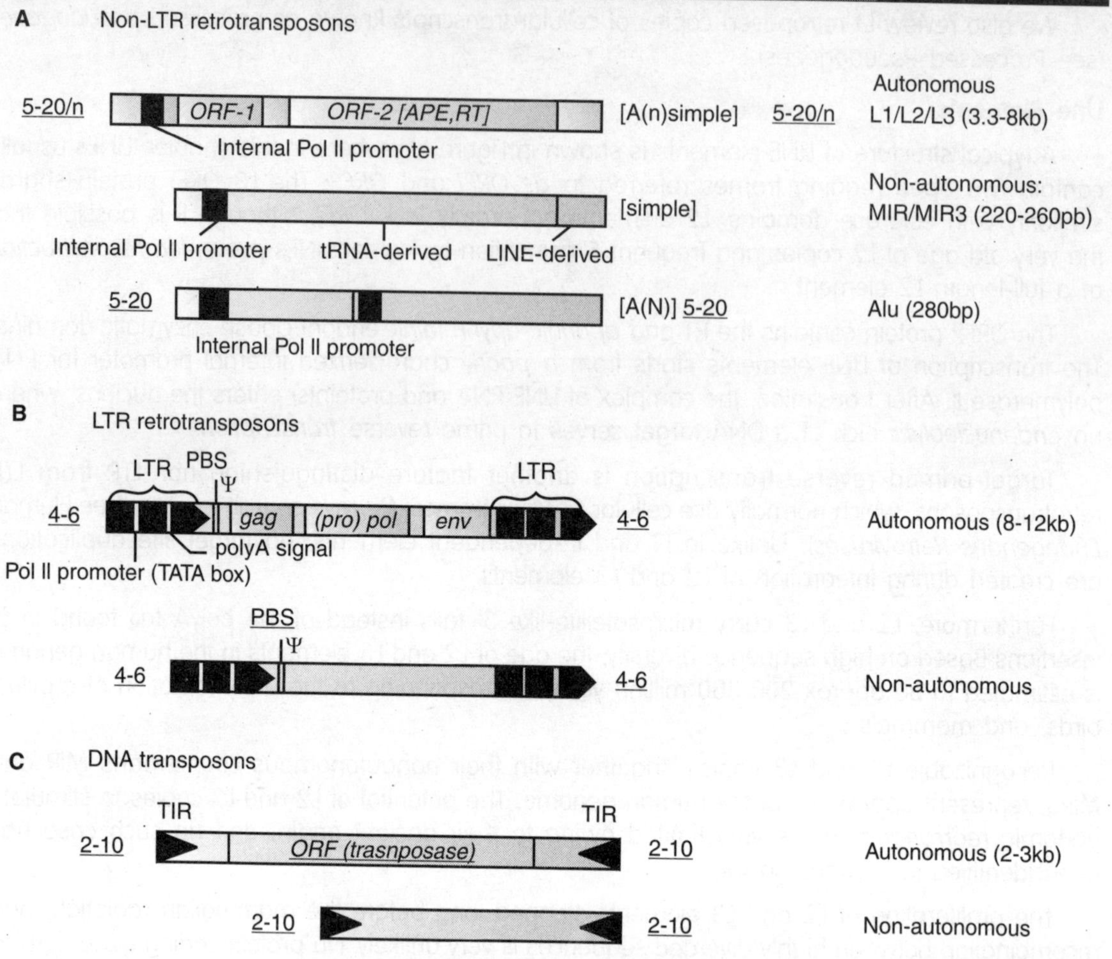

Figure 1.1: Structure of Human Interspersed Repeats

Structure of non-long terminal repeats (LTR) *retrotransposons*. The first bar shows a schematic organization of LINE elements. Human L1 and L3 elements contain two open reading frames *(ORFs)*, L2 copies apparently lack *ORF1*.

The function of the first *ORF1 is poorly* understood. *ORF2* encodes a protein with the apurinic-apyrimidinic endonuclease and reverse transcriptase (RT) enzymatic domains. L1 insertions have a 3' polyA tail and are flanked by variable long target duplications, typically 5- to 20-bp long. L2 and L3 elements lack the target site duplications, and their 3' tails are composed of simple repeat sequences. The second and third schematic bars depict organization of SINEs. L2- and L3-dependent MIR and MIR3 SINEs (middle) consist of two parts.

The 5' part (dark gray) is derived from a tRNA gene and harbors an internal pol III promoter. The 3' tails of MIRs are homologous to the 3' end of the corresponding LINE counterparts (light gray). *Alu* elements (bottom) derived from *7SL RNA* genes have a dimeric structure and contain a composite pol III promoter. SINEs share insertional characteristics with their LINE counterparts including the 3' end simple repeats in MIRs. L1-dependent *Alu* repeats contain 3' polyA tails and are flanked by 5- to 20-bp long direct repeats. (B) Structure of LTR retrotransposons.

All LTR retrotransposons contain two LTRs, which include a pol II promoter and polyA signal. The internal part of autonomous elements comprises three main open reading frames: *gag, pol,* and *env;* in some HERV families *pro* can be separate from *pol.* The 5' part of the internal sequence contains a tRNA primer binding site (PBS) for initiation of reverse transcription and also an *encapsidation* sequence (P) necessary for incorporation of retroviral RNA molecules into virions.

Internal sequences of nonautonomous LTR retrotransposons may or may not share similarity with *retroviral* ORFs, but they contain all structures required for retroviral replication and reverse transcription such as LTRs, PBS, or the encapsidation signal. Both the autonomous and nonautonomous elements are flanked by 4- to 6-bp long direct repeats. (C) Structure of DNA transposons.

DNA transposons contain terminal inverted repeats (TIRs) of variable sizes (from approx 10 to several hundred nucleotides). Autonomous elements encode the replication enzyme transposase. Nonautonomous elements share TIRs with *autonomous* copies, which are recognized by transposase. DNA transposons are flanked by 2- to 10-bp long target site duplications, characteristic for each (super)family.

Mechanism of Processed Pseudogenes

Eukaryotic genomes contain a large number of *pseudogenes,* which are homologous to known functional genes, but are apparently defective owing to the presence of various mutations, such as truncations, frameshifts, or missense changes. Pseudogenes are sometimes abbreviated using the Greek letter Ψ followed by the symbol of the original functional gene from which the pseudogene was derived.

The human genome contains two types of pseudogenes: duplicated pseudogenes and processed pseudogenes generated in germ-line cells. Duplicated pseudogenes resemble normal cellular genes and are created by direct DNA duplications. Processed pseudogenes or *retropseudogenes,* on the other hand, are *structurally* distinct.

Typically, they are characterized by the lack of both a promoter and introns, and by the *acquisition* of a polyA-like sequence at their 3′ ends. Because of co-linearity with spliced mRNAs, processed pseudogenes appear to originate by reverse transcription and integration of cellular mRNAs. The mechanism of amplification of processed pseudogenes represented a long-standing puzzle.

The presence of a polyA tail and lack of long terminal repeats indicated involvement of a non-retroviral enzymatic machinery. Furthermore, processed pseudogenes share common insertion characteristics, such as the TTAAAA insertion motif shared with LINE-L1 *and Alu* repetitive sequences, strongly pointing to active L1 elements as the donors of the RT.

Indeed, a series of *sophisticated* experiments has demonstrated reverse *transcription* and *integration* of spliced reporter mRNA by the L1 enzymatic machinery with all hallmarks of processed pseudogenes. Genomic studies of processed pseudogenes have disclosed additional characteristics of L1-mediated retroposition: integration independent of the chromosomal location of the parent gene, preferential insertions into GC-poor DNA segments (isochores), dark *Giemsa bands*

(2-0-22), as well as frequent 5' truncations and inversions. Therefore, it is generally accepted that human processed pseudogenes are copies of cellular RNA transcripts reverse transcribed and integrated into the genome by L1 machinery.

Retroposition of cellular RNAs by other RT-encoding elements such as endogenous *retroviruses* appears to be rare. The number of processed pseudogenes derived from protein-coding mRNA is estimated to be somewhere between 10,000 and 30,000, depending on detection methods and *stringency* criteria.

Some maybe misannoted as functional genes. Moreover, most pseudogenes are 5' truncated and contain only a partial 3' untranslated region and thus are not detectable by standard translated searches. he majority of processed pseudogenes preserved in the human genome were created after the split between the rodent and primate lineages. The peak of their generation roughly corresponds to the main period of *Alu* amplification 60-40 mill ion years ago, after which the frequency of retroposition has declined.

The number of processed pseudogenes depends on several properties of mRNA. Short and GC-poor mRNAs tend to produce more pseudogenes), although the significance of small size may be artificial. The number of pseudogenes is positively correlated with the breadth of expression of the *parental gene* (end, as expected, with germline/embryonic cell expression , because only germ-line *retro-transpositions* can be passed onto future generations.

The contribution of functional gene groups to the human processed pseudogene population is not random either. Although the number of processed *pseudogenes* is comparable to the number of human genes, only approx 1.0% of human genes are represented among processed pseudogene(s), and 30 human genes account for 20% of all human processed pseudogenes. *Ribosomal* protein genes, DNA/RNA binding *proteins, receptors, kinases, metabolic enzymes, mitochondrial proteins,* and housekeeping genes in general are represented by the highest number of processed pseudogenes.

In summary, it seems that genes expressed in germ-line cells, such as germ-line specific or *housekeeping* genes, have the highest chance of being retroposed by L1 elements. In addition to typical transcripts of protein-coding genes, other RNAs can also be retroposed by L1 elements. Processed pseudogenes can be derived from alternatively spliced mRNAs, *antisense transcripts,* or mRNAs derived from other repetitive elements such as endogenous retroviruses It seems that virtually any RNA including non-coding RNAs can be potentially retroposed by L1 elements.

The model of *Alu* and L1 retrotranspositions implies two principal requirements for successful retroposition: (1) the RNAs should contain a polyA-like 3' terminal sequence and (2) they should be located close to the newly synthesized L1 proteins (e.g., *cytoplasmic localization,* close to ribosomes) owing to the L1 *cis* preference. This hints that both pol I and III transcripts may be transposed as well, if they meet the aforementioned conditions.

Alu elements are a prominent example of polyAterminated processed pseudogenes derived from a pol III transcribed gene *(7SL RNA)*. Retroposition of processed pseudogenes can serve as a mechanism of gene duplication However, unless they acquire a new promoter (most pol II genes have promoters upstream of the transcription start), typical processed pseudogenes cannot be expressed. *Expression* of recently integrated pseudogenes has been reported in several cases, but it is not clear whether they encode *functional proteins.*

Nevertheless, there are a few examples of retrogenes (i.e., functional and expressed intronless genes derived from retroposed mRNAs) in the human genome. For example, *retroposition* can help to amplify X-linked genes, which have only one functional copy in somatic cells. X-linked *mutations*, thus, often result in loss of function and their transfer to autosomes or amplification on X can revert this phenotype. Several such examples of X-to-autosomes or X-to-X amplification of human as well as mouse genes have been reported.

Along the same lines, processed pseudogenes could also provide a new domain (*exon*) if they insert into introns of other genes. The only well-documented example of a successful exon shuffling in the human genome identified to date is SCAN domain-containing gene 2 *(SCAND2)*. This member of the SCAN nuclear protein family gene located on 15q25.2 harbors a C-terminal domain derived from the N-terminal part of *Clorfl2/EGLNI* (egl nine homolog 1) located on 1q42.2.

Despite the examples of functional retropseudogenes or exon shuffling, however, the vast majority of processed pseudogenes are clearly translationally incompetent (owing to truncations or stop codons), and they appear to follow mutation patterns consistent with neutral evolution. Interestingly, such expressed processed pseudogenes without protein coding capacity can act as functional non-coding RNAs. For example, *Makorin1-p1* is an expressed processed pseudogene regulating the mRNA stability of its progenitor gene *Makorinl.*

Disruption in *the Makorin 1-p 1* pseudogene results in polycystic kidneys and bone deformity in mice, demonstrating its importance for *Makorin1* function. Searches for expressed pseudogenes in the human genome indicate that this type of *regulation* may be more common than *previously* anticipated. The potential contribution of processed *pseudogenes* to human diseases is far from being understood.

Pseudogene integrations may cause insertional inactivation of genes, but, despite many reports of *Alu* and L1 insertion, no processed pseudogene-related insertional mutation has been reported to date. Potential regulatory effects of pseudogene transcripts at the RNA level are *intriguing*, particularly in the light of early *speculations* on this subject.

Finally, it should be pointed out that processed pseudogenes (especially expressed copies) may interfere with the analysis of variation of human genes, because they can be *misinterpreted* as *polymorphism* or as *mutated alleles* of the functional genes.

ENDOGENOUS RETROVIRUSES OF MAN

Retroviruses belong to a broad class of (retro)elements that replicate via an RNA intermediate and include LTRs in their DNA copies. The retroviral life cycle is characterized by reverse transcription of the retroviral RNA *genome* followed by cDNA integration into the host nuclear DNA, where they can persist in the form of a stable integrated provirus. *Retroviral* infections of early *embryonic* and germ-line cells can be inherited by subsequent *generations* and such ancient proviral relics found in the genome are called ERVs.

HERVs resemble well-known exogenous retroviruses and carry typical genes found in infectious retroviruses. The group-specific antigen *(gag)* open reading frame encodes internal structural proteins of the retroviral particle, whereas *pro* and *pol* genes encode enzymes necessary for retroviral amplification. Replication-competent retroviruses code these three enzymes: RT, integrase, which inserts retroviral DNA into host chromosome, and protease, which posttranslationally cleaves Gag, Pol, and Env polyproteins into functional proteins.

RT also performs an RNA degradation activity, encoded in a separate domain called RHase H. The last open reading frame *env* encodes surface (envelope) glycoproteins required for attachment of retroviral particles to the *cellular receptor* and *penetration* into the cell. The affinity for a particular receptor determines retroviral range of infectivity or *tropism* for cells and tissues expressing the given receptor.

The *transmembrane* subunit of the Env proteins contains a so-called *immuno-suppressive* domain, a conserved l7mer with effects on the proliferation and differentiation of lymphocytes. Interestingly, the *immuno-suppressive* domains of mammalian C-type retroviruses, including some HERV families, are similar to the analogous segments in the envelope glycoprotein of filoviruses (negative-strand RNA viruses) such as *Ebola* or *Marburg viruses*. HERVs also may contain other *auxiliary genes.*

Several class II and III families code *deoxyuridine triphosphatase* (dUTPase). Some HERV-K elements encode a 105 as functional homolog of the Rev/Rex proteins found in human immunodeficiency virus (HIV)-1 and human T-cell leukemia virus (HTLV), respectively. These proteins mediate nuclear transport of unspliced RNAs via the host factor CRM1. The structure of an integrated DNA provirus is different from the structure of its RNA genome.

During the process of *retroviral* reverse transcription two identical copies of part of the retroviral sequence, known as long terminal repeats (LTRs), are generated. LTRs are located at both ends of the provirus and flank the protein-coding internal sequence. LTRs harbor the pol 11 *promoter, enhancers,* polyA signal, and other regulatory sequences. Typical HERV LTRs have conserved 5'-TG and 3'-CA termini. During *integration,* 4- to 6-bp target site duplications (TSDs) of the host DNA are created, which flank the integrated *provirus.*

The length of TSDs is determined by the viral integrase and is characteristic of a given *retrovirus* or family of HERVs. Since the discovery of the first human endogenous retrovirus in 1981 more than 400,000 HERV fragments have been found in the human genome, contributing approx *9%* of human DNA . HERV sequences were classified using several different systems and the nomenclature used in the literature is rather confusing

The most frequently used classification is based on the binding site for the tRNA primer. For example, a recently active family of HERVs uses lysine (K) tRNA *molecules* to prime reverse transcription and is therefore known as HERVK. Although this classification was sufficient during the early years of HERV research, the system became obsolete after the discovery of many other HERV families in the human genome.

Progress in classification based on primer binding sites is hampered by the fact that the same tRNA primer can be shared by unrelated families and further complicated by the existence of *chimeric* elements composed of segments derived from different families. Such chimeras probably arise by co-packaging of two different retroviral RNA into vision and subsequent template *switching* between the RNAs during reverse transcription.

Therefore, a different, more flexible classification system of HER Vs is used in the RU database. This classification uses a combination of traditional names and numbers. RU families are primarily defined based on a substantial DNA sequence *divergence* from other retroviral families. Based on their similarity to *exogenous* retroviruses, HERV families are grouped into three classes. HERVs with homology to mammalian type C *retroviruses* (*gamma retroviruses*), such as *murine leukemia* virus (MLV), have been placed in class I.

Class I represents a highly heterogeneous group of HERV s with many different families 'This class includes numerous non autonomous families, some of which constitute a distinct subgroup called MER4. In addition, many class I elements are chimeras composed of segments derived from unrelated retroviruses. All class I families are flanked by 4-5 by TSDs. Class II consists of HER Vs related to mammalian type B and D retroviruses (*beta retroviruses*) represented by mouse mammary tumor virus.

This class is often referred as the HERVK group, because most of the families use lysine (K) tRNA as the primer for reverse transcription. The HERVK elements are flanked by 6 by target site duplications. In 1995, a new group (class III) of endogenous retroviruses similar to human foamy viruses *(Spumaviridae)* was discovered. HERVL is a prototypic member of this group. With the exception of HERV 18 and HERVL66, all other class III families lack env-like ORFs.

In addition to several families with similarity to retroviral proteins, this group contains a large number of *nonautonomous* elements known as mammalian apparent LTR retro-transposons (MaLRs) and THEI. All class III elements are flanked by 5 by TSDs. No endogenous counterparts of *exogenous lentiviruses* such as HIV are known in the human genome. All retroviruses including HER Vs belong to the *copia* superfamily of LTR-retrotransposons; no other superfamilies of LTR-retrotransposons have been detected in the human genome.

Surprisingly, however, the human genome contains eight cellular genes with significant similarity to gypsy superfamily of LTR retrotransposons, probably derived from very old *gypsy* elements, whose remnants cannot be otherwise detected in genomic DNA. Endogenous retroviruses, although still active in mice, are nearly extinct in the human genome. However, some class I elements contain potentially functional ORFs.

Class III elements are found in all placental mammals, and some lineages including simians and mice (but not rats) exhibit sustained activity of these elements. In the human genome, on the other hand, class III elements lost their activity long ago and their age ranges from approx 20 to 150 million years *(K.* Class II embraces the most recent retroviruses found in the human genome (HERVK10), many of which are human-specific, and the youngesl elements appear to be only approximately I million years old.

Only three elements in the human genome seem to harbor all full-length ORFs. The most preserved HERVK element *potentially* capable of reinfections is known to be *polymorphic* in human populations. The genomic distribution of HERVs is a result of many factors including the integration preference, intensity of *recombination*, selection, and so on. Although there are no new HERV insertions reported, HERV integration preferences can be approximated from exogenous retroviruses such as HIV and MLV.

HIV preferentially integrates into gene- and GC-rich regions. MLV also targets genes, especially positions around transcription start sites. In contrast to these exogenous retroviruses, HERVs are under-represented in GC-rich regions and especially in genes. probably owing to interference of retroviral polyA signals with gene transcription if they are inserted in the sense orientation relative to transcription. Recombination between two direct LTRs of a provirus leave a single copy of the LTR (solo LTR) in the genome, and the remaining LTR with the internal sequence is discarded in the form of a circular episomal DNA. Approximately 90% of HERV genomic elements are such solo LTRs. Full-length elements are mostly found in regions with low recombination rates, such as AT-rich regions or chromosome Y.

Thus, the HERV distribution is *strikingly* different from the insertional pattern of exogenous retroviruses and mostly reflects post-insertional processes. The extent of HERV contribution to human diseases is a subject of considerable debate in the field. On the DNA level, HERV-mediated re-arrangements are relatively rare as HER Vs are essentially extinct in humans, and no HERV-related insertion is reported in *mutational* data-bases.

Because of the ancient origin of most HERV copies and the existence of relatively low copy number families, the presence of highly similar, closely spaced HERV copies prone to recombination is infrequent. Indeed, recombination between two HER Vs is a sporadic source of genomic *instabilities.* One example involves recurrent deletions of Y-linked azoospermia factor a *(AZFa)* gene by *homologous recombination* between two HERV 15 copies, associated with male infertility.

Recombination between solo LTRs has also contributed to allelic variation of the *HLA* locus in the human population. Expression of endogenous retroviruses has been implicated in the etiology of various human diseases including cancer and *autoimmune* disorders. We should stress, however, that the potential contribution of HERVs to these diseases is a highly controversial issue and despite two decades of research on the *pathogenic potential* of *endogenous retroviruses,* convincing evidence linking HERVs to diseases is still lacking.

Aside from the negative outcomes of retroviral insertions and recombinations on the host cell, retroviral regulatory elements and proteins could serve as a new source of material for *evolutionary* experiments. Indeed, there are several examples of HERV-derived promoters, polyA signals and other *regulatory* signals *recruited* by the host genome.

One famous example represents a HERVE insertion into the promoter of the amylase gene cluster during primate evolution, which stimulated rearrangements of this locus accompanied by the emergence of amylase expression in *salivary glands.*

Also, genes of *protein-coding* repetitive elements including HERVs may occasionally be recruited by the host genome as functional cellular genes. Notably, two *env* genes from the HERVW/HERV 17 and HERV-FRD/MER50 families seem to serve as functional human genes coding for the *syncytin* proteins responsible for cell fusion during differentiation of the *syncythiotrophoblast* in the *human placenta.* The extent of such contributions to cellular functions is unclear and the number of *unequivocal* cases of HERVs benefiting the host is very limited.

ANCIENT COPY OF DNA

Approximately 5% of the human genome is derived from ancient copies of DNA transposons. Preserved copies of DNA transposons are flanked by terminal inverted repeats (TIRs) of variable length. The main enzyme *encoded* by autonomous DNA transposons is called transposase. Trasposase has several activities: it specifically binds to TIRs, excises the *integrated* copy and pastes it to another place in the genome (cut-and-paste process). After insertion, 2- to 10-bp *duplications* of the target site are created.

DNA transposons amplify by co-replicating with the host DNA as they *preferentially* excise from already replicated DNA and reinsert into nonreplicated segments. This *asymmetric replication theoretically* increases the number of DNA transposons by a factor of 1.5, though the real efficiency of this process is-probably lower.

Human transposons represent ancient genomic fossils hardly recognizable at the DNA level. Consequently, the first evidence *disclosing* the presence of DNA transposons in the human genome was not obtained until large amounts of human *genomic* sequence became available. To date, six transposon superfamilies have been found in humans: Mariner/Tc1, hAT, piggyBac, MuDR, Harbinger, and P-like elements.

Without exception, all these families became extinct in the past and no transposition-competent element is preserved in humans. As in the case of other very old repeats, the pathogenic potential of human DNA transposons is probably very limited. *Mariner* elements were detected in a large number of genomic duplications including those linked to the Charcot-Marie-Tooth disease type 1A (CMT1A) duplication, and to *hereditary neuropathy* with liability to *pressure palassies* (HNPP) deletions.

However, the proposed involvement of the Mariner *transposase* in *generating* double stranded breaks stimulating the rearrangements seems *speculative* given the fact that no translation-competent Mariner copy is found in the human genome. Yet, it should be noted that the human genome contains nearly 30 genes that were derived from DNA transposons and appear to be preserved between mammals and many *vertebrates*.

Well known recruited *transposases* include: RAG1, a part of V(D)J recombinase complex producing functional *immunoglobulin* and T-cell receptor genes in developing *lymphocytes, centromere* protein CENP-B, an the more recently described MER53; Jerky, and HARBI1. Interestingly, the *majority* of known human genes derived from TEs evolved from DNA transposons.

This is probably related to the fact that transposases are extremely diverged, as there is practically no detectable similarity between transposases from different superfamilies. Transposases, thus, provide a wide repertoire of sequence *variants* compared to more conserved retrotransposon proteins (A) Although the function of the transposon-derived genes is mostly unknown, it is likely that they are involved in DNA recognition and rearrangements. Consequently, they may contribute to pathological genomic rearrangements.

Notably, it has been shown that RAGI nicking may stimulate homologous recombination (lam) and also chromosomal translocation both in vivo and in vitro. Finally, human cells seem to be in general permissive to DNA transposons and these elements can, thus, be used as vectors for gene therapy.

CONCLUDING REMARKS

The human genome has preserved a substantial fraction of ancient DNA transposons, MIR SINEs, L2, and L3 LINEs, that have limited potential to stimulate genomic rearrangements. Some HERVs have amplified more recently and sporadically stimulate genetic instabilities. Several processed pseudogenes, DNA transposons, as well as both ancient and recent LTR-retrotransposons were recruited by the host genome as new genes.

Particularly *intriguing* is the presence of many genes derived from transposases of DNA transposons. Given the *endonucleolytic* activity of the *transposases*, it is possible that these recruited genes are involved in DNA processing.

Chapter 2

Bacterial Genomic Screening

Traditionally, *antibiotics* have been identified by exposing bacteria to novel compounds and assaying their bacteriostatic or bactericidal activity. Once *antibacterial* activity was confirmed, a lengthy series of studies would determine their mode of action (MOA) with regard to which essential cellular function was being inhibited.

While this approach was highly successful for several decades, only one new class of antibiotic has been *commercialized* in the last 20 yr. It is a *reflection* of the methods and the relatively limited chemical diversity employed, that currently available *antibiotics* target a small number of *cellular* processes involved in DNA, RNA, protein, and cell-wall *biosynthesis*.

Although antibiotic discovery declined during the 1980s and early 1990s, the same cannot be said for the ability of pathogenic bacteria to develop resistance. The appearance of high-level *vancomycin*-resistant *Staphylococcus aureus* (VRSA) (land vancomycin-resistant *Enterococcus faecium* (VREF), along with multiple drug resistance in many other *clinically* important *bacteria*, has made it essential that new antibiotics be found.

We discuss here two key strategies utilizing interchangeable technologies that have been applied to the discovery of novel *antibiotics*. The first strategy is target-based screening, where *comparative* genomics and *bioinformatic* analysis are used to select targets with the appropriate *spectrum* and selectivity.

Genetic technologies integral for the success of this approach, such as *essentiality* testing, are also described. An *unprecedented* number of novel targets have been discovered via this approach, and a number of examples are discussed. This section concludes with the case history of a target successfully progressed through identification by genomics, high-throughput *screening* (HTS), and proof of concept in curing *experimental* infections

The second approach is to screen for compounds that possess *antibacterial* activity, and then employ *genomic* and *bioinformatic* tools to identify the molecular target. The advantage of this approach is that the compounds already possess *antibacterial* activity, which is a property often difficult to engineer into molecules identified by enzyme-based screening approaches.

The recent development of novel *biochemical* and genomic *technologies* that facilitate identification and characterization of the MOA of these agents have made this approach as attractive as the genomic target-based screening strategy. Both of these approaches have advantages and unique challenges associated with them, but genomics and genomic *technologies* are integral to the success of either approach.

RESEARCH SCREENING

Screening

Pointing Agents

Comparison of the many bacterial genome sequences available allows the identification of targets that are present in most, if not all, clinically relevant bacteria. Further *comparison* with the genome sequences of eucaryotes allows the selection of targets that are sufficiently different or unique that they would be targets to inhibit bacteria. Indeed, this is particularly relevant given the recent publication of the human genome sequence.

Moreover, subset comparisons among different bacterial species suggest the possibility of more targeted chemotherapy approaches that would allow the development of *antibacterial* agents selectively active against only *Grampositive* or Gram-negative bacteria, or even against individual *pathogens* (e.g., *Helicobacter pylori, Chlamydia)*. There is an enormous benefit in knowing the *phylogenetic* distribution of potential target systems early in the drug-discovery process.

It provides some predictive value of the potential spectrum of activity that is achievable and, perhaps more importantly, allows targets of defined *spectrum* to be sought. Targets specific to only one or a few bacterial species would allow narrow-spectrum agents to be created, thereby limiting the side-effects, often inherent in the use of broad-spectrum agents, that are often caused by altering normal *bacterial flora*.

Validating Antimicrobial Targets

Valid antimicrobial targets are those that are essential for cell viability during growth and infection. To determine the role of various gene products in cell viability and *pathogenesis*, a number of techniques have been developed for testing essentiality both in vitro and in infection models. For example, *random mutants* can be generated in which growth is conditional. In one case, a *signature-tagged* transposon-based system has been developed to identify genes that are essential for infection, but which are dispensable for growth in vitro and thus encode *virulence* factors.

Temperature-sensitive mutants can also be generated in which in vitro growth is conditional, but these require rather *tedious* procedures to identify the *responsible allele*. In the past, due to the paucity of genetic information in pathogenic bacteria, these random *mutagenic* methodologies were the only way to identify genes vital for cell viability. Such approaches are limited by the fact that they do not generate truly random libraries of mutants. Transposons have preferred sites of integration, and there are regions of *chromosomes* into which they apparently do not insert. Equally, conditional *lethal* mutations cannot be generated in all essential genes.

Table 2.1: Publicly Available Bacterial Genomes.

Organism	Domain	Contigs	Genes	Total bases	Source
Actinobacillus actino-mvcetemcomitans HK1651	Bacteria	9	1988	2,095,439	University of Oklahoma
Aeropyrum pernix K1	Archaea	1	1840	1,669,695	NCBI
Agrobacterium tumefaciens C58	Bacteria	4	5299	5,673,563	NCBI
Agrobacterium tumejaciens C58 (Dupont)	Bacteria	4	5402	5,674,064	NCBI
Aquifex aeolicus VF5	Bacteria	2	1553	1,590,791	NCBI
Archaeoglobus fulgidus DSM4304	Archaea	1	2420	2,178,400	NCBI
Bacillus anthracis	Bacteria	2	226	277,885	NCBI
Bacillus anthracis Ames	Bacteria	1	5287	5,227,297	TIGR
Bacillus halodurans C-125	Bacteria	1	4066	4,202,353	NCBI
Bacillus stearotherinophilus 10	Bacteria	380	3342	3,269,999	University of Oklahoma
Bacillus subtilis 168	Bacteria	1	4112	4,214,814	NCBI
Bordetello pertussis Tohama I	Bacteria	1	3892	4,086,186	Sanger Center
Borrelia burgdorferi B31	Bacteria	22	1637	1,519,856	NCBI
Brucella melitensis 16M	Bacteria	2	3198	3,294,931	NCBI
Buchnera sp. APS	Bacteriu	3	574	655,725	NCBI
Caenorhabditis elegans	Eukaryota	0	22,407	0	NCBI
Campylobacter jejuni NCTC 11168	Bacteria	1	1633	1,641,481	Sanger Center
Caulobacter crescentus	Bacteria	1	3737	4,016,947	TIGR
Chlamydia muridarum	Bacteria	2	916	1,076,912	NCBI
Chlamvdia trachomatis D/UW-3/Cx	Bacteria	1	921	1,042,519	NCBI
Chlamydophila pneumoniae AR39	Bacteria	1	1109	1,229,853	NCBI

(Table Contd.)

Organism	Domain	Contigs	Genes	Total bases	Source
Chlamydophila pneumoniae CWL-029	Bacteria	1	1052	1,230,230	Incyte
Chlamydophila pneumoniae J138	Bacteria	1	1069	1,226,565	NCBI
Chlamydophila pneumoniae L2	Bacteria	2	1042	1,234,390	TIGR
Chlorobium tepidum TLS	Bacteria	24	3178	2,196,918	TIGR
Clostridium acetobutylicuin	Bacteria	1	3672	3,940,880	TIGR ATCC824
Clostridiur perfringens 13	Bacteria	2	2723	3,085,740	NCBI
Corynebacterium diphtheriae NCTC 13129	Bacteria	1	2128	2,488,600	Sanger Center
Corynebacterium glutamicum	Bacteria	1	2989	3,309,400	EBI
Deinococcus radiodurans RI	Bacteria	4	3103	3,284,156	TIGR
Desulfovibrio vulgaris	Bacteria	1		3,571,425	TIGR
Enterococcus faecalis V583	Bacteria	4	3148	3,359,973	TIGR
Enterococcus faecium	Bacteria	365	3309	2,928,706	DOE/JGI
Escherichia coli K-12	Bacteria	1	4279	4,639,221	NCBI
Escherichia coli 0157:H7	Bacteria	1	5361	5,498,450	NCBI
Escherichia coli 0157:H7 EDL933	Bacteria	1	5324	5.528.445	NCBI
Fusobacterium nucleaturn ATCC 25586	Bacteria	1	2067	2,174,500	NCBI
Haemophilus influenzae KW20	Bacteria		1 1714	1,830.138	NCBI
Halobacterium sp. NRC-1	Archaea	3	2605	2.571.010	NCR!
Helicobacter pylori 26695	Bacteria	1	1576	1,667,867	NCBI
Helicobacter pylori J99	Bacteria	1	1491	1,643,831	NCBI
Lactococcus lactic suhsp.	Bacteria	1	2266	2,365,589	NCBI
*Listeria innocua Clip*11262	Bacteria	1	2968	3,011,208	NCBI
Listeria monocytogenes EGD	Bacteria	1	2846	2,944,528	NCBI
Mesorhizobiurn loti	Bacteria	3	7281	7,596,300	NCBI
Methanobacterium thermo- autotrophicurn delta H	Archaea	1	1869	1,751,379	NCBI

(Table Contd.)

Organism	Domain	Contigs	Genes	Total bases	Source
Methanococcus jannaschii DSM 2661	Archaea	3	1770	1,739,927	NCBI
Methanopyrus kandleri AVI9	Archaea	1	1687	1,694,969	NCBI
Methanosarcina acetivorans	Archaea	1	4540	5,751,492	NCBI
Mycobacterium leprae	Bacteria	1	2157	3,268,203	Sanger Center
Mycobacterium *tuberculosis* CDC1551	Bacteria	1	4187	4,403,836	NCBI
Mycobacterium tuberculosis H37Rv (lab strain)	Bacteria	1	3927	4,41 1,529	Sanger Center
Mycoplasra genitalium G-37	Bacteria	1	484	580,074	NCBI
Mycoplasma pneumoniae M 129	Bacteria	1	688	816,394	NCBI
Mycoplasma pulmonis UAB CTIP	Bacteria	1	782	963,879	NCBI
Neisseria gonorrhoeae FA 1090	Bacteria	1	2129	2,146,879	University of Oklahoma
Neisseria meningitidis MC58	Bacteria	1	2025	2,272,351	TIGR
Neisseria meningitidis Z2491	Bacteria	1	2065	2,184,406	Sanger Center
Nostoc sp. PCC 7120	Bacteria	7	6129	7,211,789	NCBI
Pasteurella multocida PM70	Bacteria	1	2014	2,257,487	NCBI
Porphyromonas gingivalis W83	Bacteria	1	1777	2,343,478	TIGR
Prochlorococcus marinus MED4	Bacteria	26	1716	1,674,813	DOE Joint Genome
Pseudomonas aeruginosa PAOI	Bacteria	1	5565	6,264,403	University of W
Pyrohaculum aerophilum Micr	Archaea	1	2275	2,222,890	UCLA Dept.
Pyrococcus abyssi	Archaea	1	1765	1,765,118	NCBI
Pyrococcus furiosus	Archaea	1	2208	1,908,253	Utah
Pyrococcus horikoshii OT3	Archaea	1	2058	1,738,505	NCBI

(Table Contd.)

Organism	Domain	Contigs	Genes	Total bases	Source
Ralstonia solanacearum	Bacteria	2	5116	5,810,922	NCBI
Rhizobium sp. NGR234	Bacteria	1	417	536,165	NCBI
Rickettsia conorii	Bacteria	1	1374	1,268,755	NCBI
Rickettsia prowazekii Madrid E	Bacteria	1	834	1,111,523	NCBI
Saccharomyces cerevisiae S288C	Eukaryota	16	6261	12,057,849	NCBI
Salmonella typhi CT 18	Bacteria	3	4633	5,133,712	Sanger Center
Salmonella typhimurium LT2 (strain AZ1516)	Bacteria L 111	2	4553	4,951,371	Washington
Shewanella putrefaciens	Bacteria	2	4221	5,131,063	TIGR
Sinorhizobium meliloti 1021	Bacteria	3	6205	6,691,694	NCBI
Staphylococcus aureus EMRSA- 16	Bacteria	1	2679	2,902,619	Sanger Center
Staphylococcus aureus MW2	Bacteria	1	2632	2,820,462	NCBI
Staphylococcus aureus Mu50	Bacteria	2	2748	2,903,147	NCBI
Staphylococcus aureus N315	Bacteria	2	2625	2,838,294	NCBI
Staphylococcus aureus NCTC 8325	Bacteria	2	2631	2,836,373	University of Oklahoma
Staphylococcus epidermidis	Bacteria	367	2920	2,701,014	GSK
Staphylococcus epidermidis RP62A	Bacteria	2	2444	2,646.310	TIGR
Streptococcus mutans UA159	Bacteria	1	1871	2,030,921	University of Oklahoma
Streptococcus pneumoniae R6 hex	Bacteria	1	2043	2,038,615	NCBI
Streptococcus pneumoniae M 1 GAS	Bacteria	1	1697	1,852,441	University of Oklahoma
MGAS8232		1	1845	1,895.017	NCB1
Sulfolobus solfataricus P2	Archaea	1	2977	2,992,245	NCBI
Sulfolobus tokodaii	Archaea	1	2826	2,694,765	NCBI
Synechocystis sp. PCC6803	Bacteria	1	3169	3,573,470	NCBI

(Table Contd.)

Organism	Domain	Contigs	Genes	Total bases	Source
Thermoanaerobacter tengcongensis MB4T	Bacteria	1	2588	2,689,445	NCBI
Thermoplasma volcanium	Archaea	1	1499	1,584,804	NCBI
Thermotoga maritima MSB8	Bacteria	1	1846	1,860,725	TIGR
Treponema pallidum Nichols	Bacteria	1	1031	1,138,011	NCBI
Ureaplasma urealyticum serovar 3 (Uu)	Bacteria	1	611	751,719	NCBI
Vibrio cholerae N 16961	Bacteria	2	3835	4,033,464	TIGR
Xanthomonas axonopodis citri str. 306	Bacteria	1	4312	5,175,554	NCBI
Xanthomonas campestris ATCC 33913	Bacteria	1	4181	5,076,188	NCBI
Xylella fastidiosa	Bacteria	3	2831	2,731,750	NCBI
Yersinia pestis CO-92 Biovar *Orientalis*	Bacteria	4	4083	4,829,855	Sanger Center

With the availability of genome sequence data, it is now possible to identify every potential gene in the chromosome of a bacterium. Armed with this information, one can perform a systematic evaluation of the *essentiality* of each and any predicted gene. Indeed, a consortium of European laboratories is now attempting to evaluate the importance of every gene in the *Bacillus subtilis* genome.

A number of methods are available for knocking out genes in a directed fashion. The simplest form is *plasmid insertion mutagenesis*. In this method, a small defined internal fragment (typically 500 bp) of the target gene is cloned in a nonreplicative *plasmid harboring* a selective marker that is expressed in the host organism.

This plasmid is introduced into the bacterium and grown under selective conditions. Maintenance of the plasmid DNA requires *homologous recombination*, which results in the insertion of the plasmid sequence into the target gene. The method is quick and simple and, together with the availability of the sequence of the chromosome, has allowed a rapid genome-wide identification of essential genes in *Streptococcus pneumoniae*. Interpretation of the data however, presents several problems.

Because efficient recombination requires hundreds of base pairs of homology, a significant portion of the gene may remain intact after recombination, raising the issue of partial gene activity being expressed from the intact portions of the disrupted open reading frame. Partial activity of an essential gene may be sufficient for cell viability, and thus single crossover insertional inactivations may misrepresent genes that are truly essential.

Moreover, insertional *mutagenesis* requires integration of the entire vector into the chromosome, often altering the expression of genes downstream of the target (i.e., polarity effects). As a result, a gene may be considered essential when, in fact, the effect derives from a downstream gene vital for cell growth (i.e., *falsepositive analysis*).

An improved *knockout* approach achieves removal of the gene to ensure its *inactivation* and replacement with a small resistance determinant designed not to introduce polarity effects This *allelic* replacement method requires a double *recombination* event using *homologous* sequences that flank the target gene. The gene is deleted and replaced by a selective marker.

The *procedure* identifies genes that are likely to be essential, and very effectively defines those genes that are not. The results can be used to focus attention on targets critical for bacterial functions. Perhaps the best approach to confirm that a gene is essential for cell *viability* is to generate a conditional mutant. One method of achieving this is to place the target gene under the control of an inducible *promoter.*

This method provides quantitative information useful for evaluating potential targets for drug discovery because, in contrast to knockout methods that result in the complete loss of activity of the gene product, promoter control systems allow the level of expression to be regulated often in a titratable manner, ranging from no expression (thereby testing the *essentiality* of the gene) to overexpression (in which the cellular *concentration* of the target exceeds that normally found in the wildtype strain).

In this way, it is possible to determine the level of a particular protein necessary for *survival/ growth.* A number of inducible promoter systems have been developed and used in both *nonpathogenic* and *pathogenic* bacteria.

Bacterial strains carrying these regulated gene systems can be used to confirm the MOA of an antibacterial agent; if regulation of the levels of a particular gene target gives rise to a selective and *concomitant* change in bacterial sensitivity to an inhibitor of the target, this provides *corroboration* of the mode of antibacterial action.

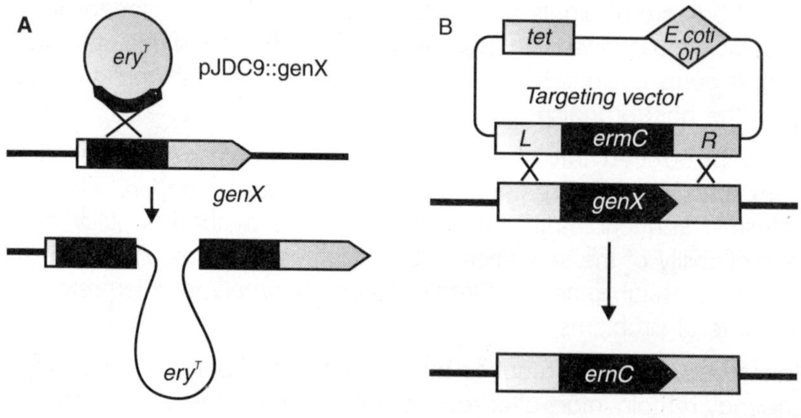

Figure 2.1: Laboratory testing. (A) Plasmid insertion mutagenesis. (B) Allelic replacement mutagenesis.

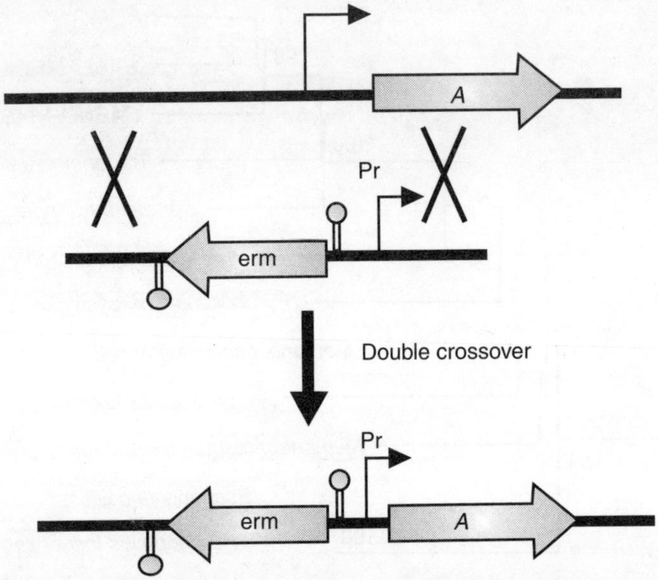

Figure 2.2: Regulation of genomic expression. Gene A is placed under the control of an inducible promoter.

Examples

In the following subsections, we discuss some of the targets progressed via the target-based screening *strategy*, and focus on the role played by genomics in identifying and *progressing* these targets. In some of these examples genomics has contributed to the identification of the target, while in others it has provided new insights into an established target area, which has enabled better *exploitation* of the target.

Aminoacyl tRNA Synthetases

Aminoacy1 tRNA synthetases catalyze the *amino acylation* of amino acids with their cognate tRNA prior to incorporation of the amino acid into a growing peptide chain. Consequently, there are 20 different tRNA *synthetases* in Gram-negative bacteria (but only 19 in Gram-positive bacteria), and each is essential for the survival of bacterial cells. The potential of these enzymes as *antibacterial* targets is *demonstrated* by the antibiotic *mupirocin*, which is a potent and selective *inhibitor* of *isoleucyl* tRNA synthetase (IRS).

Prior to 1995 these tRNA synthetases had not been well characterized in different pathogens. Genomics, however, has allowed determination of the sequences of tRNA synthetases from multiple bacterial *pathogens*, and comparative analysis of these sequences has enabled valid consideration to be given to the *spectrum* and selectivity of each enzyme. Moreover, this *analysis* can *demonstrate* whether a specific enzyme may have use as a broad- or narrow-spectrum target.

For example, as there are differences between the *methionyl* tRNA synthetases (MRS) of Gram-positive and Gram-negative bacteria, this enzyme may be an appropriate target if a selective anti-Gram-positive or anti-Gram-negative agent were required.

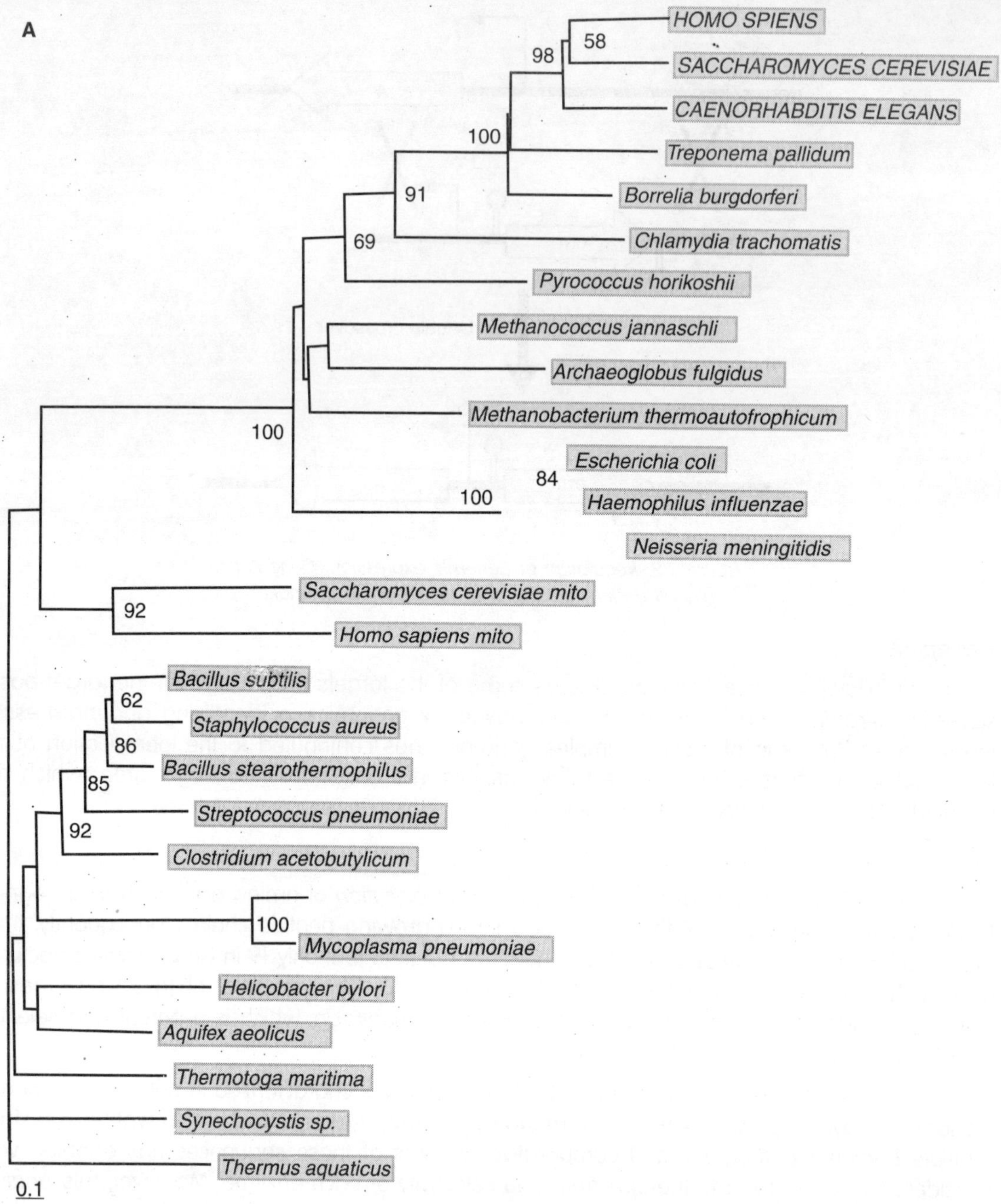

Figure 2.3: Neighbour-joining (NJ) trees of (A) methionyl-tRNA synthethetase and (B) glutaminyl- and glutamyl-tRNA synthetases, labeled as G1nRS and GluRS, respectively. The NJ tree was constructed using the program NEIGHBOUR of the PHYLIP 3.57c package (http//evolution.genetics.washington.edu/phylip.html). In order, the numbers at nodes represent the percent occurrence of nodes in 1000 bootstrap resamplings of the NJ tree. Different type faces are used to distinguish species of the Bacteria (italics), Eucarya (upper-case, italics) and Archaea (italics,

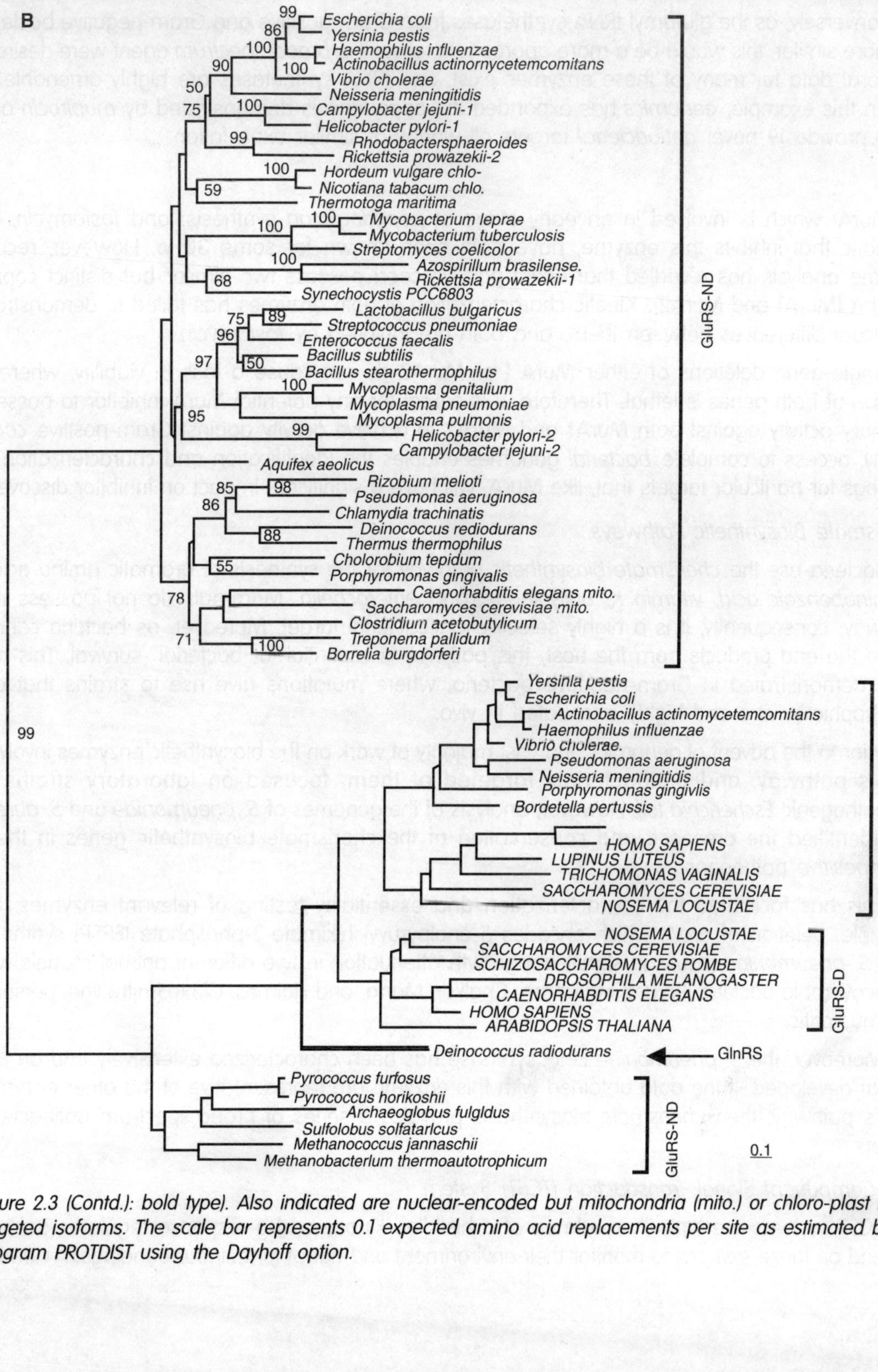

Figure 2.3 (Contd.): bold type). Also indicated are nuclear-encoded but mitochondria (mito.) or chloro-plast (chlo.) targeted isoforms. The scale bar represents 0.1 expected amino acid replacements per site as estimated by the program PROTDIST using the Dayhoff option.

Conversely, as the glutamyl tRNA synthetases from Gram-positive and Gram-negative bacteria are more similar, this would be a more appropriate target if a *broad-spectrum* agent were desired. Structural data for many of these enzymes exist, and tRNA synthetases are highly amenable to HTS. In this example, *genomics* has expanded the opportunities demonstrated by *mupirocin* and IRS to provide 19 novel *antibacterial* targets all worthy of similar exploitation.

MurA

MurA, which is involved in an early stage of *peptidoglycan* synthesis. and *fosfomycin*, an antibiotic that inhibits this enzyme, have both been known for some 30 yr. However, recent genome analysis has revealed that Gram-positive cocci possess two similar but distinct copies of MurA (MurA1 and MurA2). Kinetic characterization of both enzymes has failed to demonstrate significant differences between them, and both are inhibited by *fosfomycin*.

Single-gene deletions of either MurA l or MurA2 do not cause a loss in viability, whereas *deletion* of both genes is lethal. Therefore, it is critical for any potential MurA inhibitor to possess inhibitory activity against both MurA1 and MurA2 to achieve activity against Gram-positive *cocci*. Clearly, access to complete *bacterial* genomes enables the identification and characterization of paralogs for particular targets that, like MurA, will have a significant impact on inhibitor discovery.

Chorismate Biosynthetic Pathways

Bacteria use the *chorismate biosynthetic* pathway for the synthesis of aromatic amino acids, *p-aminobenzoic acid, vitamin K, ubiquinone,* and *enterochelin.* Mammals do not possess this pathway; consequently, it is a highly selective antibacterial target. Moreover, as bacteria cannot obtain the end products from the host, this pathway is essential for bacterial -survival. This has been demonstrated in Gram-negative bacteria, where mutations give rise to strains that are auxotrophic in vitro and highly attenuated in vivo.

Prior to the advent of genomics, the vast majority of work on the biosynthetic enzymes involved in this pathway, and antibacterials targeted at them, focused on laboratory strains of nonpathogenic *Escherichia toll.* However, analysis of the genomes of *S. pneumoniae* and *S. aureus* has identified the presence and conservation of the chorismate biosynthetic genes in these *Grampositive* pathogens.

This has facilitated the characterization and essentiality testing of relevant enzymes. For example, deletion of *aroA,* which encodes 5-enolpyruvylshikimate-3-phosphate (EPSP) synthase, from *S. pneumoniae* results in significant growth attenuation in two different animal models with no recoverable bacteria (Bryant, Ingraham, Chalker, Marra, and Holmes, GlaxoSmithKline, personal communication).

Moreover, the *S. pneumoniae* EPSP synthase has been characterized extensively and an HTS format developed. If the data obtained with this enzyme are representative of the other enzymes in this pathway, then *chorismate* biosynthesis provides a series of broad-spectrum antibacterial targets.

Two-Component Signal Transduction (TCST) System

TCST systems are typically made up of a histidine kinase and a response regulator. Bacteria depend on these systems to monitor their environment and adapt to their surroundings. A number

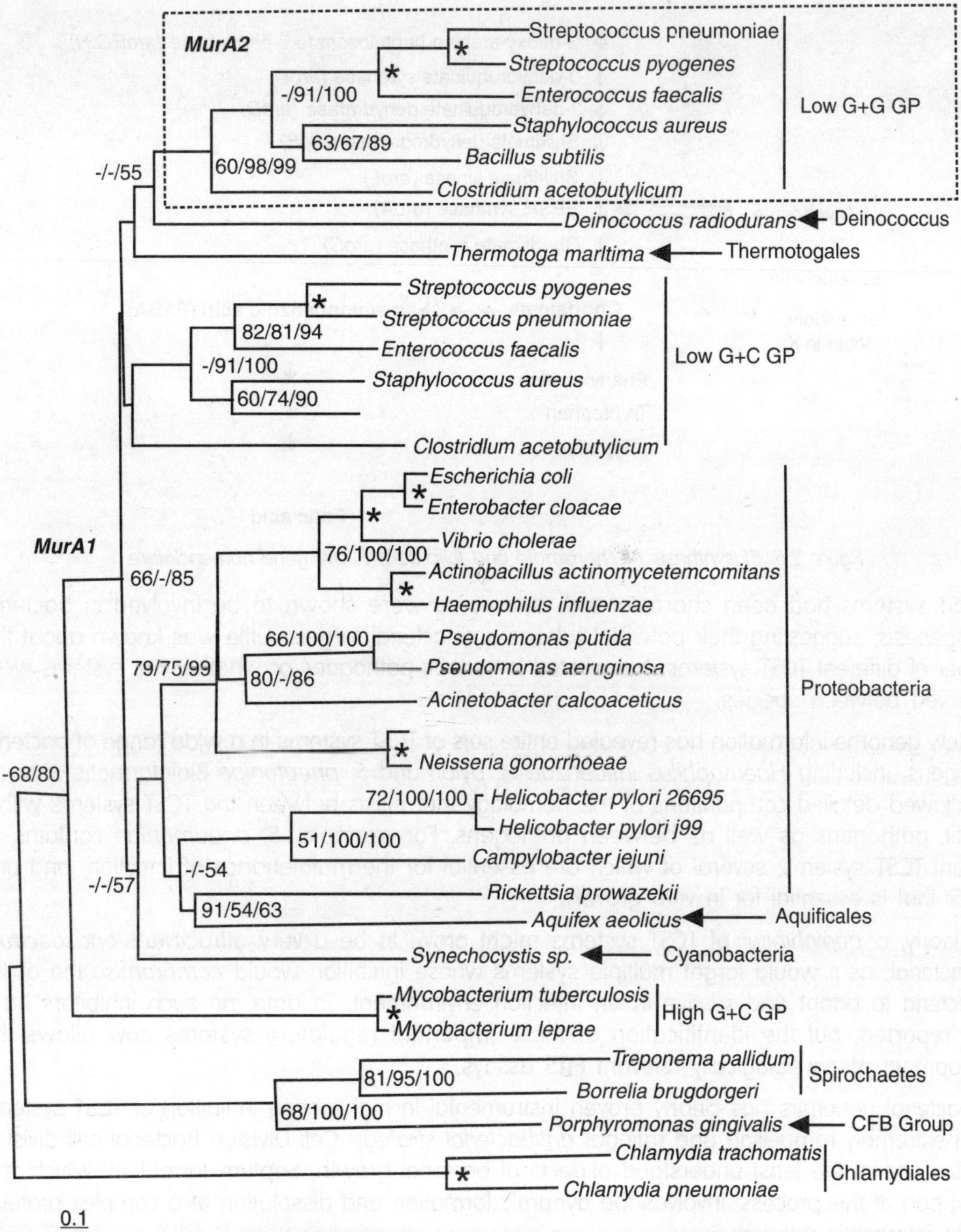

Figure 2.4: Phylogenetic tree showing divergence of UDP-N-acetylglucosamine enolpyruvyl transferase isoforms, MurA1 and MurA2.

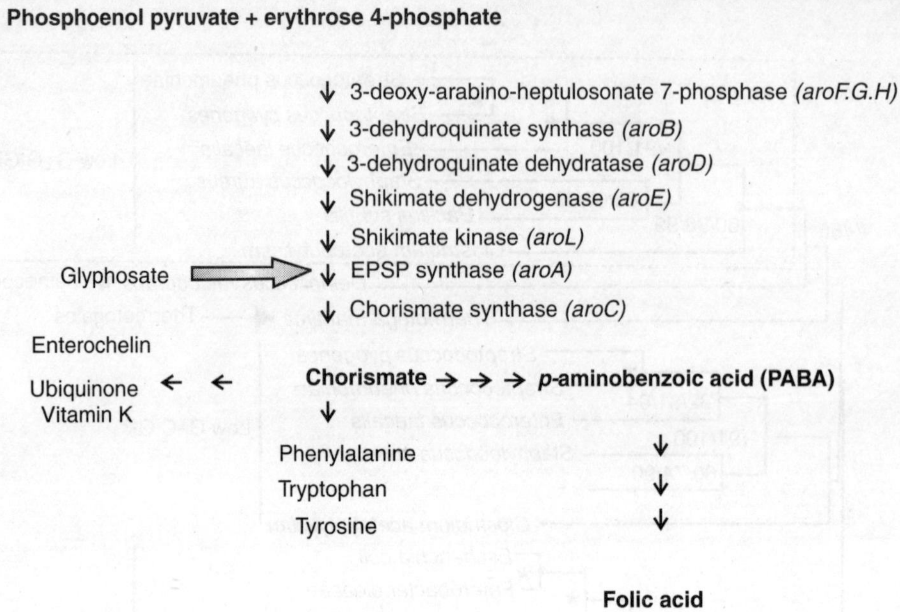

Figure 2.5: Biosynthesis of chorismate and folic acid, E. coli *gene nomenclature.*

of TCST systems had been characterized, and some were shown to be involved in bacterial *pathogenesis*, suggesting their potential role as antibacterial targets. Little was known about the number of different TCST systems that existed in various pathogens or whether the systems were conserved between species.

New genome information has revealed entire sets of TCST systems in a wide range of bacterial pathogens, including *Haemophilus influenzae H. pylori* and *S. pneutoniae* Bioinformatic analysis has allowed detailed comparisons of the homology that exists between the TCST systems within specific pathogens as well as between pathogens. For example, *S. pneumoniae* contains 13 different TCST systems, several of which are essential for the maintenance of infection; and one *(yycFG)* that is essential for in vitro growth.

Clearly, a *paninhibitor* of TCST systems might prove to be a very attractive *broadspectrum* antibacterial, as it would target multiple systems whose inhibition would *compromise* the ability of bacteria to adapt and survive in an infection environment. To date, no such inhibitors have been reported, but the identification of these important regulatory systems now allows the development of *physiologically* relevant HTS assays.

Bacterial genomirs has clearly proven instrumental in highlighting inhibition of TCST systems as an extremely rornpelling and rational *antibacterial strategy.* Cell Division Bacterial cell division remains one of the least understood aspects of bacterial growth. Septum formation, which is a critical part of this process, involves the dynamic formation and dissolution of a complex protein-protein interaction network.

FtsZ plays a central role in this network, by polymerizing into a ring structure at the mid cell, and by recruiting other proteins that play a role in the downstream events in the cell-division

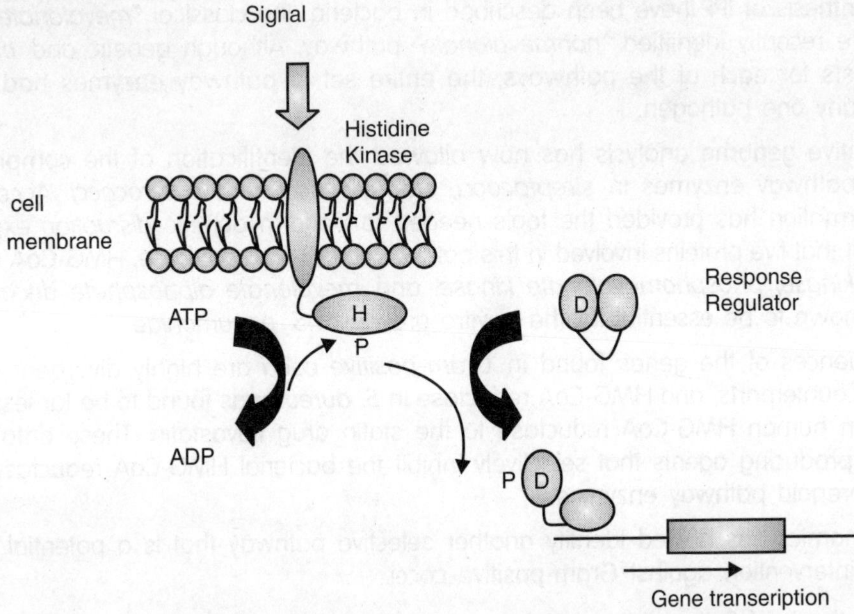

Figure 2.6: Showing the mechanism of action of a two-component signal transduction system. TCSTS are typically made up of two proteins, a histidine kinase and a response regulator. The histidine kinase generally acts as a sensor to external signals and responds by autophosphorylating on a histidine residue. This phosphate is then transferred to an aspartate residue of the response regulator, which activates the response regulator by causing it to bind to DNA and alter gene transcription. In this way, the bacterium monitors its environment and adapts to its surroundings.

cycle. A number of these proteinprotein interactions, such as the interaction of FtsA with FtsZ, are essential for cell viability and thereby represent potential *antibacterial* targets.

Analysis of the genes encoding FtsA and FtsZ proteins in a variety of bacterial genomes has demonstrated that these proteins are widely distributed and highly conserved. Furthermore, *alignment* of the sequences of FtsZ from a variety of bacteria demonstrated a conserved 10-amino-acid motif in the C-terminal region of the protein, which was proposed to play an important role in *interacting* with other *cell-division* proteins.

Detailed mutational analysis of this region of FtsZ in *S. aureus* demonstrated that a single residue (Phe376) is essential for the interaction with FtsA. Similar work with *E. coli* FtsZ has confirmed that changing single residues in this region significantly affects binding to FtsA. These data suggest that it may be possible to design small *molecules* that could either enhance or *antagonize* the interaction of FtsA with FtsZ.

Given the importance of the interaction between FtsA and FtsZ, either mechanism of action might result in antibacterial activity. An HTS assay for the FtsA/FtsZ interaction has been successfully developed, which we hope will facilitate exploitation of this novel *antibacterial* strategy.

Isoprenoid Biosynthesis

Isoprenoids, which are *ubiquitous* in nature, comprise a family of over 23,000 products, each composed of repeating 5-carbon *isopentenyl diphosphate* (IPP) subunits. Two different pathways for the biosynthesis of IPP have been described in bacteria: the classical "*mevalonate*" pathway and the more recently identified "*nonmevalonate*" pathway. Although genetic and *biochemical* evidence exists for each of the pathways, the entire set of pathway enzymes had not been identified in any one pathogen.

Comparative genome analysis has now allowed the identification of the complete set of *mevalonate*-pathway enzymes in *streptococci, staphylococci,* and *enterococci.* Access to this genome information has provided the tools needed to perform genetic *disruption* experiments, with the result that five proteins involved in this pathway (HMG-CoA synthase, HMG-CoA reductase, *mevalonate kinase, phosphomevalonate kinase,* and *mevalonate diphosphate decarboxylase*) have been shown to be essential for the in vitro growth of *S. pneumoniae.*

The sequences of the genes found in *Gram-positive cocci* are highly divergent from their mammalian counterparts, and HMG-CoA reductase in *S. aureus* was found to be far less sensitive (10^{-4}-fold) than human HMG-CoA reductase to the statin drug fluvastatin. These data raise the possibility of producing agents that selectively inhibit the bacterial HMG-CoA reductase or other bacterial isoprenoid pathway enzymes.

Thus, genomics has helped identify another selective pathway that is a potential target for antibacterial intervention against Gram-positive cocci.

Biological Synthesis of Fatty Acid

In mammals, fatty-acid biosynthesis is performed by a single multifunctional polypeptide chain, whereas in bacteria the pathway requires seven discrete enzymes. This key difference implies that bacterial fatty-acid biosynthetic enzymes are also potential selective targets for *antibacterial* drugs. The majority of the work on this pathway and its component enzymes has focused on *laboratory* strains of *E. coli.*

Table 2.2: Identification of Fab Genes in Key Respiratory and Gram-Positive Pathogens.

Target	*S. aureus*	*S. pneumoniae*	*E. faecalis*	*H. influenzae*
FabG	✓	✓	✓	✓
FabK	X	✓	✓	X
FabI	✓	X	✓	✓
FabA	X	X	X	✓
FabZ	✓	✓	✓	✓
FabB	X	X	X	✓
FabF	✓	✓	✓	X
FabH	✓	✓	✓	✓
FabD	✓	✓	✓	✓

✓ = homolog present in genome; X = homolog absent from genome.

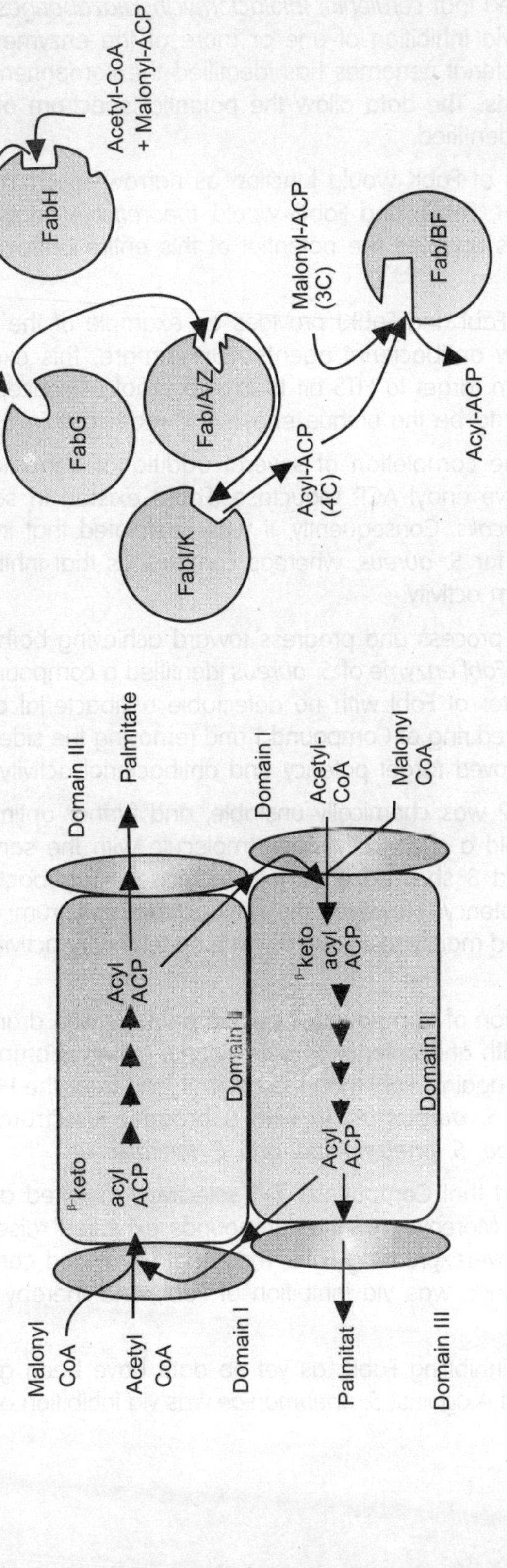

Figure 2.7: Comparison of fatty acid biosynthesis in humans and bacteria.

Early work demonstrated that *cerulenin, thiolactomycin, diazaborines,* and *triclosan* all exerted their *antibacterial* activity via inhibition of one or more of the enzymes in the pathway. Recent analysis of sequenced bacterial genomes has identified the component enzymes in a variety of important clinical pathogens. The data allow the potential spectrum of each of the component pathway enzymes to be identified.

For example, inhibitors of FabK would function as narrow-spectrum *antistreptococcal* agent, whereas inhibitors of FabH, FabD, and FabG would *theoretically* show broad-spectrum activity. Genomic analysis has thus enabled the potential of this entire pathway to be exploited as an *antibacterial* strategy.

Enoy1 ACP reductase (Fabl and FabK) provides an example of the potential of this pathway to provide targets for new antibacterial agents. Furthermore, this example demonstrates the successful progression from target to HTS hit to in vivo proof of concept. When this project was initiated, Fabl was thought to be the unique enoy1 ACP reductase in all bacteria.

However, following the completion of several additional genome sequences, it became apparent that an alternative enoyl ACP reductase (FabK) existed in some pathogens, such as *S. pneumoniae* and *E. faecalis.* Consequently, it was postulated that inhibitors of Fabl could be useful as selective agents for *S. aureus,* whereas compounds that inhibited both FabK and Fabl could have *broad-spectrum* activity.

Below we present our process and progress toward achieving both off these objectives. HTS of compounds against the *Fabl enzyme* of *S. aureus* identified a compound designated Compound 1, as a low-potency inhibitor of Fabl with no detectable antibacterial activity against *S. aureus.* Opening the sevenmembered ring of Compound 1 and removing the side chain yielded Compound 2, which had slightly improved target potency and antibacterial activity.

However, Compound 2 was chemically unstable, and further optimization of the right-hand side was performed to yield a chemically stable molecule with the same basic pharmacophore (Compound 3). Compound 3 showed a 32-fold increase in antibacterial activity and further improvements in target potency. However, the antibacterial spectrum of Compound 3 was still very narrow, being restricted mainly to *S. aureus,* with no inhibitory activity being observed against FabK.

Subsequent derivatization of Compound 3 yielded analogs with dramatically enhanced target potency and greater breadth and potency of antibacterial activity. Compound 4 was three orders of magnitude more potent against Fabl than the original lead from the HTS screen and possessed exquisite activity against *S. aureus* along with a broader spectrum of antibacterial activity encompassing *H. influenzae, S. pneumoniae,* and *E. faecalis.*

MOA studies confirmed that Compounds 2-4 selectively inhibited acetate incorporation in *S. aureus* and *H. influenzae.* Moreover, all the compounds exhibited raised MIC against *S. aureus* and *H. influenzae* strains overexpressing Fabl. These data provided conclusive evidence that the MOA of all these compounds was via inhibition of Fabl, and thereby facilitated a robust lead optimization program.

Despite Compound 4 inhibiting FabK, as yet no data have been generated to demonstrate that the MOA of Compound 4 against *S. pneumoniae* was via inhibition of FabK. The antistaphylococcal activity of Compound 4 is excellent (MIC_{90} 0.06 μg/mL for 22 multiresistant strains of MRSA). Furthermore, Compound 4 demonstrated excellent in vivo activity in experimental *S. aureus* infections.

These data clearly demonstrate the potential of enzymes in the fattyacid biosynthesis pathway as novel antibacterial targets, and illustrate how a successful *cascade* from novel genomic target to proof of concept in vivo is achievable.

Screening of Antibacterial Cell

The explosion in attempts to increase the chemical diversity of *antimicrobial* agents has meant that the numbers of compounds included in screening collections has increased from tens of thousands to hundreds of thousands. *Combinatorial* chemistry and new natural product sources and strategies have underpinned this expansion, with new *technologies* allowing for *antibacterial* screening in real-time assays.

Whole-cell antibacterial screening affords a direct approach to identifying novel antibacterials and is facilitated by advances in techniques for defining the molecular target(s) of novel antibacterial compounds. Although whole-cell antibacterial screening generally lacks the sensitivity of molecular target screening, it carries the advantage that all potential targets, both known and unknown, are screened simultaneously within a *physiological* context, and with an in vitro assay that has demonstrated relevance to the clinical endpoint.

Additionally, a key advantage of this approach is that the compound already possesses antibacterial activity, which is often exceptionally difficult to engineer into compounds identified from a target-based screen; past experience has shown that leads from target-based screens often possess no significant antibacterial activity.

However, a challenge to this approach is that antibacterial activity is frequently encountered as the result of nonspecific effects, often membrane derived. Consequently, assays are required to exclude advancing such compounds and to focus on compounds with truly exploitable, bacterially selective modes of action.

MOA Methodologies

Control of Gene Expression

The relationship between target abundance, especially for single-enzyme targets, and whole-cell sensitivity suggests overexpression constructs could be used as tools to investigate the targets of antibacterial compounds. For example, overexpression of a target in most cases will cause a concomitant increase in MIC.

Strategies for obtaining overexpression include both regulated and unregulated plasmid-mediated expression, and genomic essentiality data should allow preparation of overexpression constructs for all potential antibacterial targets.

Susceptibility testing of libraries of overexpression constructs together with wild-type strains allows rapid MOA screening for antibacterial compounds against a considerable number of potential targets (Huang and McDevitt, GlaxoSmithKline, personal communication). MOA hypotheses generated by these global analyses can be tested by use of over- or underexpression constructs for specific targets. Inducible promoters can be used to generate libraries of strains expres-

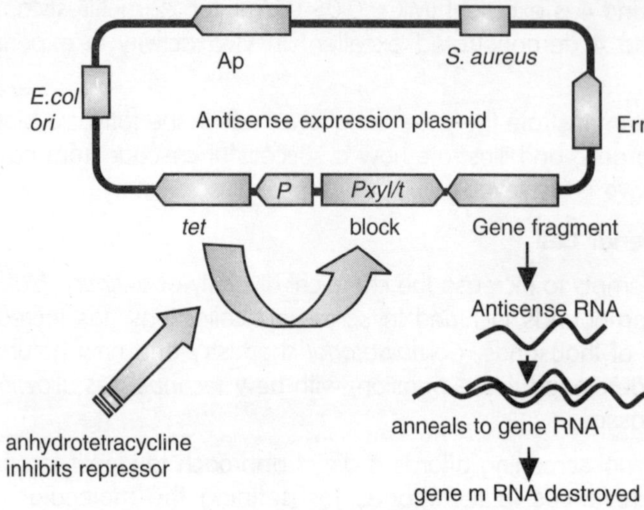

Figure2.8: The use of antisense RNA to evaluate gene essentiality.

sing antisense RNA to specific gene transcripts. In these experiments, fragments of genes are inserted into a plasmid in the *antisense orientation*, downstream of a controllable promoter. In the absence of antisense induction, cells are viable, since the normal gene is transcribed and *translated.*

However, in the presence of inducer, antisense RNA is generated, leading to downregulation of expression of the target gene and, if the gene is essential, this will result in decreased growth and/or cell death. Induction of antisense RNA to a gene that encodes an antibacterial target will result in a selective increase in sensitivity to the corresponding antibiotic. Antisense libraries encoding antisense constructs to every gene in a genome can be generated, and this enables rapid global MOA screening of compounds against a considerable number of potential targets. The utility of this procedure for both essentiality testing and MOA screening has been reported.

Translatarary Microarray

Global transcription profiling using microarrays has been used to identify antibiotic signature expression patterns in *H. influenzae* and *S. aureus.* A library of expression profiles has been established for MOA studies in *S. aureus* exposed to ciprofloxacin, triclosan, and mupirocin, with resulting inhibition of DNA replication, *fatty-acid biosynthesis*, and protein synthesis, respectively.

Cluster analysis of these profiles indicated that the modes of action of the different classes of antibiotics could be distinguished by their unique expression patterns. Therefore, the expression profile of a novel antibacterial compound can be compared with the profiles of known antibacterials to indicate its likely mode of action, although the exact molecular target is unlikely to be identified using this technique. A similar approach is possible via the generation of signature patterns using proteomics.

Selection of Target based Resistance

Many antibacterials with a specific action have the potential to select targetbased resistance, manifested either spontaneously or following *mutagenesis*. Hence, investigation of resistant mutants through library *generation* and complementation has the potential to determine the molecular target of a novel *antibacterial*. Strides made in molecular cloning technologies, especially in Gram-positive pathogens, such as *S. aureus*, have greatly facilitated this approach.

Once the MOA of an antibacterial compound has been identified, these tools then play the same role as they do in the target-based screening approach, in providing methodologies to confirm the MOA of structural *analogs* produced during attempts to optimize the properties of the initial lead compound.

Cellular Leakage and Lysis

A range of membrane assays have been used, including red blood cell lysis and intracellular K^+ leakage, to identify compounds with *nonspecific deleterious* effects on membranes as the mechanism of antibacterial activity.

The use of dyes sensitive to membrane potential, including flow cytometry applications, also offers approaches to investigating the effect of antibacterials on bacterial cell membranes (Gentry, Wilding, and Gwynn, GlaxoSmithKline, personal communication). These assays are critical for early identification of compounds that may kill bacteria via nonspecific, and thus potentially toxic, mechanisms.

Biosynthesis of Large Molecules

The effect of antibacterials on the incorporation of radiolabeled precursors into macromolecules has been used in a convenient microtiter format to define whether the primary action of a new antibacterial is on DNA replication, transcription, protein synthesis, fatty-acid biosynthesis, or cell wall biosynthesis. Collectively, the gene products of these macromolecular processes represent a significant proportion of the essential genome of pathogens.

Identification of one of these processes as the target of a novel antibacterial compound excludes the compound having a nonspecific MOA and enables more focused studies to address the specific *targeting*.

CONCLUDING REMARKS

The use of genomics has considerably expanded the number of novel and validated antibacterial targets. Soon we will know all the essential genes for key bacterial pathogens, and thus all possible antibacterial targets will be revealed. Furthermore, advances in genetic and genomic technologies means that we now have considerable resources available to identify the mode of action of novel antibacterial compounds identified from screening processes.

There is considerable synergy between the whole-cell and target-based screening approaches, as these utilize similar *technologies* and prosper from the application of *genomics*. However, a number of challenges still face this *therapeutic* area, and while genomics has facilitated the identification of new antibacterial *strategies*, potential targets vary in the ease with which they can be exploited.

For example, MurA, FabI, and the aminoacyl tRNA synthetases illustrate the importance of thorough analysis of phylogenetic distribution and conservation as a critical aspect of target

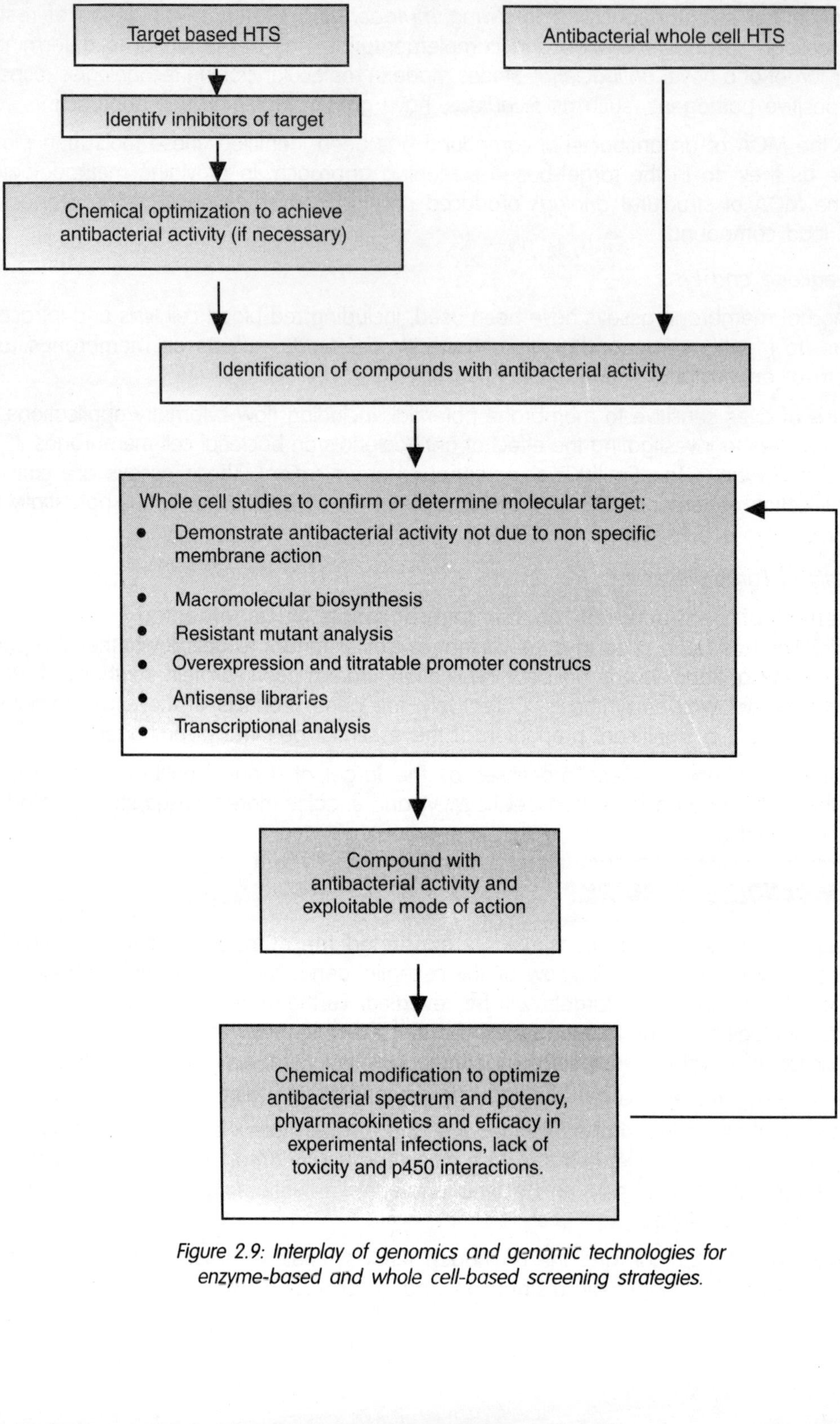

Figure 2.9: Interplay of genomics and genomic technologies for enzyme-based and whole cell-based screening strategies.

selection. In addition, target abundance is an important consideration in target selection, as a highly abundant target may be difficult to inhibit in the physiological environment of the cell.

Furthermore, the genes encoding some targets may be prone to high levels of *spontaneous* mutation, or as yet undiscovered bypass mechanisms may exist, either of which may result in resistance. However, the most substantial challenge is arguably the identification of good quality hits and leads from HTS.

It is apparent that bacteria and many of their molecular targets appear to he quite refractory to selective inhibition by small molecules; therefore, it is imperative that screening approaches utilize large collections of novel, chemically diverse, and developable compounds. Natural product strategies such as *shuffling* the genes of known secondary metabolite pathways to yield new molecules, and application of combinatorial chemistry approaches to design compound libraries especially suited to the *parameters* common to antibacterial compounds, are both welcome approaches.

New antibacterial agents that inhibit novel targets are required to help overcome the resistance problems that plague present classes of antibiotics. From this discussion, it should be clear that genomics is playing a *pivotal* role in the discovery of such agents. It is to be hoped that the continued application of this technology will result in the delivery of new *antibacterials* that are safe, *welltolerated*, and *efficacious*.

Chapter 3

Analysing Microbial Genomes

Microbial genome analysis is a growing area that is expected to lead to advances in *healthcare*, environmental cleanup, agriculture, industrial processes, and alternative energy. According to the Genomes Online Database, as of April 2007 close to 500 microbial genomes have been sequenced to date, whereas more than 1000 additional projects are ongoing or in the process of being launched.

As the genomic community is rapidly moving toward the generation of complete and draft sequences for several hundred microbial genomes, comparative data analysis in the context of integrated genome data sets plays a critical role in understanding the biology of the newly sequenced organisms. Conversely, individual organism-specific genome analysis carried out in *isolation* cannot support timely analysis of newly released genomes.

Microbial genomes are sequenced by organizations worldwide, follow an *annotation* process (gene prediction and functional characterization) that is often specific to each sequencing center, and end up in one of the public sequence data *repositories*, such as GenBank in the United States, EMBL in Europe, and DDBJ in Japan.

Genome sequence data include information on gene coordinates, transcription orientation, locus identifiers, gene names, and protein functions. Analyzing microbial genomes requires however additional functional annotations, such as *motifs*, *domains*, *pathways*, and *ontology* relationships, which are provided by diverse, usually *heterogeneous*, *data sources*, such as Pfam, InterPro, COG, CDD, KEGG, and Gene Ontology (GO).

Resources such as EBI Genome Reviews and RefSeq include such additional functional annotations, sometimes after reannotating the sequences from the public sequence data sources. These resources share common goals, but contain different collections of genomes or data with different degrees of resolution regarding the same genomes. These differences are the result of diverse annotation methods, curation techniques, and functional characterization employed across microbial genome data sources.

Comparative genome data analysis is *critical* for effective exploration of the rapidly growing number of complete and draft sequences for microbial genomes. For example, the efficiency of the functional characterization of genes in newly sequenced genomes can be substantially improved if this characterization involves methods based on observed *biological evolutionary phenomena.*

Thus, genes with related (coupled) functions are often both present or both absent within specific genomes and tend to be *collocated* (on chromosomes) in multiple genomes. The effectiveness of comparative analysis depends on the availability of powerful analytical tools and the efficiency of the *integration,* which in turn is determined by the phylogenetic *diversity* of the organisms, the quality of their annotations, and the level of detail in cellular reconstruction.

The efficiency of the integration depends on its breadth (in terms of the number of genomes it involves) and depth (in terms of different annotations it captures). Integration of available genomic data provides the context for comparative *genome analysis,* and is becoming the single most important element for understanding the biology of the newly sequenced *organisms.*

Analyzing genomes in the context of other (e.g., *phylogenetically* related) genomes is substantially more efficient than analyzing each genome in isolation. The Department of Energy's Joint Genome Institute (JGI) is one of the major contributors of microbial genome sequence data, currently conducting about 23% of the reported archaeal and bacterial genome projects world-wide.

Individual microbial genomes are sequenced and *assembled* to draft level at JGI's production facility, and finished either at JGI's production facility, Lawrence Livermore, or Los Alamos National Labs. Both draft and finished genomes pass through the automatic Genome Analysis Pipeline at Oak Ridge National Lab, which generates gene models and associates *automatically* predicted genes with functional *annotations,* such as InterPro protein families, *COG categories,* and KEGG pathway maps.

Before publication or submission to GenBank, scientific groups interested in a specific genome further review and curate the microbial genome data in *collaboration* with Oak Ridge National Lab's *Computational* Biology group and JGI's Genome Biology Program. As previously mentioned, the efficiency of microbial genome review, curation, and *analysis* increases substantially when individual microbial genomes are examined in the context of other genomes.

Providing such a *framework,* to ensure timely analysis of the genomes sequenced at JGI, is one of the main goals of the Integrated Microbial Genomes (IMG) system. IMG aims at providing high levels of data diversity in terms of the number of genomes integrated in the system from public sources, data *coherence* in terms of the quality of the gene *annotations,* and data *completeness* in terms of breadth of the functional annotations.

COMPARATIVE ANALYSIS

The IMG system provides support for comparative analysis of microbial genomes in an integrated genome data context. IMG integrates microbial and selected eukaryotic genomic data from multiple data sources. A high level of genome diversity is ensured by collecting data from public sources, such as EBI Genome Reviews, National Center for Biotechnology Information's RefSeq, and EMBL Nucleotide Sequence Database.

The data model underlying the IMG system provides the structure required for *integrating* and *managing* microbial and selected *eukaryotic* genomic data collected from multiple data sources. The system incorporates in a *coherent* biological context several data types:

(1) primary genomic sequence information,

(2) computationally predicted and curated gene models,

(3) precomputed gene relationships (which are sequence similarity based, gene context based, and so on), and

(4) functional annotations and pathway information.

The user interface is organized in a manner that allows *navigation* over the microbial genome data space along its three key dimensions representing genomes, genes, and functions, respectively. Genomes (organisms) are identified and organized either based on their taxonomic lineage (*domain, phylum, class, order, family, genus, species, strain*) or other organism specific properties, such as *phenotypes, ecotypes, disease,* and *relevance.*

For each genome, the primary DNA sequence and its organization in scaffolds or contigs, are recorded. Genomic features, such as predicted coding sequences and some functional RNAs, are recorded with start/end coordinates. Predicted genes are grouped based on sequence similarity relationships: ortholog and *paralog* gene relationships are currently computed based on *bi-directional* best hit single-linkage.

COGs provide an additional clustering of *orthologous* groups of genes in IMG. Genes are further characterized in terms of molecular function and *participation* in pathways. *Metabolic* pathways are modeled in IMG as ordered lists of reactions and consist usually of one to four reactions.

A reaction can include compounds which are reactants (*substrates*, products) catalyzed by enzymes, and physical entities such as proteins, protein complexes, electrons, and so on. *Nonmetabolic* pathways are modeled in IMG as lists of functions. Pathways are combined into networks via reactions that share common components. Networks can be further combined into more complex networks.

Note that networks are different from KEGG maps, which represent complex networks. Pathways are associated with genes via gene products that function as enzymes that serve as catalysts for individual reactions of metabolic pathways.

The association of genes with pathways in IMG is based on a controlled vocabulary of terms. IMG terms are defined by domain experts as part of the process of including IMG pathways into the system.

The IMG pathways are consistent with the BioPAX level 1 data exchange format in order to facilitate sharing these data across different systems. In addition to the IMG terms and pathways, resources, such as COG, Pfam and InterPro, are used for the functional characterization of genes.

Finally, pathways, reactions, and compounds are included from KEGG and LIGAND. The first version of IMG was released on March 1, 2005. The current version of IMG (IMG 1.4, as of March 1, 2006) contains a total of 699 genomes consisting of 395 bacterial, 30 archaeal, 15 eukaryotic genomes, and 259 bacterial phages.

ANALYSING INTEGRATED MICROBIAL GENOME

Data analysis in IMG is set in a multidimensional data space, whereby genes form one of the dimensions and are characterized in the context of other dimensions, in particular individual organisms (genomes), functions, and networks of pathways. Genes are directly associated with genomes (via *gene prediction*), as well as with functions and pathways (via functional characterization). An organism is associated with a specific function *f* or pathway *p* if its genome has a gene that is associated with *f* or *p*, respectively.

Genes can be grouped (*clustered*) in terms of their sequence similarity or associations with functions and pathways. Each dimension in the microbial genome data space is characterized by one or several category attributes whose values can be used to specify a classification hierarchy.

For example, phylogeny serves as a category attribute for organisms and is used to specify their *phylogenetic* tree classification. *Phenotypic* attributes, such as origin of the sample used for sequencing (e.g., ocean, groundwater, and so on) can also serve as category attributes for organisms.

Microbial genome data analysis operations allow navigating the multidimen-sional data space along one or several dimensions and can be set in the context of specific (i.e., subsets of) organisms, functions, or pathways. Organism (genome) selections help focus the analysis on a subset of interest, especially in terms of *phylogenetic* or *phenotypic* relationships.

For example, a set of interest may include all the *strains* within a specified species. Similarly, function selections focus the analysis on a subset of interest, such as functions involved in lipid *metabolism* pathways. Finally, gene selections reduce the scope of *analysis* to genes with certain properties, such as genes sharing a common function or genes that are *colocated* on the chromosome.

An important type of analysis operation regards examining so-called occurrence profiles of objects of interest (e.g., functions) selected from one dimension of the *multidimensional* data space, across objects (e.g., organisms) selected from another *dimension* of the data space. Consider two dimensions of the data space representing functions and organisms, respectively.

The occurrence profile for a function of interest (e.g., enzyme), *f*, shows the pattern of *f* across organisms y_1 to y_n in the form of a vector (L_1 ...,L_n) where L_i represents the set of y_i genes that are associated with *f*. Similarly, the profile for a gene, *x*, across organisms y_1 to y_n has the form of a vector (L_1, . . . , L_n) where L_i represents a set of y_i genes that are associated with *x*, where the association of yi genes with *x* is based on a specific sequence similarity method.

The number of genes in a set L_r, k_r, is called gene abundance and vectors of the form (k_1, ... , k_n) are called abundance profiles. Presence profiles are a special case of abundance profiles, whereby in each vector of the form (k_1, ... , k_n) k_i *is* replaced by either "a" (absent) if k_i is zero or "p" (present) otherwise. Figure 3.1 shows an example of abundance profiles for genes x_1 to x_4 across organisms y_1 to y_8. Profiles for objects that are aggregations (compositions) of other objects consist of all the profiles for their component objects. For example, the profile of a metabolic pathway consists of the profiles for the enzymes involved in the pathway, whereas the profile of a network consists of the profiles of its component pathways.

Analysis based on occurrence profiles usually involves:

(1) examining the profiles for objects of a given type across objects of another type; or

(2) finding objects of a given type that either have a predefined presence profile or whose presence profile is similar to the presence profile of a given object of the same type, across objects of another type.

For example, examining the profiles of the genes of a specific organism, y, in the context of other related organisms, y_1, ..., y_k allows determining what y may have in "common" with y_1, ..., y_k. Sequences with sufficient degree of similarity are deemed to encode the same gene, and accordingly are considered "common" to or "present" in selected organisms.

	y_1	y_2	y_3	y_4	y_5	y_6	y_7	y_8
X_1	2	1	1	3	0	0	1	0
X_2	1	1	1	2	0	0	1	0
X_3	0	1	1	0	0	0	0	0
X_4	1	1	1	1	2	1	2	1

Figure 3.1: Abundance profile example.

For the example shown in Figure 3.1 organism *y* has gene x_4 in "common" with organisms y_1 to y_8; and genes x_1 and x_2 have the same presence profile across genomes y_1 to y_8. Note that an organism having multiple genes corresponding to a specific gene in another organismis the result of the similarity method employed (e.g., *homology*) in computing profiles.

Finding a unique *orthologous* gene in an organism corresponding to another gene in a different organism is straightforward only for singly copy genes. For other genes, establishing orthologous relationships across organisms is complicated by the fact that most genes undergo either gene *duplications* or fusion events, with subsequent losses of some of the duplicated copies adding to the complexity of determining such relationships.

Occurrence profile operations can be used for analyzing biological phenomena such as gene conservation or gain, for a specific organism (e.g., *y*) in the context of other organisms (e.g., y_1, ..., y_k). For the example shown in Figure elsewhere in this chapter gene x_4 is conserved across y_1 to y_8, whereas gene x_3 is gained with respect to y, and y_4 to y_8.

Occurrence profiles are critical in the process of understanding the biology of the microbial genome under study. This process is based on observed biological evolutionary phenomena: genes with related (coupled) functions (1) are often both present or both absent within specific genomes that have these functions; (2) tend to be collocated (on chromosomes) in multiple genomes; (3) might be fused into a single gene in some genomes; or (4) are cotranscribed under the same regulator. Consider the example shown in Figure elsewhere in this chapter where pathway p involves reactions R_1, R_2, R_3, and R_4: genes x_1, x_2, and x_4 of genome G_1 are associated with pathway p via enzymes e_1, e_2, and e_4, respectively; genes z_1, z_2, z_3 and z_4 of genome G_2 are associated with pathway p via enzymes e_1, e_2, e_3, and e_4, respectively; if gene x_3 is similar (i.e., determined to be related via significant sequence similarity) to gene z_3, then, following the rules previously listed, x_3 may be associated with p via enzyme e_3.

For the example shown in Figure elsewhere in this chapter suppose that gene x_1 is functionally characterized, whereas x_2 is not; then the fact that genes x_i and x_2 have similar occurrence profiles across organisms y_1 to y_8, may help characterize x_2, which may participate in a similar biological process as gene x_1.

Finding objects that have a specific presence profile are used for identifying certain (e.g., unique) genes in an organism in the context of other organisms. For example, consider finding genes of a target organism in terms of *presence* or absence of homologs (or orthologs) in other reference organisms. Reference organisms can be defined based on some biological property, such as phylogenetic relationship, shared *phenotype*, or *ecological* environment.

For example, if the reference organisms are phylogenetically related then finding genes that have a specific profile could be used to identify preserved, gained, or lost genes. Although the preserved genes are shared by all organisms in a *phylogenetic* lineage and therefore are likely to be *inherited* from the last common ancestor, gene gain and loss in the target organism (or group of organisms) can be related to the specific adaptation to the ecological environment of these organisms.

A potential application of the occurrence profiles is the identification of genes and other genomic properties that can be used to distinguish between different species or strains of the same species of pathogens using a variety of *molecular diagnostics* tools. Occurrence profiles involving functions, pathways, and other genomic data are used in comparative analysis in a

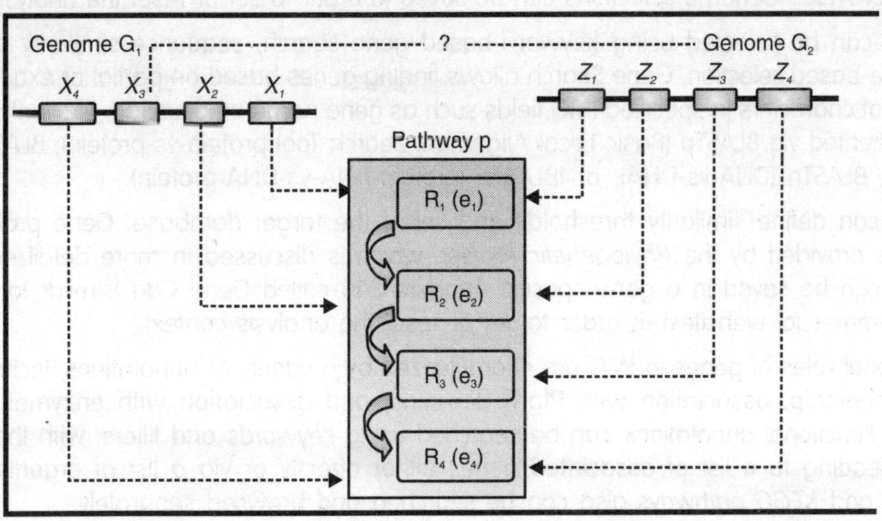

Figure 3.2: Example of functional characterization of genes.

way similar to that previously discussed for genes. For example, occurrence or *abundance* profiles of certain COGs (such as signal transduction *histidine kinase*, *serine/threonine protein kinase*, and *phosphatase*) can provide a broad overview of protein families present or absent in the genomes of interest, whereas occurrence profiles of Pfam domains found in these proteins could provide additional information on the signals sensed by the proteins.

METHOD OF ANALYSIS

Comparative genome data analysis in IMG is set in the context of integrated microbial genomes. IMG allows exploring the microbial genome data space along three key *dimensions*: genomes (*organisms*), functions, and genes. Comparative analysis for genomes is provided in IMG through a number of tools that allow genomes to be compared in terms of organism-specific summaries (*statistics*), genes, and functional *annotations*.

Next, we discuss the occurrence profile analysis tools provided by IMG in more detail. Note that all the examples provided in this section are based on IMG 1.4 (March 2006). IMG's content and user interface are extended on a regular basis, therefore these examples may be different for subsequent versions of IMG.

Profile Analysis

The context for occurrence profile analysis is defined by the set of genomes, genes, and functions of interest selected by the user. By default this context involves all the genomes, genes, and functions in the system. Genome (organism) selections provide the option of focusing the analysis on a subset of *genomes* of interest, such as strains within a specified species.

Genomes can be selected using a keyword-based Genome Search in conjunction with a number of filters, such as such as *phenotype, ecotype, disease relevance*, or *phylum*. Organisms can also be selected from an alphabetical or phylogenetically organized list available in the Organism Browser. Genome selections can be saved in order to set or reset the analysis context.

Genes can be selected using keyword-based gene search, sequence similarity search, or gene profile-based selection. Gene Search allows finding genes based on partial or exact matches to a string of characters in specified IMG fields such as gene name or locus tag. Similarity searches are implemented via BLASTp (Basic Local Alignment Search Tool protein-vs-protein), BLASTx (DNA-vs-protein), BLASTn (DNA-vs-DNA), or tBLASTn (protein-DNA-vs-DNA-protein).

Users can define similarity thresholds and select the target database. Gene profile-based selection is provided by the *Phylogenetic Profiler*, which is discussed in more detail next. Gene selections can be saved in a gene specific Analysis Cart called Gene Cart (similar to shopping carts of commercial websites) in order to set or reset the analysis context.

Functional roles of genes in IMG are characterized by a variety of annotations, including their COG membership, association with Pfam domains, and association with enzymes in KEGG pathways. Functional annotations can be searched using keywords and filters, with the selected functions leading to a list of associated genes either directly or via a list of organisms. *COG categories* and *KEGG pathways* also can be searched and browsed separately.

Function selections can be saved in a function specific Analysis Cart in order to set or reset the analysis context. In summary, the analysis context is defined by the set of genomes, genes, and functions of interest selected by the user, where the set of genomes is maintained using a genome list, whereas genes and functions are maintained using *Analysis Carts.*

Tools and Techniques

As discussed in the previous section, occurrence profiles are specified in a two-dimensional data space, where one dimension represents a set of genes or functions, x_1 to x_n whose profiles are computed in the context of the other dimension, which represents a set of organisms, y_1 to y_m. The occurrence profile for a gene or function of interest, *x,* consists of a vector of the form (L, . . . , L_n) where L_i represents the set of genes of y_i that are either (1) similar to *x* (if *x* is a gene) or (2) genes of y_i that are associated with *x* (if *x* is a function).

Occurrence profile results can be displayed as *two-dimensional* matrices or projected on a phylogenetically organized list of organisms. Next, we present several examples of employing IMG occurrence profiles in data analysis together with alternative visual *presentations* of the profile results.

COG-Based Functional Occurrence Profiles Example

The following example illustrates how functional occurrence profiles are used in comparative genome analysis. In this example, such a profile is used to examine the presence of a specific pathway (i.e., CO_2 fixation) in a set of selected organisms, namely in the archaeal class of *Methanomicrobium archaea.* These organisms can first be selected using IMG's phylogenetic-based Genome Browser as shown in Figure 3.3 and then saved in order to focus the analysis context as previously discussed.

The first step in one of the CO_2 fixation pathways is catalyzed by a CO dehydrogenase/ acetyl-CoA synthase enzyme. A keyword search on expression "CO dehydrogenase/acetyl-CoA synthase" with COG as a filter retrieves a list of five COGs corresponding to different subunits of CO dehydrogenase/acetyl-CoA synthase, as shown in Figure 3.3 III.

After these COGs are saved with the COG Cart, their occurrence profiles across the *methanomicrobia* organisms are displayed in a tabular format as shown in Figure elsewhere in this chapter, with each row *displaying* the profile of a specific COG across the selected *organisms.*

Each cell in the profile result table contains a link to the associated list of genes and displays the count (*abundance*) of genes in this list. Colours are used to represent visually gene abundance, whereby white, bisque and yellow represent gene counts of zero, one to four, and more than four, respectively.

In this example, the occurrence profile result suggests that, with the exception of one organism, *CO dehydrogenase/acetyl-CoA* synthase is present in these organisms, which means that they rely on this pathway for CO_2 fixation.

KEGG-Based Example

The next example illustrates how functional occurrence profiles can be used for comparing phylogenetically related organisms. In the example shown in Figure 3.4, occurrence profiles of the enzymes participating in nitrogen metabolism are analyzed across the organisms that belong to the family of *bradyrhizobiaceae.*

These organisms are first selected using IMG's phylogenetic-based Genome Browser as shown in Figure 3.4 (i) and saved in order to reduce the analysis context as previously discussed. Starting with the KEGG Pathway Browser, enzymes in the Nitrogen Metabolism pathway are selected with the KEGG Pathway Details as shown in Figure 3.4. A set of enzymes, including nitrogenase, different versions of nitrate reductase, and nitrite reductase, is then saved with the Enzyme Cart as shown in Figure 3.4.

The occurrence profiles of these enzymes across the bradyrhizobiaceae family are displayed in a tabular format as shown in Figure elsewhere in this chapter, with each column displaying the profile of a specific enzyme across selected organisms. Each cell in the profile result table contains a link to the associated list of genes and displays the count (abundance) of genes in this list.

Note that the occurrence profile tools in IMG provide two alternative display options (functions vs genomes and genomes vs functions) as illustrated in this and previous examples. In this example, the analysis of occurrence profiles shown in Figure 3.4 suggests that *nitrogen metabolism* may be different across these organisms.

Example of Gene *Occurrence* Profiles

The following example illustrates how gene occurrence profiles can be used to *examine metal* binding in *Shewanella.* First, metal binding-related functions are found with IMG's Function Search using Pfam or InterPro as filters. For example, Pfam 02805 is associated with a list of genes that include *Shewanella* genes that are related to metal binding.

These genes are saved using Gene Cart, as shown in Figure 3.4 (I) In this example, the presence profiles for genes are displayed in the form of vectors where each position in the vector corresponds to an organism, as shown in Figure 3.5 (II): the organisms are *phylogenetically* ordered to facilitate *comparison* of closely related organisms.

Presence of an ortholog of a gene in a given organism is indicated by a domain letter, "B" for *bacteria,* "A" for *archaea,* and "E" for eukarya, whereas the absence of the gene is indicated by a dot ("."). One can mouse over the letter or dot to see the organism name along with its *phylum.* For the example shown in Figure 3.5 III, the occurrence profiles for the *Shewanella* genomes are highlighted.

For a single gene, IMG also provides the *Phylogenetic* Distribution Viewer, which presents the abundance profile for that gene across the phylogenetically organized list of organisms. The *abundance* of the selected gene is indicated by the count of *homologous* genes at each *taxonomic* level as shown in Figure 3.5 IV.

Occurrence Profile

Occurrence profiles can be used for finding objects (e.g., genes, functions) that share a specific presence profile across a set of organisms. IMG's Phylogenetic Profiler is a tool that allows finding genes in a target organism that share the same gene presence profile, where presence or absence of genes is based on (homologous) gene similarity, with cutoffs used to define the similarity relationship.

In the example shown in Figure 3.6 the Phylogenetic Profiler is used to find genes from a

Burkholderia mallei strain that have no *homologs* in a *Burkholderia pseudomallei* strain. Similarity cutoffs can be used to fine-tune the selection. The list of genes with the specified profile is then provided as a *selectable* list as shown in Figure 3.5.

The Phylogenetic Profiler can be used, for example for finding unique, common, or lost genes in the (query) organism of interest compared to a target group of organisms. In the example shown in Figure 3.6 548 genes are found to be unique in *B. mallei ATCC 23344 (B. mallei)* with respect to *B. pseudomallei K96243 (B. pseudomallei). As* we discuss next, such gene profile-based selections provide the context for analyzing *phylogenetically* related genomes and reviewing their gene models.

Interpreting Results

Occurrence profile results involve organisms, functional roles (e.g., Pfam families, COGs, enzymes), and sets of genes, each of which can be further examined. For a set of selected organisms comparative summaries are provided using the Organism Statistics as illustrated in the left panel of Figure 3.7, where summaries for the *B. mallei* and *B. pseudomallei* strains previously mentioned are presented in the context of other related *Burkholderia* strains.

These summaries include the total number of genes and enzymes, and the number of genes with various characteristics, such as genes associated with KEGG pathways, COGs, Pfam, and InterPro domains.

Such summaries can be configured by selecting the properties that are of comparative interest. Individual organisms can be further examined using the Organism Details that includes various statistics of interest, such as the number of genes in the organism that are associated with KEGG, COG, Pfam, InterPro, or enzyme information, as shown in the right panel of Figure 3.7.

For each organism one can also examine the associated list of scaffolds and contigs: for each coordinate range, a Chromosome Viewer allows displaying genes coloured according to COG functional categories. Individual COG pathways or general categories can be examined using the COG Browser, which provides a hierarchical listing of the COG general categories (i.e., amino acid transport and metabolism) and individual pathways (i.e., arginine biosynthesis).

The COG Pathway or Category Details lists the COGs of the selected pathway/category and the number of organisms with genes that belong to these COGs. For a given COG, the "*organism counts*" are linked to a list of organisms and their associated "*gene counts.*" KEGG pathways can be explored in a similar manner using the KEGG Pathway Details. Individual genes can be analyzed using Gene Details, as illustrated in Figure 3.8.

A Gene Information table includes gene identification, locus information, biochemical properties of the product, and associated KEGG pathways. Gene Details also includes evidence for the functional prediction: gene neighbourhood, COG, InterPro, and Pfam, and precomputed lists of homologs, orthologs, and paralogs.

The gene neighbourhood displays the target gene with its neighbouring genes in a 25-kb chromosomal window, as shown in Figure 3.8, where the target gene is pointed out by an arrow.

The Gene Ortholog Neighbourhoods, also shown in Figure 3.8, includes the gene neighbourhood of orthologs of the target gene (pointed out by an arrow) across several organisms: each gene's neighbourhood appears above and below a single line showing the genes reading in one direction on top and those reading in the opposite direction on the bottom; genes with the same colour indicate association with the same COG group.

For each gene, locus tag, scaffold coordinates, and COG group number are provided locally (by placing the cursor over the gene),whereas additional information is available in the Gene Details associated with each gene.

A gene can be also examined in the context of its associated pathways, through links to KEGG maps available on the Gene Information table.

On such a map, the EC numbers are colour-coded and linked to the Gene Details for the associated genes, as illustrated in Figure 3.9, which displays the *Purine Metabolism* KEGG map for the *B. mallei* gene shown in Figure 3.9 (pointed out by an arrow).

Gene Model Validation

The following example illustrates how occurrence profile results can assist in gene model validation. Consider the B. mallei and B. pseudomallei genomes previously mentioned. The result of the Phylogenetic Profiler indicates that, although *B. mallei* is approx 20% smaller than *B. pseudomallei* (*4764 vs 5855* protein coding genes, respectively), it has 548 unique genes.

This high number of unique genes (more than 11.5% of the total number of predicted genes) suggests that a large percentage of the coding capabilities of *B. mallei* is distinct compared to *B. pseudomallei.*

However, examining these genes using IMG's Ortholog Neighbourhoods, as illustrated in Figure 3.10, suggests that most of the differences in gene content between *B. mallei* and *B. pseudomallei* are owing to inconsistencies of the gene models. Detailed analysis of these 548 genes subsequently revealed that:

1. Genes *BMA3300, BMA3308, BMA3320,* and *BMA3324* appear as unique in *B. mallei,* although each of them has an ortholog in *B. pseudomallei;* these *B. mallei* genes seem to be unique because their ortholog in *B. pseudomallei* was not identified as a valid gene.

2. Genes *BMA3286* and *BMA3303* in *B. mallei* and *BPSL0240* in *B. pseudomallei* are functional genes that were erroneously identified as pseudogenes because they supposedly contain authentic frameshifts or stop codons; analysis of their BLAST hits against orthologs in other *Burkholderia* genomes available in IMG shows that they encode full-length proteins with no frameshifts or stop codons and their identification as pseudogenes was based on the alignment to multidomain homologs-fusion proteins.

3. Gene *BMA3290* indicates a gene in *B. mallei,* which is longer than all its homologs and is likely to have an incorrect start codon; indeed, analysis of this region and its comparison to the regions of synteny in other *Burkholderia* genomes shows that the start codon of BMA3290 is incorrect; moreover, a gene in a different frame was missed as a result of erroneous prediction of the gene start.

Although Phylogenetic Profiler shows that *B. mallei* and *B. pseudomallei* have 10 different genes in this region, in fact there is only a 2-gene difference: a transposase in *B. mallei,* which is absent from *B. pseudomallei* and an ortholog of BPSL0240, which is a pseudogene in *B. mallei.*

Thus, the comparative analysis of the genes in *B. mallei* and *B. pseudomallei* indicates an up to 90 % error rate (either false-positive genes in one genome or falsenegatives in the other genome) in the results because of the difference in gene prediction algorithms used to identify coding sequences in these two genomes.

CONCLUDING REMARKS

Effective microbial genome data analysis across biological data management systems involves providing support for comparative analysis in an integrated data context. We presented the comparative analysis *capabilities* provided by the IMG system, in particular those that are based on occurrence profiles.

The comparative analysis capabilities in IMG are based on techniques that follow observed biological evolutionary phenomena regarding functional coupling of genes. Some IMG tools have similarities to analogous tools in microbial genome data analysis systems such as WIT, ERGO, MBGD, SEED, Microbes Online, and PUMA2. However, IMG has also a number of unique comparative analysis *capabilities*.

Thus, instead of restricting users to a predefined collection of metabolic pathways compiled from the literature and mostly comprising model organisms, IMG provides users with the opportunity to define their own pathways and functional categories by employing Analysis Carts regardless of existing annotations. Such user-defined pathways can be further analyzed using a variety of tools, such as COG, Enzyme, and Pfam Profiles, and the *Phylogenetic* Profiler.

These tools were specifically developed in order to enable the analysis of genomes that are poorly characterized, are phylogenetically distant from model organisms, and cannot be analyzed efficiently using traditional pathway databases. The first version of IMG was released in March 2005, followed by quarterly releases consisting of data content updates and analytical tool extensions.

A data warehouse framework was used in developing IMG, and was found to provide an effective environment for developing a system that needs to support the integration and management of data from diverse sources, where data are *inherently imprecise* and tend to evolve over time. The data warehouse environment has provided an established framework for modelling and reasoning about genomic data.

IMG data content extensions have focused on data quality in terms of the coherence of annotations, based on sound validation and correction procedures, as well as *corroboration* of annotations from other public microbial genome data resources. IMG's occurrence profile tools have proved to be effective in the *detection* and subsequent correction of annotation errors.

We plan to further enhance the occurrence profile tools in IMG. First, we plan to extend the occurrence profile based selection to include additional biological objects, such as gene clusters (e.g., COGs), enzymes, and chromosomal gene clusters. Note that unlike the profile-based selection of genes, no target organism needs to be selected for functional features such as COGs and enzymes that are common to all organisms.

To support the selection of chromosomal gene clusters, we plan to extend the content of IMG by precomputing these clusters. Second, we plan to develop improved occurrence profile viewers in order to increase their usability. For example, we are considering presenting occurrence profile results in a hierarchical (tree) phylogenetic context, which would enhance these tools' ability to support examining biological phenomena of interest, such as gene loss and lateral gene transfer. The existing phylogenetic distribution viewer lays out the taxonomy of each organism in a text-based format, which has expressivity limitations.

A more intuitive, and therefore more effective, way to represent this type of information in a phylogenetic context could be based on the 16S ribosomal RNA tree. IMG will continue to be extended through quarterly updates, whereby it aims at continuously increasing the number of genomes integrated in the system from public resources and JGI, following the principle that the value of genome analysis increases with the number of genomes available as a context for comparative analysis. IMG will also continue to address the needs of the scientific community for *comprehensive* data content and powerful, yet *intuitive*, *comparative* analysis tools.

Chapter 4

Visualizing Microbial Genome

Several circular genome visualization tools have been developed, and offer a wide variety of features. The Microbial Genome Viewer is one such online tool. Users can choose from several genomes and create plots within the web browser. It also offers a data upload facility to plot *experimental* data. However, the plot customization is tedious, and if a mistake is made it is not possible to undo and repeat without destroying the entire plot.

Search functionality is limited and the plot is not interactive enough. Genomap provides the functionality to create *circular maps* and offers a large number of *customizable* features, but little help in creating plots quickly and easily. Also, the plot *interactivity* is limited. BugView also allows some comparative analysis, but is limited to only two genomes. Though the abilities in linear comparison are useful, the circular plots are static.

Genome-Plot provides a user-friendly tab-delimited file format for easy modification by users, but the plot must be customized for each genome, and once again, no interaction is possible with the resulting plot. CGView offers much functionality, which makes it easy to create and customize the plots and provides excellent hyperlinked circular plots. But the search ability is limited, and no markup is possible on the plot after It has been created.

GenomeViz offers several advantages to the user. It uses a simple tabdelimited file format that can be readily modified by the user. It provides users with several premade files ready for beginning plotting immediately. Features like "*tagging*" provide the user with complete control over the colours of each gene. It also offers several different plotting methods for *numerical* data.

Moreover, the plot is interactive (albeit with limited zooming ability), and it is easy to locate genes in one or all the genomes plotted, and extract data from either selected regions or parts of the plot. Creating the plot is itself an interactive process providing the user with complete control over the plot appearance. The resulting figuresare publication quality.

Some scripts are also provided to make the common tasks simpler for the user. There are two types of information that a visualization program must be capable of displaying; qualitative and quantitative. It is important to be able to visualize both qualitative and quantitative data from microbial genomes. Functional classifications (like Clusters of Orthologous Groups [COGs]) and identification of *horizontally* transferred genes are examples of qualitative data.

They allow us to classify genes into different groups. Thus, it is informative to compare, for example, the distribution of potentially horizontally trans-ferred genes between two related microbial genomes. Such a comparison can provide us with clues to regions that are more prone to insertion and deletion events in the coevolution of these two genomes.

Quantitative data is simple numerical data, e.g., gene length, GC skew, GC content, conservation scores, gene expression intensity values, and so on. Quantitative data may be of two types, gene-based (gene length, expression values) or window-based (GC skew, conservation scores).

Gene-based quantitative data refers to a data where each gene is associated with a single value, e.g., gene length or fold change at one time point in a microarray experiment. Window-based quantitative data refers to values calculated for short, overlapping segments of the genome. GC content and GC skew for a genome are usually calculated in this manner.

CONCEPT OF TAGS

GenomeViz uses the concept of "tags," which may be applied to groups of genes for classification-type qualitative data. A tag is just a name given to a group of genes. It may be a short word, or a letter of the alphabet (e.g., "U" for genes with unknown function, or "CON" for genes conserved across a comparison of a few genomes). The genes of a genome may be divided into different groups and each group given its own "tag."

Tagging provides the user with the ability to change colours for entire groups easily and gain more control over the GenomeViz display All the information on the groups and tags to be used in a particular plot must be written in a tag file. A tag file is a tab-delimited text file of at least two and at most three columns. It has the "tags" in the first column, their colours in the second, and their brief descriptions in the third column.

A small two-column tag file is shown next.

```
Transcription     RED
Translation       GREEN
OtherGenes        GREY
```

The first column is the tag column. In this example, it means that we have three types of groups (and so three tags) for the genes, "*Transcription*," "*Translation*," and "Other Genes." The second column simply states the colour that should be used for colouring each group.

To change the colour of the genes involved in "Translation," simply change the text GREEN in the second column to say, BLUE. When the plot is reloaded, the new colours will be displayed.

However, a tag file may also have three columns, as shown next.

```
T    orange    transcription
R    blue      translation
M    green     cell motility
```

```
S    violet   signal transduction
-    grey     function unknown
```

The third column can be used to describe the tag if we wish. Its purpose is to provide a more informative description. It is recommended that numbers (0, 1, 2, 3 ...) not be used as tags. The character "–" can also be used as a tag. All these columns must be separated with a "single" tab character only.

When one has a large number of tags, then it is useful to have a short description of the tag. The tag file can be displayed within GenomeViz to read the *descriptions* anytime. A tag file with all the COG *categories* is provided with GenomeViz.

CONSTRUCTION OF GENOMIC MAP

The file that contains the actual data to be plotted is called the map file. This has been designed to be a simple format that can be easily edited and modified by anyone manually or with a program.

A sample map file is shown next (first few lines from the genome of the *hyperthermophilic archaeon Aeropyrum pernix* genome).

```
1669695
APE0001   -   -   213    938    hypothetical protein
APE0002   K   -   938    1276   hypothetical protein
APE0004   R   -   1260   2174   hypothetical protein
APE0006   -   +   2261   2836   hypothetical protein
APE0007   -   +   3896   5440   hypothetical protein
APE0009   P   -   5774   6091   transport protein
```

The first line of the map file contains only a single column, and a single value: the total number of bases in the genome, in this case, 1,669,695. All other lines of the map file contain six tab-delimited columns. The six columns are described next.

1. A gene identifier or a name. National Center for Biotechnology Information (NCBI) frequently uses a "Locus Tag" feature to describe bacterial gene identifiers. For example, APE0001 is the locus tag for the first gene in the *A. pernix* genome. The locus tag for each gene can be seen in the NCBI Gene database.

   There are some limitations to this identifier. First, it must be only a single word. Second, it must not be entirely a number, e.g., 1, 10, 124, are all invalid gene identifiers. Third, it must be unique for the genome the user is trying to plot. All *identifiers* for the genomes provided with GenomeViz follow these three basic rules.

2. The tag/value column. The second column contains the tag that has been applied to this gene to make it a part of a group of genes. In the example previously listed, four types of tags are visible, "K", "R", "P", and "–". The colours for these tags (and for others in the map file) must be described in the tag file.

The second column contains tags in this example because this is an example of a qualitative data file. A map file, which contained the gene lengths for example, would have, in place of the tags, integer values for each gene.

3. The strand column. This column simply denotes the strand on which the gene lies. There can be only two values for this column, "+" or "-". No other values are acceptable.
4. Gene start. This column contains the location of the start of the gene feature.
5. Gene end. This column contains the location of the end of the gene feature. Both the gene start and gene end must be valid integer values.
6. Description. The last column of the map file. It contains the description, annotation, name of the gene, and any other text information.

The only difference between a qualitative data map file and a quantitative data map file is the values in the second column. All other columns are *identical* for the same genome. If there is any line in the map file that does not have six columns in the correct format, GenomeViz will show an error, point out the incorrect line number and the column, and stop the plotting.

In such as case, one must identify the error, correct it, and redo the plot again. The map file format is easy to maintain and modify in simple text editors or spreadsheets, and the extensive format checking performed by GenomeViz before plotting helps identify and correct mistakes before they are incorporated in the plot.

The map file alone is sufficient for plotting numerical data, but both the map and tag files are needed to plot classification-type data. The type of data, qualitative or quantitative, is automatically detected from the map file.

PREPARATION OF GENOME CIRCLE

Plotting Genomic Data

It is possible to plot data in several ways with GenomeViz. Given next is a list of methods available for plotting.

1. Plotting classification style data (qualitative).
   a. Two circles (+ and – strand separately).
   b. Single circle (both + and – strands as a single circle).
2. Plotting numerical data (quantitative).
   a. Gradient style graph with two circles (+ and – strand separately).
   b. Gradient style graph with single circle (+ and – strands as a single circle).
   c: One-sided line graph (like a circular bar chart, useful for alignment data).
   d. Two-sided line graph (useful for GC content and GC skew).

Plotting Classification-Style Data

Both the tag and map files are needed to create a classification-style plot in GenomeViz. Follow the following steps to create a classification style data plot in GenomeViz.

Loading a TAG File

1. Go to File in the Main menu.
2. Select Load Tag File, and choose for which genome to be loaded a TAG file for (Genome 1, 2, 3,... 8). Choose "Genome l."
3. Browse to the location of a tag file (say the TAG file supplied with GenomeViz-tagfiles/ COGs.tag).
4. Click Open. The tag file COGs.tag is now loaded and this is displayed in a small frame below the main menu. The loaded tags are also shown in the text display area. Now follow the steps next to load a map file and create the plot.

Loading a MAP file

1. Go to File in the Main menu.
2. Select "Load Map File l."
3. Choose "Draw Two Circles."
4. Choose "Classification Style Graph."
5. Browse to the location of a map file (e.g., the map file supplied with Genome Viz for *Escherichia coli* K12 in the samples/classification-data directory - Escherichia coli K12.map).
6. Click Open.
7. The genome of *E. coli* K12 will be displayed (two circles for two strands) coloured in the COG colours (as specified in the tag file) as Genome 1.

Plotting Numerical Data

No tag files are needed for plotting numerical data. Only a map file containing quantitative data is sufficient. Follow the following steps to load a map file containing numerical data to create a plot.

1. Go to File in the Main menu..
2. Select "Load Map Filc 1."
3. Choose "Draw Two Circles."
4. Choose "One Sided Line Graph."
5. Choose "Blue."
6. Browse to the location of a map file (e.g., the gc-content map file supplied with GenomeViz for *E. coli* K12 in the gc-content-mapfiles directory).

The GC content of the *E. coli* K12 genome in the map file will be displayed as a one-sided line plot coloured in blue.

GRAPHICAL PRESENTATION OF NAVIGATION

Using Mouse Over

In all plots, Mouse Over on any gene immediately displays all the information about the gene in the display areas just below the Main menu. The line number in the map file, the gene identifier, the tag/value, strand, gene start, gene end, and description all are displayed.

Marking Genes

Clicking on any gene in the plot highlights it in a colour called the "Selection Colour." The default Selection Colour is yellow. The information on a selected gene is also displayed in a text display area on the right side of the drawing area. Right clicking on a gene deselects it.

Marking COGs

One can select COG categories directly for each genome using this menu provided they are available in the map file. Thus, Select COGs → Select COGs in Genome 3 → K-transcription, selects all genes classified in the category Transcription in the Map file for Genome 3. It is possible to select different categories in the same genome in different colours by simply changing the selection colour before selecting the category.

However, it is advisable to use a neutral background tag file, e.g., COGsGrayScale.tag, to provide a better contrast for the categories of.the user's choice. This tag file colours all COG categories in a neutral gray colour. The user may also edit this tag file to reflect any other colour as well.

Finding Desirable Genes

The complete information in the map file can be searched using the Search option. All genomes may be queried independently of one another. Go to Search → Search Genome 1 (to search in the first genome). A Search window appears. Type in the term to search, and press "search". After the search is completed, a pop-up window appears and lists how many results were found.

These results can be examined in the text display area on the right hand side of the drawing area. The search results may also be saved to a text file. In addition, all the genes that matched the search pattern are *highlighted* in the GenomeViz plot in the *"Search Colour."* Several different searches (each with a different search colour) can be run on the same genome or the plot.

In this manner, the search and highlight functionality provides one with a rapid overview of distribution of search terms over the genome. A global search function is also available, i.e., all the plotted genomes may be searched at once for a single pattern. The results are displayed genome-wise in the text display area.

Removing Unwanted Genes

If there has been an error in plotting a genome circle, this particular circle can be easily removed without affecting the rest of the plot. Navigate to Clear→Genome 1, to remove the outermost circle. Choose File → Clear All, to reset the entire plot.

Plot Summary

To have a quick overview of which files have been used to create each genome circle, one can go to Summary-Plot Details to have look at a table containing the names of all the tag files and map files being used for each genome circle in the plot.

Print Order

It is possible to create publication quality plots with GenomeViz. Once the user is satisfied with the plot created and wants to finally print it, the user can go to File → Print. A print dialog box appears with several options. Give the dialog box time to complete its rendering of the print preview plot in the small window.

Choose the paper size and choose "Print to file" option. Provide a name for an output file, e.g., myplot.ps. GenomeViz creates postscript output files that can be easily read in by standard graphics programs, and converted to a PDF if desired.

Invisible Menu

The search function provides highlighting genes based on a pattern match, and the tag file allows genes to be coloured based on the group in which it belongs. To colour genes on a numerical data plot, that do not share any common search pattern, it is not possible to colour them using these options. However, individual genes of interest in both the classification-style plots and the *quantitative* data plots are searchable and can be coloured by using the special mask genome menu.

It is somewhat like a multiple search option, but with the facility of colouring each result in a specific colour. It has a simple format, a two column tab-delimited format, as shown next. The first column is the gene of interest, and the second column specifies the colour it should be displayed in.

```
Gene1  red
Gene2  blue
Gene3  yellow
```

The Gene1 will be red, Gene2 will be blue, and Gene3 will be yellow. No format checking is performed on the mask file. It must be ensured by the user that the format is correct, all gene identifiers used are present in the map file, and that the colours are valid Tk colours.

BRING INTO FORCE

Platforms

GenomeViz has been tested to run successfully on Linux and Solaris operating systems. Unix systems that have ActiveTcl installed may also run GenomeViz but we have not tested this.

Active*Tcl*

It is required that the user install ActiveTcl distributed by ActiveState to run GenomeViz. It is recommended over any other existing Tcl installation that the user might have to run GenomeViz. Installing ActiveTcl will not interfere with the user existing Tcl installation and will have no effects on the user's Tcl programs, if the user has any.

Perl

The user will also need Perl to run the scripts that are distributed with GenomeViz. Perl is usually installed by default on Linux/Unix systems in the path/usr/bin/perl. The user can easily check this by typing the following command on the terminal.

```
which perl
The user may get
/usr/bin/perl
```

which means the user already has Perl installed, or the user may get something like

```
perl not found
```

which means the user does not have Perl and will need to install it.

If the user does need to install Perl, once again it is recommended that the user gets the ActivePerl *distribution* from ActiveState. It is easy to install and should not pose any difficulty.

POINTS TO REMEMBER

1. If the user can program in Perl, it is easy to modify the scripts provided with GenomeViz to create new programs that can compute parameters using a window-based approach, e.g., dinucleotide content, complexity, and so on.
2. A common application involves a list of genes that one would like to plot and visualize along with other data. The script *"tagit"* makes it simple for to create a file that can be plotted and visualized easily with GenomeViz. Suppose the user has a list of genes that the user is interested in. The user should provide the script with the file containing this gene list and the tag the user wants to attach to these genes.

   The user should also provide the map file to be used (currently GenomeViz provides 120 map files to choose from). The script creates a new map file, but with all the genes tagged with the designated custom tag the user provides.
3. GenomeViz and accompanying scripts and data can be download at the GenomeViz homepage (http://www.uniklinikum-giessen.de/genome/).
4. Once a plot has been made, it should be saved to a postscript file. However, when the plot needs to be recreated, one needs to use the same input files once again. Use the Summary-Plot details to save the details of the files used to create the user's plot in such a case.
5. There are many different ways to specify the colours in the tag file. The colours in a tag file may be written by their name, e.g., Red, red, or RED are all acceptable. Hexadecimal codes are also allowed. Two colour browsers are provided within GenomeViz that can help to select colours and obtain their standard names or hexadecimal codes.
6. The search box supports advanced pattern matching abilities provided by the Tcl/Tk regexp. For example, if the user wants to search for genes containing the pattern tRNA or rRNA, the user can type tRNAIrRNA, where the "I" character denotes OR. A link to a complete guide for regular expression pattern matching using Tcl can be found at the GenomeViz homepage.
7. Whole genome alignments provide us information on which regions of the genomes have been conserved and which have been subject to deletion and insertions. It is easy to get

complete genome alignments of bacterial genomes using AVID. Avid also provides a simple format for the representation of such alignments.

The script *avid2viz* can reformat genome alignment data from the AVID program to a map file format that can be plotted in GenomeViz. This map file can be used to visualize conservation data of genomes along with other data such as GC content, Basic Local Alignment Search Tool scores, and so on in GenomeViz.

8. Use the Perl programs gc2viz and gcskew2viz to compute window-based mapfiles for plotting in GenomeViz. They use only the nucleotide fasta file as input and create a mapfile that can be plotted in GenomeViz. The GC content map files supplied with GenomeViz contain only the GC content values of the actual genes themselves.

   The user can download whole genome nucleotide files for any sequenced bacteria genome from NCBI (http://www.ncbi.nlm.nih.gov/ genomes/lproks.cgi). The nucleotide sequence files on the NCBI server have the ".fna" file extension.

Chapter 5

Prokaryotic Fingerprinting

For accurate assembly of a bacterial clone reap (or contig), it is necessary to precisely detect the degree of overlap and map order of clones. One way is to generate *fingerprints* for all the clones in a given project. The overlap between clones is determined by the proportion of *comigrating* restriction fragments on either acrylamide or agarose gels, with *mulitfold* coverage of clones allowing rapid identification of *artifacts* such as rearranged or deleted clones.

There are two fingerprinting methods described here double-digest end labeling. using either a radioactive isotope, or, as described here, *fluorescent molecules* where fragments are resolved on a denaturing acrylamide gel and single-restriction. digestion-using agarose gels. It should be noted that the data produced by the two methods are *incompatible*.

The software described here for data entry and analysis are Image v3.9 and FPC v4.5. For fluorescently fingerprinted clones, the Genescan software package (Perkin-Elmer ABI, Foster City. CA) can also be used for *entering* gels.

Some of the main differences between the two techniques are summarized in Table 5.1. With fluorescent fingerprinting, an increased sample to gel ratio, and decrease in gel variation affects can be achieved by *multiplexing* finger-printed clones and a standard marker (used to assign *normalized* values to bands) in each well. Along with the *shorter run* time, this can increase the speed of data generation when using fluorescent fingerprinting.

However, once the raw data are produced it is generally quicker to enter and analyze the restriction digest agarose fingerprints. Band calling in Image is more *automated*, as the *absorption* of label and emission of signal are directly proportional to the size of the fragment (larger fragments absorb more label), a predictive decay of signal strength is seen facilitating *automatic* band calling.

Double bands can be scored due to the additive effect of the signal. Fluorescently fingerprinted clones exhibit random signal decay as dye incorporation is determined by sequence or *conformation* factors at the ends of fragments. The data produced by restriction digest agarose fingerprinting are more amenable to the automated algorithm routines used in Fingerprinting

Contigs (FPC). The size information attached to the restriction digest agarose fingerprints facilitates more accurate positioning of clones within contigs, especially when picking sequence clones.

The Genome Sequencing Center (GSC) at Washington University School of Medicine (St. Louis MO) have undertaken to fingerprint the human male RPCI- I I BAC library (6), using the *HindIII* restriction digest agarose gel method. If incorporation of data with the GSC is required, it is necessary to fingerprint novel clones using this method.

The *HindIII* restriction digest method described here matches closely the GSC protocol (Marco Marra and Tammy Kucaba, personal communication, GSC, Washington University School of Medicine) especially with respect to digestion conditions and marker used and produces compatible data.At the Sanger Centre, Carol Soderlund has written a script called get __GS'C, which will extract chromosome specific clones from the Genorne Sequencing Centre's database of fingerprints. These can then be imported into FPC where they are binned and ordered.

Method of Southern Blotting

The aim of Southern blotting is to transfer DNA from one medium, the gel, to a stable medium which is suitable for subsequent hybridizations. DNA transfer is achieved by the upward movement

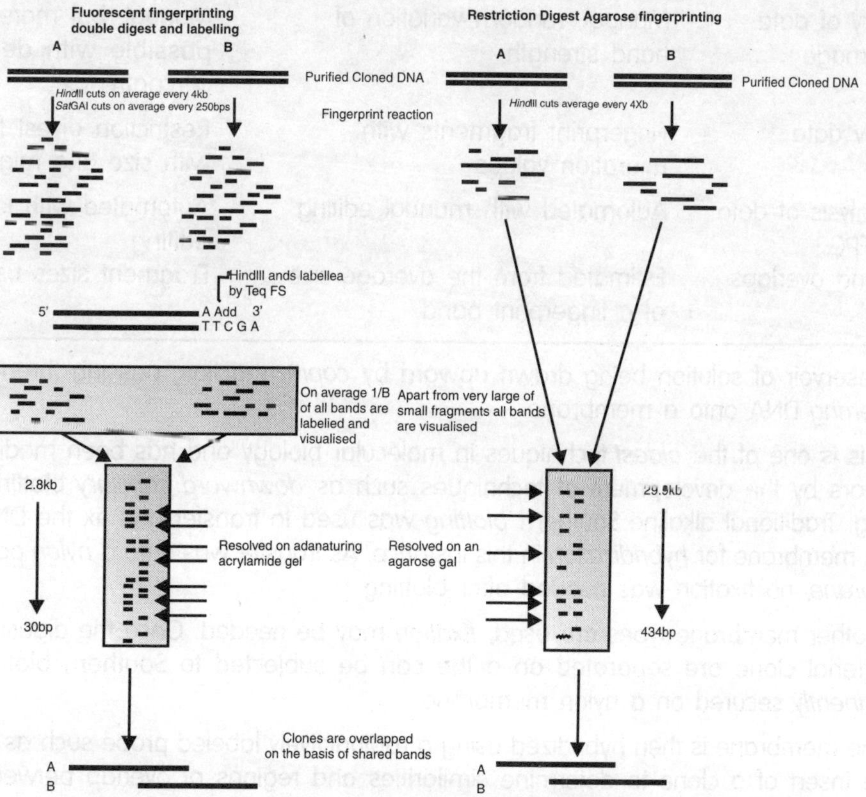

Figure.5.1. A schematic of the two types of fingerprinting described here. Although the digestion, labeling, and fragment separation techniques differ significantly for both methods, and the data produced are incompatable. the same results will usually be seen after analysis of clones.

Table 5.1: Comparison of Fluorescent End Labeling and Restriction Digest Fingerprinting.

	Fluorescent fingerprinting	*Restriction digest agarose fingerprinting*
Digestion	*HindIII* and San3AI I h then 1.5 It precipitation	HindIII 2 h
Labeling used	Incorporation of fluorescent dideoxy adenosine triphosphate molecule at 1-findIII ends	DNA stained in gel using an intercalatary dye (e.g., Vistra Green)
Resolution of fingerprints	Acrylamide gels 4.5 h run tune, mulitplexed clones and standard marker. increases throughput	Agarose gels 16 h runtime. no multiplexing, post run stain 1.5 h and accuracy
Entry of data in image	Manual, random variation of hand strength	Manual, but more automation possible with decaying band >trcnath
Raw data	Fingerprint fragments with migration values	Restriction divest fragments with size and migration values
Analysis of data in FPC	Automated with manual editing	Automated with less manual editing
Sizing overlaps	Estimated from the average size of a fingerprint band	Fragment sizes used

of a reservoir of solution being drawn upward by *capillary* action, passing through the gel and *transferring* DNA onto a membrane.

This is one of the *oldest* techniques in molecular biology and has been modified throughout the years by the development of techniques such as *downward capillary* blotting and vacuum blotting. Traditional alkaline Southern *blotting* was used to transfer and fix the DNA froth the gel onto a membrane for *hybridization* in this instance. As transfer was onto a *nylon* positively charged *membrane,* no fixation was needed after blotting.

If other membrane types are used, *fixation* may be needed. Once the digestion products Of a bacterial clone are separated on a the can be subjected to Southern blotting, and *semi-permanently* secured on a nylon membrane.

The membrane is then hybridized using a radioactively labeled probe such as an end product or the insert of a clone to determine similarities and regions of overlap between clones. Both small regions of overlap (5 kb) and large regions of overlap can be determined by this method. In this chapter we describe a hybridization protocol using a DNA polymerase based strand labeling, with a *Rediprime* labeling kit.

This results in uniform labeling throughout the probe and high specific activity. Here, random oligo-nucleotide labeling was used. whereby all possible 9-mer *nucleotides* are present in a reaction mixture *comprising* of a buffered solution of dATP. dGTP. dTTP, and exonuclease-free Klenow enzyme.

These random primers bind to the DNA, and the Klenow fragment of DNA polymerase I adds both labeled and unlabeled dNTPs. Uniform labeling of the probe is achieved throughout the probe due to all *combinations* of nonamers being present, and the random binding of these to the target DNA.

In this experiment, [α-^{32}p] dCTP was used, resulting in the newly synthesized probe being *radioactively* labeled throughout (9,10). Competition DNA was added to the probe as placental DNA to block out any repetitive signals that could be falsely identified as positive signals. Relatively low *stringency* was used in these experiments to allow overlapping and similar clones to be easily identified. This was achieved by using a standard 65°C temperature for *hybridization* and *posthybridization* washes, and relatively high *concentrations* of NaCl in the posthybridization washes.

BASIC REQUIREMENT

Isolation of DNA for Fingerprinting

1. 2X TY (per liter): Add 16 g Bacto-Tryptone. 10 g yeast extract, and 5 g NaCl. Dissolve in dH_2O, adjust pH to 7.0 with 5 *N* NaOH, adjust to final volume with dH_2O.
2. Combitip repeat dispenser (Eppendorf. The Netherlands).
3. 1 mL sterile deep-well microtiter plates (Beckman Instruments Inc., Fullerton, CA).
4. Deep-well microtiter plate caps (Beckman Instruments, Fullerton, CA).
5. U-bottom sterile microtiter plates and lids (Greiner Labortechnik).
6. 0.2 µm Filter bottom plates (Millipore, cat. no. MAGVN2250).
7. Multichannel pipels 5 50 µL and 50-300 [AL and tips (Lab Systems, Finnpipet, Life Sciences International).
8. Sterile *cocktail* sticks or a 96-pin replicating tool (Denley, Labsystems Instruments).
9. RNase A (Ribonuclease A, Sigma Chemical, St. Louis, MO).
10. Orbital shaker (New Brunswick Scientific G24 environmental incubator, Edison, NJ).
11. Vortex genie fitted with a microtiter plate hold (Scientific Industries, Bohemia, NY).
12. Centrifuge with microtiter plate holding rotor (Sorvall RT7, du Pont Co., Sorvall, DE).
13. Class 11 microbiological safety cabinet (Walker Safety Cabinets Limited, UK).
14. Appropriate selective agents, e.g., chloramphenicol. kanamycin or ampicillin (Sigma Chemical).
15. Solution I (GTE): 4.504 g glucose (BDH Laboratory Supplies, Poole. UK), 10 nit, 0.5 14 EDTA, 12.5 mL 1 *M* Tris-HC1, pH 8.0. Make up to 500 mL with double-distilled H_2O and filter

sterilize and store at 4°C.

16. Solution II (make fresh each time): 8.6 mL double-distilled H2O, 1 mL 10% SDS. 400 μL *5M* sodium hydroxide (BDH Laboratory Supplies).
17. Solution III: 3 M potassium acetate. pH 5.5. Store at room temperature.
18. Ethanol (70% with water) and isopropanol (BDH Laboratory Supplies).
19. Plate sealers (Costar, Corning. Corning, NY).

DNA Hybridization

1. LB broth: Yeast extract 5 g, tryptone 10 g. NaCl 10 g, made up to 1 L with deionized distilled water and autoclaved.
2. 20 μL, 200 μL, and 1000 μl. pipets with tips (Gilson Medical Electronics, France).
3. 50 mL sterile falcon tubes (Greiner Labortechnik).
4. Sterile cocktail sticks.
5. Appropriate selective agent, normally kanamycin for PACs (Sigma Chemical).
6. Orbital shaker (Lh fermentation).
7. Centrifuge (Beckmann J2-21 and J-6B, Beckman Instruments).
8. Solution 1: 15 mM Tris-HCl. pH 8, 10 mM *ethylenediaminetenraacetic* acid (EDTA), filter *sterilized* and stored at 4°C.
9. Solution II (make fresh each time): 0.2 M NaOH, 1 % (w/v) sodium dodecyl sulfate (SDS). Stored at room temperature.
10. Solution III: 3 M potassium acetate, pH 5.5 (chill before use).
11. Glass Corex tubes (du Pont Instruments).
12. Parafilm (American National Can).
13. Isopropanol and 70% v/v ethanol (BDH Laboratory Supplies).
14. Deionized distilled water containing 180 μg/mL RNasc A (Sigma Chemical).
15. Phenol chloroform isoamyl alcohol 25:24:1 (Sigma Chemical).
16. 20. 200, and 1000 μL pipets and appropriate tips (Gilson).
17. 3 M Sodium acetate, pH 5.3.
18. 100% and 70% v/v ethanol (BDH).
19. 1.5 ml- Eppendorfs (Eppendorf, The Netherlands).
20. -70 C freezer (Forma Scientific, Canada).
21. Microcentrifuge (MSE, Sanyo. UK).
22. Deionized distilled water.
23. Vortex (Genetic Research Instrumentation, Essex, UK).

Fluorescent Fingerprinting

1. AB1377 Sequencer (Perkin-Elmer Applied Biosysters, Foster City, CA).
2. Lambda DNA (500 ng/μL, New England Biolabs).
3. Ethanol (96% and 70% with water [BDH]).
4. 0.3 *M* sodium acetate.
5. Matrix standards for ABI; NED matrix standard (cat. no. 402996 Perkin-Elmer Applied Biosystems) and Fluorescence Amidite matrix standard kit, with tetracycline (TET), HEX, and ROX *dichlororhodomide*-based dyes, (cat. no. 401546 Perkin-Elmer Applied Biosystems).
6. Fluorescently tagged dideoxy adenosine triphosphates. e.g., ddA-TET, -HEX and -NED Taq FS (cat. no. 4306379C, Custom Fingerprinting Kit, 1 mL each of at 10 1 *μM* NED, TET, and HEX, 3 mL Taq FS. Perkin-Elmer Applied Biosystems).
7. Fluorescently tagged dideoxy cytosine triphosphate, e.g.. ddC-ROX 5.08 μM (cat. no. 402118, Dye Terminator Cycle Sequencing Core Kit, Perkin-Elmer Applied Biosystems).
8. 37°C incubator (Economy Incubator Size 1, Gallenkamp).
9. 80°C oven (MIND/18/CLAD, Philip Morris Ltd., UK).
10. Centrifuge with microtiter plate holding rotor (Sorvall RT7, du Pont Sorvall).
11. Benchtop centrifuge (Eppendorf 54150, The Netherlands).
12. Blue dextran formamide dye: 9.8 ml., deionized formamide, 200 lit, 0.5 M EDTA, 0.01 g blue dextran dye.
13. 66-Well square tooth combs 0.2 mm (cat. no. 402183, Perkin-Elmer Applied Biosystems).
14. 10X (Tris-borate-EDTA) TBE buffer (cat. no. EC860. National Diagnostics. Atlanta, GA).
15. HindIll (20 U/μL, New England Biolabs), NEB2 buffer (New England Biolabs), Sau3AI (50 U/μL, Amersham Life Sciences), BsaJI (2.5 U/μL. New England Biolabs).
16. Hamilton repeat dispenser (Hamilton. Reno, Nevada).
17. Plate sealers (Costar, Corning, Corning, NY).

Agarose Fingerprinting

1. 1 % agarose (SeaKem LE. FMC Bioproducts, Roekland. ME) gel, made with 1 X Tris-acetate-EDTA (TAE).
2. Gel tanks, Gator Wide Format System model A3-I (Owl Scientific).
3. 121-Well coin b (Marco A. Marra, personal communication, Washington University School of Medicine).
4. *HindIII* (40 U/ μL, Roche Molecular Biochemicals, Switzerland, formally Bochringer Mannheim. Germany).
5. Loading dye for marker (6X Buffer II for 10 rill,. 1.5 g Ficoll. 0.025 g *bromophenol blue.* 0.025 g xylene cyanol. 10 mL sterile water)- Loading dye for sample (for 10 mL: 0.1 g Orange G, L5 g Ficoll, 10 rill- sterile water).

6. Marker mix: Analytical Marker DNA, wide band (Promega); DNA Molecular Weight Marker V (Roche Molecular Biochemicals, Switzerland. formally. Boehringer Mannheim, Germany).
7. Plate Sealers (Costar, Corning).
8. Bench top centrifuge (Eppendorf 5415C).
9. Combitip repeat dispenser (Eppendorf).
10. Hamilton repeat dispenser (Hamilton).
11. Cold room regulated to 4°C. (or recirculation system can be used).
12. Orbital platform shaker, Belly Dancer (Storyall Life Sciences, NC).
13. Vistra Green intercalating stain (Amersham Life Sciences).
14. FluorImager SI Vistra Fluorescence (Molecular Dynamics. Sunnyyale, CA).

Restriction Digests of Clones

1. EcoRI (10 U/μL), *NaI* (10 U/μL), *SaI* (10U/L) and buffer D (all Promega).
2. Deionized distilled water.
3. 20, 200, and 1000 μL pipets and suitable tips (Gilson).
4. 0.5 mL Eppendorfs (Eppendorf).
5. 37°C incubator (Digi-Block, Laboratory Devices).

Electrophoretic Method

Electrophoresis for EcoR/ Not/ and Sal/ Digest

1. 0.8%l, agarose (0.5% agarose for check gels). (Amresco. OH), made with 1 X TAE (40 *mat* Tris Acetate. 2 mM Na_2 EDTA.2 H_2O).
2. Ethidium bromide 10 mg/rill. (Sigma Chemical) added gel to give a final concentration of 0. 1 μg/μL.
3. Gel tank, casting tray 27 x 20 cm for overnight electrophoresis and 24-well comb (Hybaid. Hampshire).
4. One kilobase DNA ladder (Gibco-BRL Life Technologies, Paisley, UK).
5. Loading dye (15% w/v Ficoll and Orange G dye).

Gel Electrophoresis

1. 140 × 120 cm casting tray (Bio-Rad, Hercules, CA).
2. 100 mL 1% low inciting point (LMP) gel (Gibco-BRL Life Technologies) made with 0.5X TBE (89 mM Tris base, 89 mat boric acid, 1 mM EDTA).
3. Cold room at 4°C.
4. PFGE equipment (Bio-Rad, CHEF DRII, CA).
5. 24-Well comb (Hybaid).
6. Low-range PFGE marker (New England Biolabs).

7. Lambda *HindIII* (Promega).
8. Loading dye.

Southern Blotting

1. 3MM blotting paper (Whatmann. BDH).
2. 0.4 *M* NaOH.
3. 0.25 *M* HCl.
4. Fluorescent ruler (Promega).
5. Blotting, towels (Kimberly Clark, Kleenex).
6. Hybond N+ nylon membrane (Amersham Life Sciences).
7. Saran Wrap (Dow).

Hybridization

1. Hybridization glass bottles (Hyhaid).
2. Hybridization oven set at 65°C (Hybaid).
3. Nylon meshes (Hybaid).
4. Hybridization solution (10% dextran SO_4, 10X Denhardts v/v, 50 mM Tris-HCl 6X standard saline citrate (SSC) v/v, I% sarkosyl).
5. Rediprime labeling kit (Amersham Life Sciences).
6. [α-^{32}P] dCTP (Amersham Life Sciences).
7. Boiling water bath (Grant, Cambridge, UK).
8. Pipets and tips (Gilson).
9. 37°C incubator (Digi-Block).
10. Competition DNA, 10 μL placental DNA, 11 mg/mL (Sigma), 7.5 μL deionized distilled water, 32.5 μL 20X SSC, pH 7.
11. Shaking incubator set at 65°C (Innova 4080 Brunswick Scientific, NJ).
12. Posthybridization wash 1.0.1% w/v SDS. IX SSC, wash 11.0.1% w/v SDS, 0.5X SSC, both preheated to 65°C.
13. Saran Wrap (Dow).
14. Cassette (G.R.I Hi Speed X, Essex, UK).
15. Film (Sterling Diagnostic Imaging Newark. DE).

PROCEDURES

Preparing DNA for Fingerprinting

Prior to the generation of fingerprint data it is necessary to extract cloned DNA from a bacterial host. The same DNA preparation procedure is used for both methods described here. The procedure is based on an *alkaline lysis* method, updated first to a 96-well format, then improved by *incorporating* a filter-bottomed plate step.

1. In a deep-well *microtiter plate* dispense 500 µL of 2X TY plus an appropriate selective agent.
2. Using either a 96-well inoculating tool, or *sterile* cocktail sticks, inoculate each well from one of the clone glycerol stocks. After *sealing* the wells, grow the mix for 18 h in a 37°C incubator with shaking.
3. From the overnight growth transfer 250 µL to a clean *microtiter* plate and centrifuge at 1550g for 4 min.
4. Remove the supernatant and resuspend the pellet by vortexing gently (vortex genie set to 4-5). A cocktail stick can be used for *resuspending* pellets still attached to the plate.
5. To the resuspended pellet add, 25 µL solution I, tap the plate gently, and add 25 µL of solution II. Mix by tapping and leave at room temperature for 5 min. The solution should clear as the bacterial *clones* lose.
6. To each well add 25 µL solution III and mix by gentle tapping and leave at room temperature for 5 min. Seal the plate and vortex gently for 10 s.
7. Tape a 2-um filter-bottomed plate to the top of a microtiter plate which contains 10 µl. of isopropanol. Add the total well volume to the. filter-bottomed plate and centrifuge at 1550,8 for 2 min at 20°C.
8. Remove the filter-bottomed plate and leave the microtiter plate at room temperature for 30 min. *Centrifuge* at 1550g for 20 min at 20°C.
9. Remove the supernatant and dry the DNA by inverting the plate and placing it on a clean tissue paper, being careful to avoid any disruption to the pellets.
10. Prepare RNase A: 10 µL of 1 mg/mL RNase A/1 mL of T01E.
11. To the dried DNA add 100 µL of 70% ethanol, mix by gentle tapping.
12. Centrifuge at 1550g for 10 min at 20°C.
13. For restriction digest agarose fingerprinting, repeat above wash.
14. Remove the supernatant and dry as before, ensure that the pellet is transparent and add 5 µL of $T_{0.1}E$ with RNase A. Store samples at -20°C.
15. Check the preparation by separating the DNA on a small 1 % agarose gel. Typical DNA yield should be between 20-30 ng/µL.

Preparing DNA for Southern Blotting and Hybridization

In the case of PAC clones, which are used in these experiments, the DNA is inserted into a pCYPAC2 vector, which is based upon the PI artificial chromosome (PAC) and Bacterial Artificial Chromosome (BAC) vector systems of cloning DNA.

The clones contain, on average, approximately 120 kb of insert. A rapid alkaline lysis prep is used to isolate DNA, with average yield of 20-30 ng/µL. The *protocol* detailed in under the heading isolating DNA can be scaled up or down depending on the yield desired.

This is then followed by a standard phenol-chloroform-isoamyl alcohol cleanup of the DNA to remove protein carry over from the prep, which may interfere with subsequent digestions.

Isolating DNA

1. Inoculate 20 mL LB containing 250 µg *kanamycin* with a single *bacterial* colony.
2. *Incubate* overnight (approx 16 h) in an orbital shaker set to 250 rpm. 37°C.
3. Centrifuge (*Beckmann J2-21*) for 10 min at 21,000g to pellet bacterial cells.
4. Discard supernatant and resuspend in 3 mL of Solution 1.
5. Add 3 mL of Solution II and allow to stand at room temperature for 5 min. Dur ing this time, the solution should change from being fairly turbid to become clear as the bacterial cells lyse.
6. Slowly add 3 mL of Solution III and place on ice for at least 5 min. Solution III precipitates the *Escherichia coli* DNA and proteins.
7. Transfer to 30 nil. Corex tubes and centrifuge (Beckmann .1-6B) at 14,500 foi·l 0 min to precipitate the E. *coli* DNA and proteins.
8. Avoiding white precipitate, remove supernatant containing PAC DNA, and add to 8 mL of ice cold *isopropanol* to precipitate the DNA.
9. Stand on ice for at least 5 min. At this stage, tubes may be left at 20°C over-night. Alternatively, if a lot of white precipitate is removed with the *supernatant*. step 7 can be repeated.
10. Centrifuge tubes (Beckman J-6B) for 15 min at 14,500g.
11. Without disturbing the pellet, remove the isopropanol and wash pellet with 5 mL 70%. ethanol.
12. Centrifuge (Beckmann J-6B) for 5 min at 14,500g. 13. Repeat 70%k ethanol wash (optional).
14. Allow pellet to air dry until translucent and resuspend in 200 µL of deionized distilled water containing RNaseA. DNA can be stored at -20°C until needed.
15. Io determine concentration, digest 5 µL of DNA with *EcoR*I and run on a 0.8% agarose gel in 1X TBE buffer.

Cleaning DNA

1. To DNA add an equal volume of phenol-chloroform-isoamyl alcohol. Vortex briefly to mix.
2. Centrifuge at 10,000g for 15 s and remove supernatant containing DNA into a fresh Eppendorf tube.
3. Add 1/10th volume of sodium acetate, vortex to mix, and 2X volume of cold 100% ethanol to precipitate the DNA.
4. Place at -70° C for 15 min or at -20°C for 30 min.
5. Centrifuge at 10,000g for 5 min.
6. Without disturbing the pellet, remove supernatant and wash pellet in 70% ethanol.

7. Centrifuge at 10,000g for 5 min.
8. Repeat ethanol wash (optional).
9. Resuspend pellet in 50 μL of deionized distilled water.

Fluorescent Fingerprinting

1. Prepare a 5%c *denaturing acrylamide gel*. (20 mL acrylamide, 140 μL 10% *ammonium persulphate*, 28 μL *N,N,N',N'*-tetramethylethy-lenediamine [TEMED]).
2. Set up three digest premixes. one for each fluorescent label in three 1.5 MLA microfuge tubes labeled TET, HEX, and NED.
3. For one 96-well plate add to each tube, 25.5 μL $T_{0.1}E$, 24.5 μL NEB2 buffer, 5 μL *Hind*III (20 U/μL), 8 μL Taq FS (32 U/μL) and 3 μL Sau3AI (30 U/μL), then to the appropriate tube add 4 μL of one of the three ddA dyes .
4. Mix using a vortex, and spin in a benchtop centrifuge.
5. To the first third of the prepped DNA plate (A 1-H4) add 2 μL of the TET premix using a Hamilton repeat dispenser, to the second third (A5-H8) add the HEX premix, and the to last third (A9-H12) the NED premix; cover with a plate sealer.
6. Mix the reaction by gentle agitation on a vortex. and spin the plate to 150g for 10 s.
7. Incubate the reaction for 1 h at 37°C,
8. Using a multichannel pipet, add 7 μL of 0.3 *M* sodium acetate and 40 μL of 96% ethanol to each well. Pool samples labeled with different dyes.
9. To column I add columns 5 and 9 and mix, to column 2 add columns 6 and 10 and mix, to column 3 add columns 7 and 1 1 and mix, and to column 4 add columns 8 and 12 and mix.
10. Incubate at room temperature for 30 min in the dark.
11. Pellet the DNA by spinning the plate at 1550g for 20 min at 20°C.
12. Discard the supernatant and dry the pellet by tapping the plate face down onto tissue paper.
13. Wash the pellet by adding 100 pL of 70% ethanol to each well and mix by gentle tapping.
14. Spin at 1550g for 10 min at 20°C.
15. Discard the supernatant and dry the pellet.
16. Resuspend the DNA in 5 μL $T_{0.1}E$.
17. To a 1.5-mL microfuge tube add 70 μL $T_{0.1}$ 1E. 10 μL NEB2, 6 μL lambda DNA (500 ng/μL), 6 μL BsaJ1 (2.5 U/μL), 4 μL Taq FS (32 U/μL) and 4 μL ddC-ROX.
18. Incubate for 1 h at 60°C.
19. To the reaction mix add 100 μL 0.3 M sodium acetate and 400 μL 96% ethanol.
20. Leave at room temperature in the dark for 15 min, then at -20°C for 20 min.
21. Spin in a benchtop centrifuge at maximum ..speed for 2O min.

22. Discard the *supernatant* and dry the pellet by tapping the plate gently onto a tissue paper.
23. Wash the pellet by adding 200 µL 70% ethanol and spin in a benchtop centrifuge at maximum speed for 5 min.
24. Discard the supernatant and dry pellet as before, this time ensuring that the DNA is clear.
25. Resuspend the pellet in 120 µL $T0.1_E$ and 120 µL blue dextran formamide dye. 26. Prior to loading, add 2 µL of the marker mix to each sample using a Hamilton repeat dispenser.
27. Spin the plate at 150g in a centrifuge.
28. Denature the DNA in the plate for 10 ruin at 80°C and load 1.25 µL of each sample on a AB1377 sequencer.
29. Use ABI Prism Collection Software v 1.1.
30. After data collection, transfer the gel image to a UNIX workstation for entry into Image.

Agarose Fingerprinting

1. Prepare a 1 % agarose gel in IX TAE. (for a Gator Wide format gel tank, to 4.5 g agarose add 450 ml- IX TAE).
2. Set up one reaction mix to digest a 96-well plate of DNA in a 1.5-mL microfuge tube. Add 231 µL H_2O, 99 µL buffer B, and 55 µL *Hind*III.
3. Mix using a vortex, and spin in a benchtop centrifuge at maximum speed.
4. To each well add 4 µL of the reaction mix and cover with a plate sealer.
5. Mix the reaction by gentle agitation on a vortex, and spin the plate to 150g for 10 s.
6. Incubate at 37°C for 2 h.
7. Terminate the reaction by adding 2 µL of loading dye using a. Hamilton repeat *dispenser*, and seal the plate. Either load immediately or store at 4°C.
8. Before loading the samples, make the marker in a 1.5-mL microfuge tube by adding (per gel) 19.2 µL_ $T_{0.1}F$, 1.5 *µL,* analytical marker DNA wide range, 0.2 µL molecular weight marker V and 4.2 l- 6X loading; dye.
9. Load 0.8 µL of the marker in the first well and then every fifth well.
10. Load 1 µ.L of sample between the marker lanes.
11. Run the gel at 4°C in a cold room for 15 h at 90 V.
12. Prepare a fresh stain mix for the gel by mixing 1 M Tris-HCl, 0.5 mL 0.1 M EDTA and 50 µL Vista Green. Make up the vol to 500 ml, with H_20. Store at 4°C and use within 48 h.
13. Once run, cut the gel down so the length is 19-20 cm and stain for 30-45 min on a shaker.
14. Briefly wash with H_2O to remove excessive stain.
15. Scan the gel on a Fluor Tmager SI. For best results, use a 530-nm emission filter.
16. For best results use the following parameters: 100 µm pixel size: normal detection

sensitivity; 16-bit digital resolution; single label dye; 488 nm excitation filter: 530 nm Em filter 1, and *photomultiplier* tube (PMT) voltage is 800.

17. Transfer the gel image to a UNIX *workstation* for entry and analysis.

Entry of Fingerprint Data Using Image

Raw fingerprinting data is processed in the program Image, this was first developed in the late 1980s and upgraded at The Sanger Centre by Darren Platt and Friedemann Wobus, current development and curation is performed by Jim Mullikin and John Attwood (see http:// www.sanger.ac.uk/ Software/Image). Using a package of analysis algorithms, Image converts raw data to a set of normalized band values and gel images.

This is achieved by interactively stepping through three main windows. The first of these is lane tracking, lines matching the fingerprint lanes are traced along the length of the gel. The next step is band calling; the presence and fine position of *authentic* bands are edited while *artifacts* or *nongenuine* bands are removed.

To analyze the clones, each fingerprint band must have a normalized integer value attached to it, this is achieved in the final step of Image, called standard marker locking. Here, the band call in the standard lane is aligned against a known standard pattern and this is used to normalize fingerprint band values.

1. Download Image via File Transfer Protocol (FTP) by selecting the appropriate file for your system.
2. Make a directory for the each project. To import a raw gel image into Image prior to data entry. use the command, imimport <project name> <gel name> <image file>.
3. Set up default parameters need to be set for each project in Image. If all the gels in one project are to have the same characteristics, then the parameters need only be set once. *(see* http://www.sanger.ac.uk/Software/image/help/).
4. Enter the gel number. Once the defaults are set for a project they may need to be slightly altered for individual gels, e.g., the number of sample lanes may differ between gels, however, the main defaults should remain constant for a given project.
5. Enter the FIRST CLONE NAME. and the GEL NUMBER; this can be alphanumeric. Each window has the same basic layout: scroll bars at the top and right-hand side of the image. a zoom function (400% for band calling and marker locking appropriate) and a ,'ray ramp tool. Use the arrow in the top right hand corner of the window to negotiate through the different stages.
6. Clip the image and remove unwanted portions of the gel.
7. Edit Lanetracking: The lines superimposed on the image represent the lanes on the gel, these lines must accurately delineate the lanes. On agarose-digest gels, there are two types of lines; markers (red) and samples (yellow). Moving the marker lines interpolate :all :ample lint in between to their regular space with regard to the *neighbouring* marker line. For fluorescent gels, there is only one type of line, as the marker is *multiplexed.*

There are two main ways to alter the position of the lines to match the lanes:

a. Whole lines can be moved by activating (they will go black) the circular check box(es) at the left-hand end of the screen.

b. Smaller regions can be adjusted by highlighting (they will go green) the box(es) running down the length of the line. For whole lines and groups of highlighted boxes, use the cursor keys to adjust the position, for single boxes use the mouse to drag up or down.

8. Edit Bandcalling: Genuine, informative bands should be tagged and partials or gel artifacts removed. For fluorescent gels the marker and the three clones present in a lane will he deconvoluted. On the band-calling screen, the first clone is selected and highlighted red, the top half of the screen shows an x -y plot representation of the *pixel* values scanned along the middle of the lane.

   Bands are represented by points on the plot and by triangles on the lane strip, an active hand is red, nonactive ones are green. One lane is active at a time, move up or clown lanes using the *cursor keys* or. F and B.

   Remove deleted clones (missing insert DNA few fingerprint hands) or partially cut clones (numerous fingerprint bands of equal intensity) using the *middle mouse button* on the undesirable lane.

   Add hands by using the *felt mouse button* on the t-c· plot or 11, Move forward by using f or *space bar,* and back using *b.* Use *d* to delete a band and *n or middle mouse button* to remove one. Reposition bands by using the *left* or *right cursor keys,* if a band has been positioned on the side of a peak *r,* will roll the band to the nearest peak.

   At this point, clone names can be added by activating the box to the side of a lane using the *left mouse button,* then typing the name (this can he alphanumeric). All hands below a certain height can be removed by using the *left mouse button* to draw a box in the plot window and all bands within that section can be deleted.

9. Edit Standard Marker lock: The display and functions are similar to *step 3.* however, this time only the marker lanes are shown along with bands, from the standard file. To accurately interpolate the position of sample hands the standard file hands and the marker lane bands must line up. Use the ,v—v plot to check if the two match.

   To increase the match, pick a lane and move through each band checking the pattern. Once this is locked use *relock on lane.* This will apply the position of the current lane's bands onto the bands in all other marker lanes.

   Finally, go through each marker lane and check that each band is correctly positioned at the top of appropriate peaks. To disregard any changes made by the locking module, use *rein clock on whole*: the gel is then ready to finish.

   This will create a <gelname> bands and a <projectname; <gelname>.gel files, for restriction digest agarose gels an additional <gelname>. sizes file will be written.

The program "imtransfer," which comes with the Image package, will copy the finished files to the appropriate FPC directories if the name of the image project is the same as the FPC project, so, if in *step 2* the project was called 9GSC, then:

Image directory 9GSC		FPC project 9GSC
<gelname>.hands	will be copied to	/Image
<projectname>.<gelname>.gel	will be copied to	/Gels
<gelname>.sizes	if present will be copied to	/Sizes

Analyzing Fingerprirnts Using FPC

Once processed, the data are analyzed in another program developed at the Sanger Centre: FPC (FingerPrinting Contigs). FPC is an interactive program that can operate on a selfcontained basis or interact with other programs particularly ACcDB. *It* provides rapid comparisons between fingerprinted clones using a probability-of-coincidence equation, and an interactive graphical display allowing user interface and *manipulation*.

Each clone is represented by a set of integer band values. Overlap between any two clones are identified by the similarity of these values, within a set tolerance. Once overlap is established, clones are organized into *contigs,*coordinates are used to represent the clones' bands and the degree of overlap equates to the number of shared bands.

1. Download FPC along with a postscript manual, FPC.ps, by anonymous FTP from. ftp.sanger.ac.uk.puh/fpc. A User's Guide is available at http://www.sanger.ac.uk/ cari/ fpc__faq.shtml.
2. Create a project directory and sub-directories called Image, Bands, and Gel. When using restriction-digest agarose gels, a subdirectory called Sizes is also needed.
3. In Image the following output files and destinations can be found:

   1 Image →* bands →* /Image fluorescent fingerprinting

   2 Image →**.bands →* /Image restriction digest fingerprinting

   .sizes →/Sizes where migration values are used for analysis

   3 Image →*.sizes →* /Image restriction digest fingerprinting

   .bands → /Bands where size values are used for analysis

   In each case:

   Image →* .gel → /Gel

   where * is the name of the gel.

   In *step 2,* the /Sizes are used only when comparing two clones and in the size calculator (written by Ken McDonald at the GSC).

   In 3, the /Bands are used with the gel data.
4. Start *FPC*: From *File...* drop-down menu, go to *Create new project* and name the project. This will write a <projectname>.fpc and a <projectname>.cor file.
5. To read the file(s) in the Image directory into FPC. select *Update cor,* this will add the clones to the database and move the <gelname>.bands to the Bands directory.
6. To achieve the best results in FPC, the parameters must be set correctly, set these in the

Main Analysis window. or *Contig Analysis* window.

a. Tolerance: This determines the variation allowed between the position of any two bands that can be said to overlap. Set too high and spurious overlaps will be seen, too low and genuine matches will be missed. When fingerprinting human bacterial clone libraries we have found that for fluorescent gels or digest agarose gels using migration values from the <gelname>.bands set it to 7.

   For restriction digest agarose gels using fragment sizes from the <gelname>.sizes, use a variable tolerance (0.07% along the length of the gel). This is changed in the *Configure* menu. Migration values are generally used for both *fluorescent* gels and *restriction* digest agarose gels, this gives a more informative visual display. It is important not to change the tolerance once set as this figure is used in the overlap equation.

b. Cutoff: This is used in the overlap equation, the higher the exponent the. greater the *stringency* applied to the probability that two clones will overlap. The following factors are taken into account: the total number of bands in each clone, the *tolerance*, and number of matching bands between two clones.

   The number of bands in a clone is proportional to the size of the insert (given an even distribution of restriction sites), therefore it is sometimes necessary to vary the cutoff when comparing clones with different-sized inserts.

   When fingerprinting human bacterial clone libraries, we have found that for fluorescent gels a cut off of between le-06 to le-08 is appropriate, and for restriction digest agarose gels le-10 is used. This should be set in the *Main Analysis* or *Contig Analysis* windows.

c. Vector file: Cloned fragments are maintained in their bacterial hosts by ligation to vector DNA, which contain resistance genes allowing selection of transformed bacterial cells. Vector DNA will normally produce at least one fingerprint band, usually present in each clone, which should be removed prior to analysis. To identify the vector bands enter at least 25 clones into FPC that do not overlap.

   Assemble the clones In one fingerprint and gel-trace window. and look for hands that are common between all of the clones. Find the migralion values for these hands from the bands file, and the sizes, if available, from the sizes tiles.

   Take the mean value of each vector band from the 25 clones. It is these values that are added into the Vector File, e.g., for the RPCI-11 BAC library restriction *digest agarose* fingerprinted.

d. Migration values: bacfilter

   *

   1387

   3695

   3766

If a band lies within +/-7 it will be removed.

e. Sizes: bacfilter

*

6524

510

452

If a band lies within 7% of the value it will be removed where * is the prefix used for each clone from a library.

Only one band will be removed, so if a genuine band lies in the same position as a vector, and they were scored as a double in Image, it will remain. Enter the Vector file name in the Configure menu.

f. Bury: A facility exists to bury clones within another, this can be achieved manually using the *Edit clone* function. However FPC will automatically bury clones and the stringency for this should be set in the *Main Analysis* window. A 90% match of bands between any two clones in normally used, therefore set the *Bury* to 0.10.

A clone will become buried in another if it shares 90% of its bands with the parent (or canonical) clone. As well as facilitating manual manipulation, this can also help identify potentially deleted climes and confirm the legtimacy of the canonical clonel.

7. Once the parameters are correctly set the fingerprint data can be analyzed. There are two train ways large sets of data can he analyzed. *(a) CB map,* or the more interactive *(b) Ketiset → FPC.*

a. The *CB map* will group together clones which share a high probability of overlap (i.e., they have a probability of coincidence score that is below the Cutoff). It uses a "greedy" algorithm, which cart produce an suboptimal solution, so it generates N solutions (default is 10) and uses the best one.

The fingerprint bands for a contig are arranged to fit the physical order in which they lie. This is a powerful and quick tool for contig assembly. Although it can be used for fluorescent data, it is particularly useful for restriction digest agarose fingerprinting.

From the *Main Analysis* window either select *Build Contigs,* this will kill all contigs (unless a sequence clone is present and the *Use Seq Ctgs is off)* and use all the singleton data and rebuild the entire data set and construct new contigs, or *Incremental Build Contigs,* which will take only data that were added after the last build. A contig with a *Score* approaching or at 1 M O is optimal.

The Q column indicates the number of clones that, although they overlap significantly with at least one other clone in the contig, could not be *accurately* aligned within the contig. This can be caused by poor fingerprints, repeat elements within clones producing similar fingerprint patterns, or sub-optimal solution as a product of the *algorithm*. In the latter. manual manipulation may be the only option.

For contigs with a low score it is advisable to rerun the *CB map* by selecting *Calc* on the *Contig Analysis* window. An example is shown elsewhere in this chapter Even after this point, fine manipulation of clones may be necessary to the refine the order and optimize overlap of clones especially minimum sets of sequence clones.

b. Using *Keyset ® FPC* offers a less-automated approach to contig building. First, the type of keyset required should he selected as follows. using the mouse select *Clones* in the *Main Menu,* then select the *Search Commands* option. Clones can be selected on the basis of their gel, the (late created or last modified, singleton status, contig number, or on the basis of a remark added to the clone. The selected clones are then compared to the database.

If there are no contigs in the database then the output will be:

>> bA88716 ctg0 41 h → Fpc (Tol 7, Cutoff le- 10)		
Ctg0	bA60019 36h	25 le-12
Ctg0	bA6lH8 33b	26 3e-15
Ctg0	hA215015 42b	34 le-20
Ctg0	bA31713 17 37b	27 2e-14

To form a contig Edit one of these clones so that their *Contig* does not equal 0. Select *Accept.* This number will appear in the *Clone* window. Select it with the *left mouse bottom,* to open the contig. On the *Contig Analysis* window select *Compare Keyset* then *Next* then *Add.* The first clone (in this case. bA60019) should appear in the contig. Highlight the new clone and define the overlap of the two clones using the *Clone*1-* *Ctg Prob.* in the *Contig Analysis* window and position it using the *Snap Left/Snap Right* facility in the *Edit Contig* window. Repeat the procedure for each new clone.

If there are already contigs in the database the output window is similar to that produced after *CB map is* run. Select contigs that have clones overlapping with them and follow the same procedure of *Compare Keyset then Next then Add.* Selecting *Auto add* in the *Main Analysis* window will automatically add clones that overlap with existing contigs, so only the positioning need be done.

8. Once contigs have been constructed, it may be necessary to selecl a minimum set (i.e., tiling path) of clones to sequence. The extent of sequence overlap desired depends on the sequence protocol. For full (10X) sequence clone coverage, it is critical to obtain minimally overlapping clones with as little redundancy as possible for the project to be economic. For draft (5X) sequence clone coverage, the restrictions for minimum set construction may be slightly relaxed, as two significantly overlapping clones will give full coverage. The selection criteria to follow include:

   a. The clone should be a true representation of the genomic DNA it contains, avoid potentially deleted clones, e.g., clones which are missing hands thatarc clearly present in overlapping clones in the contig, or clones with too many extra bands.

   b. The bands in a sequence clone should he contained within at least one other clone in the contig; avoid end clones, as these may contain rearranged inserts, represent mixed library wells or even be chimeric.

c. When identifying minimally overlapping clones apply all of these criteria as well obtaining the least redundant overlap, be careful not to miss any intervening contig bands between two sequence clones.

Of invaluable use when identifying sequence clones are the gel traces of clones. Use these and the fingerprint window to identify sequence clones and minimum sets. For the latter it is often useful to use the CB map again. From the *Contig Analysis* window select, *Fp Order.* Manual contig manipulation is often necessary when using clones of very different length or poor fingerprints as a result of chemistry or inaccurate gel entry. Using the GSC BAC data, we are currently defining the amount of automation that can be achieved in FPC (paper in progress).

Clones for Southern Blotting

Following is the description for restriction digests of three enzymes, *EcoR*I, *Not*I. and Sal1, and their double digests. These three enzymes produce digestion fragments that are one of three different types:

1. Fragments containing a vector only.
2. Fragments containing an insert from the clone only.
3. Fragments that contain the end fragment, from the clone and vector.

Comparison between clones to detect possible overlap is based on fragments containing an insert only. Clones that have similar fingerprints for insert-only digestion fragments may be overlapping. For a precise overlap to be determined, it is necessary to determine the fragments that contain the ends of the clone and a vector end. This is possible by the digestions performed in this experiment.

The pCYPAC2 vector in which the clones are inserted contains two RNA polymerise promoters T7 and SP6. The insert, usually approximately 120 kb (14), is cloned into the vector between these two promoters. *NotI* digests the PAC vector twice, at both the T7 and SP6 ends of the vector and releases the insert that can then be sized by pulse field gel electrophoresis (PFGE). *SalI,* however, only cuts at the SP6 end of the vector.

By electrophoresis of the digestion products it is possible to determine a T7 end fragment by comparison of the *EcoRI* and *EcoRI/NotI* and *EcoRI/SalI* digests. As only *Not!* cleaves at the T7 end, there will be a fragment present in the *EcoRII* digest and the *EcoRI:Sal1* digest that is absent in the *EcoR*I/NotI digest. Likewise. it is also possible to determine the SP6 end fragment by examining the digestion fingerprint.

As both *NotI* and *SalI cut* at the SP6 end, there will be a fragment present in the EcoRI-only lane that is absent in the *EcoRI/NotI* and *EcoRI/SalI* digests. By studying these fragments it is also possible to determine in which orientation the insert has been cloned into the vector. *NotI is* a rare cutting enzyme, with a GC-rich recognition site, 5'-GC GGCC GC-3'.

Such sites are not common within the genome and are indicative of potential CpG islands where the dinucleotide *CpG is* present at higher than expected frequencies and are frequently associated with genes. Determining the length of the *NotI* fragments separated by PFGE allows

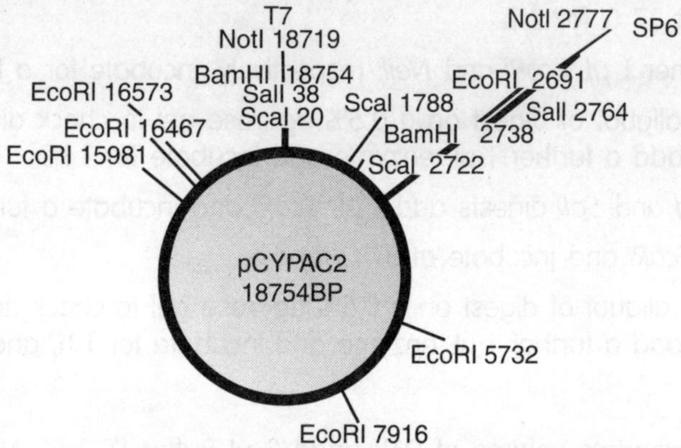

Figure 5.2: Restriction sites within the pCYPAC2 vector (accession no. U09128). Restriciton sites were determined using GeneJockeyII for Macintosh. During cloning, sequential digestions of Scat *and* BamHI *are used to removethe stuffer fragment. This produces four products, one the cloning vector, with* BarnHI *ends, two stuffer fragments with* Sca 1 *ends, which are incompatable with BamHI and therefore will nut be cloned back into the vector, and two oligonucleotide linkers with* BamHI/ScaI *ends, which are removed prior to the cloning in of the insert. The* EcoRI NotI, *and* SalI *digests determine at which orientation the insert is cloned into the vector (see text for details) and identifies* EcoRI *fragments that are vector only within the digests and will be common to all clones.*

us to estimate the length of the cloned insert with high accuracy.

Clone size data are important for the construction of clone contigs and physical maps.

EcoRI Digest

1. To the appropriate volume of DNA (500 ng), add 3 µL Buffer D, 1 µL *EcoRI* enzyme and deionized distilled water to make up to 29 µL total reaction volume.
2. Incubate at 37°C for 2 h.
3. Add 1 µL of *EcoRI* and incubate a further 2 h.
4. Run a 3-µL aliquot of digest on a 0.5% agarose gel to check digestion. If digestion is not complete, add a further 1 µL of enzyme and Incubate for 1 h. and recheck on a gel.

Sal/ Digests

1. Double the amount of DNA and add 6 µL buffer, 1.5 µL *SalI* and deionized distilled water to make up to 51 µL total reaction volume.
2. Incubate at 37°C for 2 h.
3. Add 1 µL enzyme and incubate a further 2 h.
4. Run a 2 L aliquot of digest against undigested PAC DNA on a 0.5% agarose gel to check digestion. If digestion is not complete, add a further 1 µL enzyme and incubate for 1 h, and recheck on a check gel.
5. Remove the 10 µL digest and reserve for the PFGE. Split the remaining digest into two. To the first, add 1 µL *EcoRI,* and to the second 1 *µL NotI.*

6. Incubate at 37°C for 2 h.
7. Add a further 1 *μL EcoRl* and *Neil,* respectively, incubate for a further 2 h.
8. Run a 2-L aliquot of digest on a 0.5% agarose gel to check digestion. If digestion is not complete, add a further 1 µl, enzyme and incubate for 1 h.
9. To the *Nod* and *Sall* digests add 1 µl, *EcoRl* and incubate a further hour at 37°C.
10. Add 1 µL *EcoRl* and incubate at 37˚C for t h.
11. Run a 2-4, aliquot of digest on a 0.5% agarose gel to check digestion. If digestion *is* not complete, add a further 1 µL enzyme and incubate for 1 h, and recheck on a check gel.

***Not/* Digests**

1. To the appropriate volume of DNA, add 3 µL Buffer D, 1 µL *Nod* and deionized distilled water to make up to 37 µl., total reaction volume.
2. Incubate at 37˚C for 2 h.
3. Add 1 µl, enzyme and incubate a further 2 h.
4. Run a 2-µL aliquot of digest against undigested PAC DNA on a 0.5% agarose gel to check digestion. If digestion is not complete, add a further 1 µL enzyme and incubate for 1 h, and recheck on a gel.
5. Remove 10 µL of digest for PFGE. To the remainder, add 1 µL EcoRI and incubate for a further hour.
6. Add 1 *μL EcoRl* and incubate at 37°C for 1 h.
7. Run a 2-µL aliquot of digest against undigested PAC DNA on a 0.5% agarose gel to cheek digestion. If digestion is not complete. add a further 1 µL, enzyme and incubate for 1 h, and recheck on a gel.

Electrophoretic Method for Southern Blotting

To resolve digestion products *EcoRl, EcoRI/Noti, EcoRI/Sall* and *EcoRI/Notl/Sall,* the digests should be run on a 0.8% TBE 27 x 20 cm get. A more detailed restriction map can be produced by knowing the sizes of the clones and determining any internal *Notl* or *Sall* sites. However fragments from these digestions will be large and need to be resolved at high resolution to accurately determine size.

To produce a more detailed restriction map. *Notl. Still* and *Notl* and Still double digest should be performed, and PFGE need to be used to resolve the large DNA fragments. PFGE is suitable for electrophoresis of large fragments of DNA, as fragments are viewed with greater resolution than conventional gel electrophoresis. This method is also used to obtain an insert for hybridization experiments as the insert and vector band of 16 kb can be seen as distinctly separate bands, lessening the chance of contamination by vector during hybridization.

Gel Electrophoresis

1. Prepare a 500-mL, 27 x 20 em, 0.8% agarose. 1 X TBE gel (4 g of agarose in 500 mL 1 X TBE, 10 µL ethidium bromide) with a 24-well comb. Prepare 2 L 1 X TBE for the running buffer.

2. Load samples onto the gel. with a marker in the first and last lanes.
3. Run the samples out of the wells and into the gel, 90 V for approximately 15 min. 4. Run gel at 45 V for 20 h.
5. Remove gel and photograph with a fluorescent ruler to facilitate band identification on the autoradiograph.

Inserts and Internal Not/ and Sal/ Sites

1. Prepare a 1 %% agarose, 0.5X TBE gel (la agarose to 100 ml, 0.5X TBE) with a 2a-well comb.
2. To each 10 µL digest to be loaded, add 3 µL loading dye.
3. Place PFGE marker in the first and last lanes of the gel to enable sizing of bands.
4. Carefully remove casting tray and place gel in tank. Cover with 2 µL of 0.5X TBE buffer.
5. Load samples, including 500 ng of Lambda *Hind*lll.
6. Running conditions: 6 V/cm, with a pulse time of 0.1-10 s, running time 14 h.
7. After 14 h running, stain gel with ethidium bromide (250 µg ethidium bromide in 500 mL of 0.5X TBE) for 50 min and photograph with fluorescent ruler to facilitate identification of bands on the autoradiograph. The gel should he handled gently. as it is fragile and will break easily.

Methods for Southern Blotting

1. Fill a large container with 0.4 *M* NaOH. Place a glass sheet or similar on top of the tray as a platform, and place two 3MM filters over the glass so that the ends are in the 0.4 *M* NaOH. This forms the wick of the blot.
2. Allow the wick to become saturated with 0.4 *M* NaOH; this wlll lake approximately 30 min. This starts the capillary action, which will draw the 0.4-*M* NaOH upward through the gel, transferring DNA onto the membrane.
3. Cut the nylon membrane and three sheets of 3MM to the size of the gel.
4. After photographing the gel with fluorescent ruler, place in a *0.25-M HCl* solution so that gel is immersed in liquid. Shake gently at room temperalure for 30 min. This step causes partial depurination and strand cleavage of the DNA. resulting in shorter strands that are easier to transfer onto the membrane.
5. Discard 0.25-M HCl and rinse gel briefly in deionized distilled water.
6. Wash gel in 0.4 M NaOH for *20 min*, shaking gently at room temperature. This denatures the DNA resulting in transfer of single-stranded DNA to which the single-stranded probe may bind during hybridization.
7. Invert the gel and place it on a prewetted wick. Ensure that there are no air bubbles between the wick and the gel, which could cause patchy transfer of DNA.
8. Seal the edges of the wick that are not covered by the gel using Saran Wrap to ensure that the 0.4 M NaOH is drawn up only through the get and onto the membrane and to

prevent evaporation.

9. Place membrane on top of the gel, ensuring that there are no air bubbles and full contact between the gel and the membrane to ensure even transfer. Once membrane is in place, do not move, as some DNA transfer will occur very quickly with movement of filter resulting in smeared bands on an autoradiograph.
10. Prewet three sheets of 3MM in *0.4-M* NaOH and place on top of the membrane again ensure that there are no air bubbles between the membrane and the 3MM. If this occurs, remove the 3MM and replace with a fresh sheet.
11. Stack the blotting towels on top of the 3MM, ensuring that the gel is well covered by the towels. The stack should be approximately *4* in. high. The 3MM and towels will cause the capillary action.
12. Place a weight (approximately 0.4 kg) on top of the blotting towels and leave to blot overnight.
13. After blotting, dismantle the blot and mark the positions of the wells on the membrane.
14. To neutralize the membrane, wash at room temperature for 30 min in 2X SSC.
15. Blot dry with 3MM, wrap in Saran Wrap and store at room temperature until needed.

Hybridization

Probe Preparation

1. Digest PAC to be used with *Not*I to release the insert. To 500 ng–1 Vg DNA add 3 µ.L of buffer D and 2 µL of *Not*I. Make up the vol to 30 µL with deionized distilled water.
2. Incubate at 37°C for 2 h and add a further 2 µL of *Not*I.
3. Electrophorese through a 0.5% 1X TBE check gel against undigested DNA to check digestion. If digestion is not complete. add a further 1 µL enzyme and incubate for 1 h. and recheck on a gel.
4. Run on a PFGE gel and develop as described in elsewhere in this chapter.
5. Excise insert hand removing as little as possible agarose, and store at 4°C until needed.

Probes Labeling

These steps are common to both labeling for an insert and a whole PAC. Determine concentration of DNA. 50 ng is used per labeling.

Labeling Whole PAC DNA

1. When labeling a whole PAC, take 50 ng of DNA and make up to 50 µL with demonized distilled water.
2. Boil solution containing DNA for 5 min in a pierced Eppendorf tube, resulting in single-stranded DNA to which probe synthesis can occur.
3. Microcentrifuge for 15 s at 10,000g
4. Place tube on ice for 1 min.

5. Remove 45 µL and place in a Rediprime labeling tube.
6. Add 5 µL of [α-^{32}P] dCTP to the tube. Flick gently to mix until the blue pellet has dissolved.
7. Incubate at 37°C for 15 min.

Labeling of DNA within a Gel Slice

1. Weigh gel slice and add 3 mL of deionized distilled water per gram of gel.
2. Boil gel slice for 5 ruin in a pierced Eppendorf tube, to melt the gel and denature DNA.
3. Remove volume of DNA corresponding to 50 rig and if necessary: add deionized distilled water to give a final volume of 45 µL in the Rediprime labeling tube.
4. Add 5 µLL of [α-32P] dCTP radiolabel to the tube. Flick gently to MIX until the blue pellet has dissolved.
5. Incubate at 37°C for 30 min.

Prehybridization

1. Place membranes between two nylon meshes and roll into a cylinder.
2. Place into a hybridization bottle and add 30-50 mL of hybridization solution.
3. Incubate the cylinders while rotating them in the hybridization oven at 65°C.

   There should be no air bubbles between the filter and the side of the bottle that may prevent the probe reaching the membrane in that region.
4. Prehybridize for a minimum of 0.5 h upward.

Competition of the Probe and Hybridization

1. Add competition DNA mixture to the tube and boil for 5 min. This denatures both the probe and competition DNA to single strands, allowing them to reanneal to each other.
2. Incubate tube at 65°C for 30 min. The competing DNA will bind to repetitive areas of the probe blocking them and preventing binding to target DNA.
3. Mix labeled probe with 5 mL hybridization mix and add to bottles containing prehybridization solution and membranes.
4. Place in rotating hybridization oven overnight (for 16-24 h).

Washing

1. Preheat shaking incubator to 65°C.
2. Remove membrane from hybridization bottle and place in wash 1 solution.
3. Wash in shaking incubator for 30 min.
4. Pour off wash 1 solution and replace with wash 11 solution.
5. Incubate for 15 min in a shaking incubator.
6. Discard wash 11 solution and replace with fresh wash solution 11.
7. Incubate for 15 min in the shaking incubator.

8. Discard wash and place membrane between two sheets of 3MM and gently blot to remove excess liquid.
9. Wrap blot in Saran Wrap to prevent filters drying out. If filters dry out while radioactive, the radiation sticks to the membrane and is very difficult to remove.
10. Place in an autoradiography cassette with film and leave at -70°C overnight or less, depending on the signal strength on the membrane.
11. Develop the film and if necessary, relay for a shorter or longer exposure.

POINTS TO REMEMBER

1. Southern blots may be reused and probed with several different probes on different occasions as the DNA is fixed to the membrane. Between hybridizations, the membrane may be stored safely and allowed to decay naturally, or the probe can be removed by washing membrane in a boiling solution of 1% SDS, allowed to cool and monitored for signal.
2. The dichlorhodamine dyes used for fluorescent fingerprinting are extremely photolabile: they should be stored in a covered box at -20°C. any solutions containing dyes should be kept in the dark as much as possible.
3. Each new hatch of fluorescent dyes should be tested before use to find the optimal amount to be added to each reaction, as batch concentrations can vary. Fingerprint a number of clones that are known to give varying strengths of signal with differing amounts of dyes. While fingerprinting human bacterial clone libraries, we have found that undiluted TET dye and 1:8 HEX and 3:4 NED give best results.
4. Care should be taken that the gel does not come into contact with TBE, ethidium bromide, or lint from tissues as these emit at the same wavelength as vistra green and will therefore affect results. It is important to use the same chemicals (especially agarose and acetic acid) as the GSC if incorporation with their data is desired. It is possible that slight variations in separation will be seen if chemicals are riot duplicated.
5. RNase A can be stored for up to 4 mo at -20°C, as long as it is stored in I nil aliquots in 1.5 mL microfuge tubes to prevent excessive freeze-thaw cycles.
6. Prepped DNA can bee stored at -20°C for up to I wk: after that there may he degradation in the quality of fingerprints
7. Tracking gels in Image; for 64-lane fluorescent gels put the LANE NUMBER to 70 and delete the extra tracks. With 121 lane digest gels put the MARKER REPEAT number to 120, then adjust the two outside marker lanes to match the ends of the gel, go back to the ENTER GEL NUMBER window and put the MARKER REPEAT number back to 5.
8. For consistent results when hand calling, it is best to move through each clone one at a time, evaluating the authenticity and fine positioning of each band in the clone. Scan in to 400% and use the gray ramp tool. especially at the top and bottom of the lane. The same approach should be adopted when standard marker locking.

9. You should aim to finish a full agarose digest gel (96 clones) in about 2-3 h, and a full fluorescent gel (192 clones) in 4-5 h. The time taken to enter a gel varies with user experience, data quality, and computing power.
10. The complexity of the clone library screened can be used as a guide to establish a suitable cut off; on the *Main Analysis* window, enter a clone name and select -**FPC*, you can adjust the cut off and tolerance until the clone has an appropriate number of matches. For example, for a five-hit library;

>> bA202A15 → Fpc (Tol 7, Cutoff le-10)		
Ctg0	hA214G9 40b	35 7e-22
Ctg0	bA335N 18 50b	33 Se-IS
Ctg0	bA343J 19 38b	35 le-23
Ctg0	hA3971 12 39h	37 4e-26

11. When assembling contigs it is important to take into consideration the possible presence of mixed clones. i.e. clones containing more than one DNA insert. This can occur at library construction, through cross contamination of library wells, or mixing during prepping and fingerprinting. These clones may give misleading fingerprint overlaps.

    Look for large clones with up to 50% of their fingerprint bands not contained in the other clones in the contig. Supporting evidence such as, STS data, end-sequence hits, mapping information, fluorescence *in situ* hybridization data, and electronic PCR should be taken into account when merging-contigs. If there is data indicating two contigs may join, it is advisable to lower the *Cutoff* to assess the fingerprint overlap.
12. It is advisable to run a check gel of all the digests together on a check gel after digestion with *EcoRI*. This allows a direct comparison of concentrations to be made, and alterations in the amount loaded per lane on the large gel to be made to ensure that each lane contains the same amount of DNA. This is also a good last-minute check for partial digestions.
13. LMP gels are more fragile than conventional agarose gels, but have the advantage that bands may be excised and used directly in hybridizations without first cleaning. If no hands are to be excised from the gel, it is better to run a converttional agars gel that is more robust.
14. Instead of adding RNasel to the resuspension media during the PAC preps. it can be added to the Solution I at 110 µg/mL. However. some PACs give lower yields, which can result in a pellet that is difficult to see after the precipitation in isopropanol. By not removing RNA, the size of the pellet is increased and is therefore more visible, and the RNase also has a longer period of time to act upon RNA in the sample.
15. Labeling time can be as little as 15 min at 37°C or alternatively, the labeling reaction can be left overnight at room temperature depending on the time available.
16. Depending on the repetitiveness of the region, time taken for competition to bind to the labeled DNA at 65°C can be increased to 1 h.

17. Hybridization of probe to target DNA normally occurs within this time period even for low copy numbers of target DNA However, hybridizations may be left for longer without harming.
18. Depending on the incorporation of label into the probe and the amount of target DNA. the membrane may still be very hot after these washes (200 counts or greater). It may he necessary to increase the stringency of the washes by performing another wash with 0.1 X SSC and monitoring at 10 mire intervals until the counts drop, or by continuing to wash at the same stringency, again with regular monitoring.

Chapter 6

Sequencing Bacterial Genomes

The world of bacterial genome sequencing "changed utterly" when the first two complete bacterial genome sequences were published by scientists at The Institute for Genomic Research (TIGR) in Maryland in 1995. Sequencing of the *Haemophilus influenzae* genome was followed quickly by that of *Mycoplasma genitalium-both* using a whole-genome shotgun sequencing strategy that shattered the dogma that genomes should be sequenced using a top-down approach, where detailed mapping and creation of an ordered library had to precede genome sequencing.

Consequently, aside from a handful of projects already underway in 1995, nearly all other bacterial genome sequencing projects have relied on the quick, easy, and efficient approach of making a random shotgun library, automatically sequencing tens of thousands of clones, and then using computer power to assemble this molecular jigsaw puzzle into a handful of fragments ("*contigs*"), followed by a "*finishing phase*" to close every gap and remove every ambiguity.

The success of whole-genome shotgun sequencing has been stupendous-at the time of writing over 60 bacterial genome sequences have been completed and over 200 more are underway (*see* http://wil.integratedgenomics.com/ GOLD/). The genome of every significant bacterial pathogen of humans, plants, and animals has already been or soon will be sequenced, including, in many cases, the genomes of multiple strains, pathovars, or serovars from the same species.

PURPOSE OF SEQUENCING GENOMES

The clinical bacteriologist may well ask, "What's the use of a genome sequence?", to which the scientist might reasonably reply with a quote variously attributed to Faraday or Franklin: "What's the use of a baby?" Although a completed genome may consist simply of millions of As, Cs, Ts, and Gs, genome sequencing has already amply justified its existence as an enabling technology in the research laboratory, energizing novel, hypothesis-driven research on bacterial pathogenesis and physiology, and spawning a whole range of high-throughput, genomic-based technologies that fall under the umbrella term "functional genomics". For example, efforts are already underway to make knockout mutants, express proteins, survey protein-protein interactions, solve structures, and measure gene expression on a whole-genome scale in several different microorganisms.

And thanks to the synergies between genome sequencing, the Internet, and the polymerase chain reaction (PCR), any scientist who wants any gene from any organism can have it within a few days for the effort of a few Web searches and for the price of a pair of oligonucleotides.

Genome sequencing has transformed the research landscape! One overarching theme to emerge from the genome-sequencing revolution is how little we still know about bacteria. In every new genome sequence, we usually find that at least 30-50% of the genes have no known function-even in the laboratory workhorse *Escherichia coli K* 12, at least 20% of the genes are without a functional assignment.

For organisms that we cannot grow in the laboratory (or cannot grow easily), such as *Treponema pallidum* or *Mycobacterium leprae,* genome sequencing is the only way to gain insights into their biochemistry, physiology, and virulence mechanisms. In addition, functional genomics approaches are allowing us to probe aspects of physiology and assign functions to genes that may not be obvious when bacteria are grown under laboratory conditions on agar plates.

Also, genome sequencing allows a complete "naming of parts"-once we know every gene in the organism, finding all the genes associated with a particular pathway, structure, or *phenotype* becomes a finite, *tractable* task. One early example of this was the identification and characterization of all lipopolysaccharide biosynthesis genes in *H. influenzae* from scrutiny of the genome.

On a more applied note, a major driving force in the genomics revolution has been the desire of the pharmaceutical industry to find novel targets for *antimicrobial* drugs. Comparing *microbial* and animal genomes allows the identification of genes or pathways that are present in pathogens but absent from humans, i.e., plausible targets for compounds showing selective toxicity.

Similarly, genome sequences have been used as the starting point for finding new vaccine targets. The best example of this was a stepwise sieving of genes from the *meningococcal* genome, starting with several hundred *surface-associated* proteins identified through *bioinformatics,* through to a handful of highly credible vaccine candidates.

ANTIMICROBIAL RESISTANCE

Aside from the indirect effects that might flow from *vaccine* and *drug* discovery, genome sequencing has already had some impact in the clinical microbiology laboratory, most *noticeably* in the field of *epidemiological* typing, and, to a lesser degree, in diagnostic bacteriology and the study of antimicrobial resistance.

Applications of Genomic Sequences

The least *adventurous* use of genome sequences in molecular epidemiology is an adjunct to established methods. For example, having a genome sequence can assist in setting up a multilocus sequence-typing method for a new species. In a similar vein, Desai and colleagues reported using the genome sequence of *Campylobacter jejuni* strain NCTC 1.1168 to develop a fluorescent amplified fragment length polymorphism (fAFLP) typing method for this organism, with the aim of

Table 6.1: Landmarks in Bacterial Pathogen Genome Sequencing

Year	Landmarks
1995	*Haemophilus influenzae*
	Mycoplasina genitalium
1996	*Mycoplasma pneumoniae*
1997	*Helicohacter pylori*
	Borrelia burgdorferi
1998	*Mycobacterium tuberculosis*
	Treponema pallidum
	Chlamydia trachomatis
	Rickettsia prowazekii
1999	A second *H. pylori* genome
	Chlainydia pneumoniae
2000	*Campylobacter jejuni*
	Neisseria meningitidis serogroup A
	Neisseria meningitidis serogroup B
	Two more *C. pneumoniae* genomes
	A second *C. trachomatis* genome
	Vibrio cholerae
	Pseudomonas aeruginosa
	Ureaplasma urealyticum
2001	*Escherichia coli 0157:H7 (two* strains)
	Mycobacterium leprae
	Pasteurella multocida
	Streptococcus pyogenes
	Staphylococcus aureus (two strains)
	Streptococcus pneumoniae (two strains)
	Rickettsia conori
	Yersinia pestis
	Salmonella typhi

(Table Contd.)

Year	Landmarks
2002 (to August)	*Salmonella typhimurium* *Listeria monocytogenes* *Brucella melitensis* *Clostridium perf'ringens* *Streptococcus pyogenes* (two strains) *Fusobacterium nucleatum* A third *S. aureus* strain *Bacillus anthracis* (two strains)

establishing a molecular epidemiological database.

More exciting is when genome sequencing challenges an accepted dogma of epidemiological typing. Studies in Wren's group arising from the genome sequence of *C. jejuni* NCTC 11168 showed that the established Penner serotyping method for this bacterium depended not on lipo-oligosaccharide determinants, but instead on variation in polysaccharides from a previously unsuspected capsule, overturning a decades-old preconception.

Yet, more innovative is the use of complete genome sequences to prime novel methods that would have been impossible in the pregenomic era. The most commonly used method for this "*genomotyping*" is hybridization to a microarray. Here, oligonucleotides or PCR products representing all genes in the genome of a given *bacterium* are arrayed as *microscopic* spots on a solid surface (typically a glass slide).

Fluorescently labeled genomic DNA from a test strain is then hybridized to the array, and a *confocal* microscope is used to determine the pattern of *hybridization*. Spots that do not show hybridization represent genes that are present in the genome-sequenced strain used to create the array, but absent from the test strain. Several groups have used this approach to look at strain-to-strain variation:

- Dorrell et al. performed whole-genome *microarray* comparisons of 11 *C. jejuni* strains of diverse origin to show that at least 21 % of genes in the sequenced strain were absent or highly *divergent* in one or more of the *isolates* tested. This study also provided further evidence that the capsule accounts for Penner serotype specificity and could potentially lay the foundations for novel *epidemiological* typing methods for *C. jejuni.*
- Workers in Mekalanos's laboratory exploited a microarray built from the genome sequence of the *Vibrio cholerae* El Tor 01 strain N16961 to compare the gene content of N16961 with that of other *V. cholerae* isolates. Although they found a high degree of conservation among the strains tested, they were able to identify genes unique to all pandemic strains as well as genes specific to the seventh-pandemic El Tor and related 0139 serogroup strains.
- Hakenbeck and colleagues used a microarray to investigate genomic variation in *20*

Table 6.2: Genome Sequencing Projects on Multiple Strains in the Same Species.

Species	Completed projects	Underway or planned
Bacillus anthracis	Two strains	Several more
Bacillus cereus		Two strains
Campylobacter jejuni	One strain	One strain
Chlamydia pneumoniae	Three strains	
Chlamydia trachomatis	Two strains	
Clostridium perfringens	One strain	One strain
Escherichia coli	Two strains of 0157:H7 One strain of K12	At least five pathovars
Haemophilus influenzae	One strain	Two strains
Helicobacter pylori	Two strains	
Listeria mononcytogenes	One strain	One strain
Mycobacterium aviurn		Two strains (includes paratuberculosis)
Mycobacterium tuberculosis	Three strains (includes *M. bovis*)	One strain (BCG)
Neisseria meningitidis	Serogroup A strain Serogroup B strain	Serogroup C strain
Salmonella enterica	Serovar typhi Typhimurium	>10 more serovars
Staphylococcus aureus	Three strains	At least six strains
Staphylococcus epidermidis		At least two strains
Streptococcus pneumoniae	Two strains	At least two strains
Streptococcus pyogenes	Two strains	
Tropheryma whippelii		*Two* strains

Streptococcus pneumoniae isolates representing major antibiotic-resistant clones and 10 different capsular serotypes. They found that up to 10% of genes varied between individual isolates and the reference strain.

Among the variable loci were mosaic genes encoding antibiotic resistance determinants and gene clusters related to bacteriocin production. They also performed genomic comparisons between *S. pneumoniae* and commensal *Streptococcus mitis* and *Streptococcus oxalis* strains.

- Fitzgerald and colleagues used a microarray to study genomic diversity in *Staphylococcus aureus.* They concluded that the most prevalent clone *of S. aureus* producing toxic-shock *syndrome* toxin evolved long ago and that the epidemic of toxic-shock syndrome that arose in the 1970s was caused by a change in the host environment rather than by rapid *intercontinent*al dissemination of a new hyper-virulent clone.

Unfortunately, all of the previous studies have been performed on archived strain collections and none have used microarrays to type strains prospectively, in real time, "in anger." Although it is only a matter of time before microarray-based typing methods reach the clinical bacteriology laboratory (or at least the reference laboratory), one study, yet to be definitively published, stands out in its timely use of microarrays in an outbreak investigation.

Shafi and coworkers at the University of Leicester used a whole-genome *Mycobacterium tuberculosis* microarray to detect differences between the outbreak strain from the recent Leicester Crown Hills Community College outbreak, the genome-sequenced strain H37Rv, and other *M. tuberculosis* isolates.

Having quickly defined which genes were missing in the outbreak strain using the microarray, they then went on to design a rapid PCR-based method to distinguish the outbreak strain from other strains prevalent in Leicester in time to influence management of the outbreak.

One problem with microarray-based approaches to the study of genornic diversity is that they can tell you only what is absent from the test strain compared with the genome-sequenced strain, rather than what new genes might be present. In my own laboratory, we have been exploring the use of long PCR to determine differences between strains of *E. coli.* So far we have successfully amplified fragments of up to 32 kb in length, and have detected large-scale genomic differences in a handful of *chromosomal* regions.

In the near future, we plan to scale up to achieve genome-wide coverage. The ultimate in molecular *epidemiological* typing is to sequence the entire genomes of the strains one wants to type. Already multiple strains have been sequenced for several bacterial species. With the price of genome sequencing falling *dramatically*-around $0.10 per base pair for a fully finished and annotated sequence (i.e., approx $200,000 for an average genome)-and now clearly within the resources of a research project grant, it is easy to envisage this approach becoming ever more popular over the coming years, and it becoming routine to sequence a *particularly* tricky outbreak strain of a hospital or community *pathogen.*

As is often the case with technical innovation, the most futuristic molecular epidemiological use of genome sequencing per se-rather than derivative technologies such as microarrays—has come from defense applications. Read et al. sequenced the entire genome of a *Bacillus anthracis* strain isolated from a victim of a recent *bioterrorist* anthrax attack and compared it with that of a recently sequenced reference strain.

Their study revealed 60 new epidemiological markers that included single nucleotide *polymorphisms*, inserted or deleted sequences, and tandem repeats. Selected markers were tested on a collection of anthrax isolates and were used to show that the *"bioterrorism"* strain was indistinguishable from laboratory isolates of the Ames strain, but distinct from the only other known natural occurrence of the Ames genotype (a goat isolate).

Comparative Genomes

Just as comparative genome sequencing can reveal differences useful for typing strains within a species, it can also be used to provide molecular diagnostic targets that can distinguish closely related species, subspecies, or pathovars, or to find novel targets for immunodiagnosis.

The close similarity between *Mycobacterium avium* subsp. *paratuberculosis* and other mycobacterial species has hindered the development of specific diagnostic reagents. A comparison of an incomplete *M. avium* subs *p. paratuberculosis* genome sequence with that of *M. avium* subs *p. avium* revealed 21 genes that were present in all *M. avium* subs *p. paratuberculosis* isolates and absent from all other mycobacterial species tested, thus providing novel targets, e.g., for diagnostic PCR.

Raczniak and colleagues used genome sequences to classify organisms according to the class of lysyl-tRNA synthetase that they employed. This led to the realization that the Lyme disease agent, *Borrelia burgdorferi,* possessed a class of enzyme distinct from the human host and from almost all other bacteria, thus providing a potential new target for either DNA-based or immunologically based diagnosis.

Determinants

Genome sequencing has the potential to reveal novel resistance determinants or new arrangements of known resistance genes. Researchers in Hiramatsu's laboratory compared the genome sequence of a methicillinresistant *S. aureus* (MRSA) strain isolated in 1982 with that of an MRSA strain exhibiting low-level vancomycin resistance (minimum inhibitory concentration, MIC, 8 μg/mL) isolated in 1997.

They reported that most of the antibiotic resistance genes were carried either by plasmids or by mobile genetic elements including a unique resistance island. They subsequently sequenced the genome of an apparently highly virulent, community-acquired MRSA and reported that the *mecA* gene, which encodes the novel penicillin-binding protein (PBP2') responsible for methicillin resistance, was carried in this strain by a novel form of Staphylococcal chromosomal cassette *mec (SCCmec),* and that the strain carried 19 additional virulence genes.

Fitzgerald and colleagues used a microarray to study genomic diversity in *S. aureus,* including MRSA. They discovered that MRSA strains fell into five or more highly divergent groups that differed by as much as several hundred genes. They therefore concluded that MRSA strains have arisen independently by lateral transfer of the *SCCniec* element into methicillin-susceptible precursors.

A highly novel use of whole-genome microarrays for determining which genes might contribute to resistance has recently been reported by Gill et al. Their strategy was to create a genomic library of *E. coli* sequences expressed at high copy number in a wild-type *E. coli* host, then to select for clones that showed increased resistance, and finally to use DNA microarrays *simultaneously* to isolate and identify the enriched gene inserts.

They recovered 19 genes conferring tolerance to the cleaning agent Pine-Sol, and 27 genes conferring sensitivity, and suggested that the method could have wide applications in identifying resistance determinants.

CONCLUDING REMARKS

The exploitation of genome sequences in the clinical laboratory remains in its infancy. Two impediments limit progress here:

1. The very success of traditional agar-based methods. However, the ability of molecular, genome-based methods to compete successfully with established approaches will improve as automation and molecular methods make inroads into the laboratory, so that the marginal costs of setting up a new molecular assay steadily decrease.
2. The lack of individuals dually trained in clinical *bacteriology* and molecular biology/ genomics. Would anyone interested in doing an MD or PhD or *collaborating* in this area please contact the author!

Perhaps the ideal outcome of genome *sequencing* in the clinical laboratory would be the development of simple biochemical tests that could assist in diagnosis and typing. For example, something as simple as the use of sorbitol MacConkey agar and beta-glucuronidase fermentation to distinguish *E. coli* 0157:H7 from harmless commensals.

Although in this case the frame-shift mutation that inactivates the *beta-glucuronidase* gene in *E. coli* 0157:H7 was identified without using the genome sequence, detailed comparisons of genome sequences from *closely* related species or *pathovars* might reveal similar kinds of differences that could then be detected in the *clinical* laboratory by simple biochemistry.

In closing, one can conclude that, although the genomics revolution has already had a *decisive* impact on microbiology research, it has yet to have a major effect in the clinical laboratory. However, it is only 7 yr since the first bacterial genome was sequenced.

One is reminded of the dialogue between US President Nixon and the Chinese Prime Minister Chou En-Lai during Nixon's historic visit to China. Nixon asked Chou, a keen student of history, what impact he thought the French Revolution had had on Western Civilization. To which Chou replied "It's too early to tell!"

Chapter 7

Prokaryotic Genomes

SRNAs are small, noncoding RNA species that have been shown to regulate diverse cellular processes in a number of *prokaryotes.* Most sRNAs are 100-200 nt in length and, thus, are difficult to identify by traditional functional approaches such as random *transposon mutagenesis.* Furthermore, many sRNAs are significantly less abundant than other RNA species such as rRNAs, making their physical *isolation* from total cellular RNA difficult.

Finally, because they do not encode proteins, sRNAs are difficult to identify based on their primary sequence. Owing to the challenges of identifying sRNAs by traditional functional and bioinformatic *approaches,* until several years ago only 10 bacterial sRNAs had been identified. These bacterial sRNAs had been discovered *serendipitously,* as a result of their high cellular abundance, their inadvertent *insertion* into a *multicopy plasmid* along with other genes of interest, or their detection in the study of operons.

In 2001, the development of several *bioinformatic* approaches ushered in a new era in the study of bacterial sRNAs, one in which *fortuitous* discovery was replaced by accurate genome-wide predictive searches. In these computational studies, putative sRNAs encoded in the intergenic regions (IGRs) of the *Escherichia coli* genome were predicted based on the *colocalization* of genetic features commonly *associated* with previously characterized sRNAs.

These features included predicted *Rho-independent* transcriptional terminators, putative promoters, sequences conserved among closely related species, and regions predicted to encode conserved secondary structure. A number of the putative sRNAs predicted in these studies were subsequently confirmed by Northern analysis. Of the approx 120 bacterial sRNAs known to date, the large majority has been identified using one or more of these integrative bioinformatic approaches.

Although bioinformatic approaches have proven successful in identifying *E. coli* sRNAs, their implementation has presented significant *computational* challenges that have severely limited their utilization. A genome wide search for putative sRNAs can include tens of thousands of individual predictive features such as *terminators, promoters,* and regions of conserved sequence.

Integrating these thousands of individual features to identify the instances in which they are colocalized in IGRs has, in the past, required either the use of inefficient noncomputational methods, severely limiting the rate at which searches could be conducted, or the *de novo* development of a computational tools, necessitating a level of computer expertise not possessed by most biological researchers.

Thus, although it is widely accepted that sRNAs are encoded by most if not all prokaryotes, until recently genome-wide annotations of sRNAs had been conducted in only 3 of the over 270 sequenced bacterial species. To facilitate the efficient annotation of bacterial sRNAs, the author developed and made publicly available a program called sRNAPredict that flexibly integrates different combinations of individual sRNA predictors to rapidly identify putative sRNA-encoding genes in the IGRs of any annotated bacterial genome.

By searching for putative transcriptional terminators encoded *downstream* of regions of sequence conservation, sRNAPredict identified 104 candidates for novel sRNAs in *Vibrio cholerae* IGRs. Nine predicted *V. cholerae* sRNAs were subjected to experimental verification by Northern analysis; five were confirmed.

A similar search was conducted in the *opportunistic* Gram-negative pathogen *Pseudomonas aeruginosa,* leading to the identification of 34 previously unannotated putative sRNA-encoding genes. Of these, 31 were experimentally tested by Northern analysis; 20 were confirmed. In addition, we developed an improved version of sRNAPredict2, sRNAPredict2, which we used to identify more than 2700 previously *unannotated* putative sRNA-encoding genes in the genomes of 10 other species of bacterial *pathogens.*

An sRNAPredict2 search is conducted in four main stages. In stage I, primary data files such as databases of annotated ORFs, tRNAs, rRNAs, previously annotated sRNAs, and riboswitches, as well as genome sequence files are obtained from websites such as NCBI and TIGR. In stage II, the IGRExtract program is used to convert the genome of the species of interest to a database of *intergenic* sequences.

In addition, primary ORF databases are converted to a format that can be used by TransTerm in its prediction of intergenic Rho-independent terminators. In stage III, the individual predictive elements of sRNAs such as regions of conservation and *putative* terminators are identified using several publicly available UNIX-based programs. If a database of putative promoters or transcription factor binding sites (TFBSs) is available, this database must be converted to the appropriate format by the user.

In stage IV, the names of the databases obtained in stage I and of the output files and databases created in stage III are entered into the Initial Input File along with the values of various search parameters. This Initial Input File is then entered into sRNAPredict2 and the predictive search is conducted, producing an *annotated* database of intergenic putative *sRNA-encoding* genes.

PROCEDURES

The methodology below describes how to conduct sRNAPredict2 searches on the Mac OS X operating system. Although sRNAPredict2 can be compiled and executed on other operating systems, some of the more detailed instructions (such as "drag and drop") are based on Mac OS X and may not be *applicable* to other operating systems. The instructions are written with the *assumption* that the reader has some basic *experience* using Unix-based programs.

Computerized Preservation

The sRNAPredict2 directory can be downloaded at http://www.tufts.edu/ sackler/waldorlab/ sRNAPredict.html and contains the sRNAPredict2 source code file (sRNAPredict2.cpp), executable (sRNAPredict2), as well as sample input and output files. These sample input files can be used to test sRNAPredict2.

The source code for sRNAPredict2 is written in C++. The sRNAPredict2 executable was compiled on Mac OS X using the gcc compiler. To run sRNAPredict2 on a different operating system, the source code will need to be recompiled. Double-clicking the sRNAPredict2 icon should automatically open in the "*Terminal*" application on Mac OS X.

If doubleclicking on the executable does not launch sRNAPredict2, open a new terminal window, enter "chmod 777," drag and drop the executable file, press return, then try to launch the executable again. If there is an error message while compiling the source code, enter "chmod 777" then drag and drop the source code file before attempting to recompile.

Observing sRNAPredict2

1. Ensure all sample input files are located in the same directory in which the sRNAPredict2 executable is located.
2. Launch sRNAPredict2 and at the "Enter the number of searches you would like to conduct:" prompt, enter 1.
3. At the "Enter the name of input file #1:" prompt, enter "Sample initial_input.txt." 4. The output file created will be named "Test output.lxt." This file should be identical to file "Sample output.txt" found in the sRNAPredict2 directory.

Stage 1

Storing Cenome Sequence Files

1. From NCBI:
   (a) Go to ftp://ftp.ncbi.nih.gov/genomes/Bacteria/.
   (b) Follow the link for the species of interest and partner species.
   (c) Download files with ".fna" extensions.
2. From TIGR:
   (a) Go to http://pathema.tigr.org/tigr-scripts/CMR/shared/Genomes.cgi.
   (b) Follow the green "T" link corresponding to the species of interest and partner species.
   (c) Download files with ".icon" extensions.
3. From the Sanger Institute:
   (a) Go to (http://www.sanger.ac.uk/Projects/Microbes/.

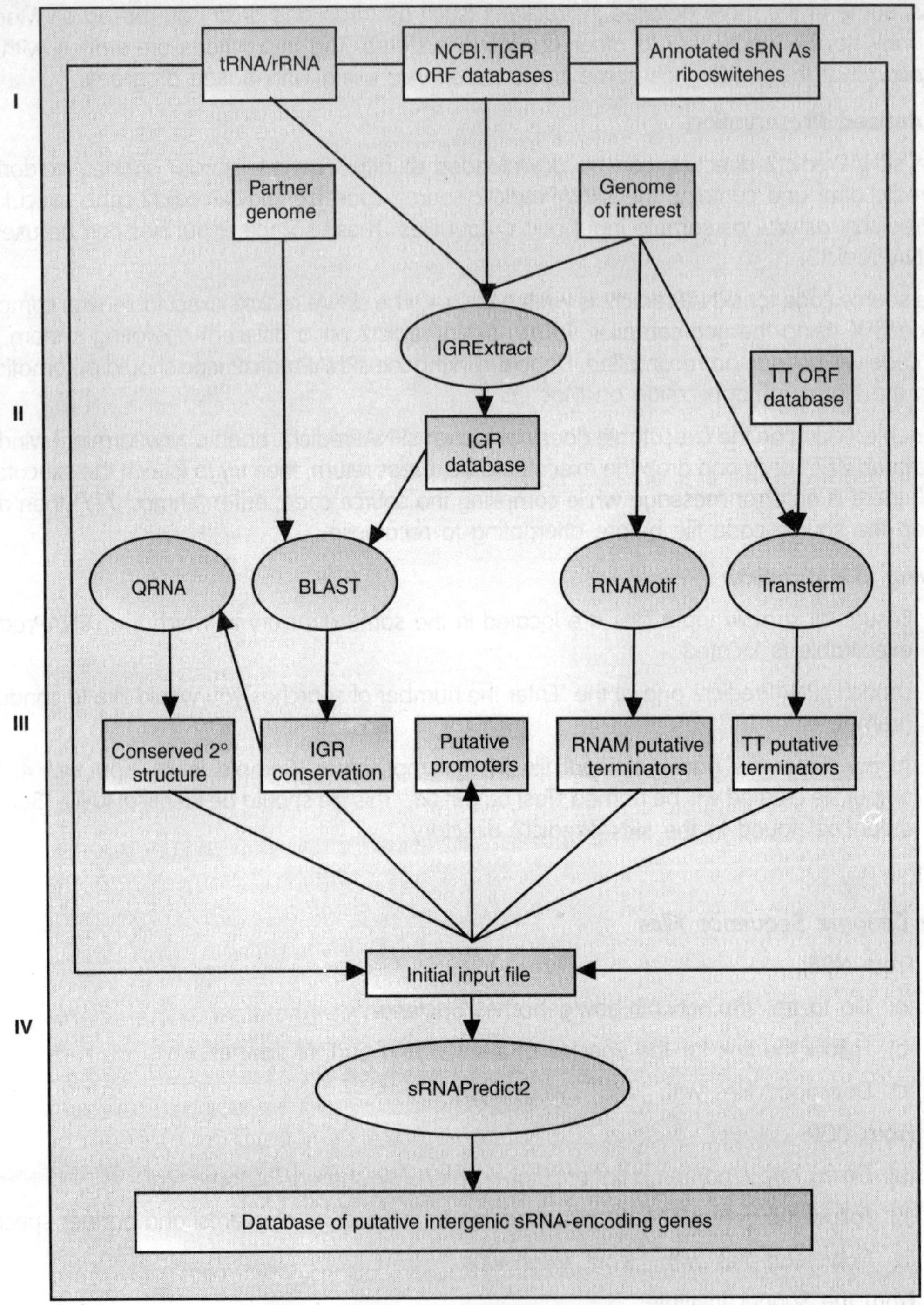

Figure 7.1: Schematic of an sRNAPredict2 search for putative sRNAs using a single combination of predictive features. The number of each stage of the search is denoted on the left.

(b) Follow the link for the species of interest and partner species.

(c) Follow the FTP link.

(d) Download files with "albs" extensions.

Storing ORF Databases

sRNAPredict2 is designed to utilize NCBI and TIGR ORF databases. Databases available at NCBI contain the coordinate positions, strand orientations, and, usually, the *locus* and product names of all ORFs. They do not include *annotated* genes encoding frame-shift mutations.

Databases available at TIGR contain the locus names and coordinates of all annotated ORFs and of annotated genes encoding frame-shift mutations.

The strand orientations of ORFs in these databases are automatically inferred by sRNAPredict2 based on the order in which the start/end coordinates are listed.

All ORFs found in the TIGR database that are not found in the NCBI database are assigned the locus and product name "TIGR." TIGR ORF databases are not available for all sequenced genomes. If no TIGR database is available for the species of interest, sRNAPredict2 will utilize only the NCBI database.

1. Storing from NCBI:

   (a) Go to ftp://ftp.ncbi.nih.gov/genomes/Bacterial.

   (b) Follow the link for the organism of interest.

   (c) Download file(s) with ".ptt" extensions.

2. Storing from TIGR:

   (a) Go to http://pathema.tigr.org/tigr-scripts/CMR/shared/Genomes.cgi.

   (b) Follow the green "T" link corresponding to the organism of interest.

   (c) Download file(s) with ".coords" extensions.

Storing tRNA

1. Go to http://pathema.tigr.org/tigr-scripts/CMR/shared/MakeFrontPages.cgi?page =rna_list.
2. Follow the link of the organism of interest. This will go to a page that, in most cases, is split into three sections: tRNAs, rRNAs, and sRNAs.
3. Follow the "download" link found next to the heading of the tRNAs and rRNAs sections. This will open a new window containing the names and *coordinates* of putative tRNAs and rRNAs, respectively.
4. Download then copy/paste the database of tRNAs into an excel spreadsheet. Repeat with the rRNAs.
5. Delete all columns in the excel spreadsheet except those containing the coordinates of the tRNAs or rRNAs.
6. Copy/paste and save these coordinates in a text file.

Compiling Annotated sRNAs and Riboswitches

1. Using Internet Explorer, go to http://www.sanger.ac.uk/cgi-bin/Rfam/genome_dist. p1.
2. Follow the link for the species of interest.
3. Select all Rfam families that are not classified as tRNAs or rRNAs.
4. Click the "Selected families" button.
5. Copy/paste the list of putative sRNAs into a text file. Delete the headings.
6. To include genes not included in the Rfam database, add them to the end of this text file.

Stage 11

Developing IGR Databases

In BLAST comparisons between the *complete* genomes of two closely related species, short stretches of *intergenic* sequence conservation are often not identified because of the relatively large amount of ORF conservation.

Thus, to increase the sensitivity of BLAST searches for sRNA *conservation,* comparisons should be conducted using a database containing only the intergenic sequences of the genome of interest. To facilitate conversion of a genome sequence to a FASTA-formatted IGR database, the author has developed a C++ program named IGRExtract.

The IGRExtract directory, which contains both the IGRExtract executable (compiled using gcc in Mac OS X) and the IGRExtract source code, is available for download at http://www.tufts.edu/sackler/waldorlab/sRNAPredict.html. Follow the instructions in elsewhere in this chapter to download and launch IGRExtract.

1. IGRExtract takes as input (1) a file containing the genomic sequence (in FASTA format), (2) an ORF database(s) (NCBI and/or TIGR), and (3) a tRNA/rRNA database formatted as described above. When prompted, enter the name of these files.
2. Enter the size of the genome.
3. Enter how much sequence upstream/downstream of the IGRs are to be extracted. To extract only IGRs enter "0;" to exclude the ends of IGRs adjacent to their flanking ORFs, enter a positive number; to include ORF sequences flanking the IGRs, enter a negative number.
4. IGR sequences are extracted in FASTA format. Verify that each FASTA name in the output file begins with "*IG*" followed by the coordinates of the IGR flanked by parenthesis. This format is critical for conversion of IGR coordinates to chromosome coordinates by sRNAPredict2.

Developing a TransTerm-Compatible Input Files

Running the terminator-predicting program TransTerm requires 2 input files, one containing the genome sequence and the other containing ORF coordinates. The header of the genome sequence file must conform to format ">SpeciesName_id#." This header is followed by the genome sequence.

The authors has developed a C++ program, TT_ORF, that converts NCBI or TIGR ORF databases to a TransTerm-compatible ORF database. This program is available for download at http://www.tufts.edu/sackler/waldorlab/ sRNAPredict.html. Follow the instructions given in computational preservation. to download and launch TT_ORF. TT_ORF automatically assigns 1 as the id no. for all ORFs; thus if TT_ORF-produced databases are used, ensure 1 is entered as the id no. in the sequence file header.

Stage III

ntergenic Conservation

BLAST comparisons should be conducted using the WU-BLAST 2.0 program (available at http://blast.wustl.edu/licensing/). BLAST comparisons should be conducted between the IGRs of the species of interest and either the entire genome or the IGRs of the partner species.

1. BLAST 2.0 supports databases in XDF (eXtended Database Format). Thus, before conducting a BLAST comparison, drag and drop the xdformat program icon (located in the BLAST 2.0 directory) into the Terminal window, type "-n," then drag and drop the IGR database file.
2. Once XDF formatting has been completed, drag and drop the blastn alias icon (located in the BLAST 2.0 directory) followed by the IGR database file and the genome sequence file or IGR database of the partner species into the terminal window. BLAST output file names can be *assigned* by entering "-o" followed by the desired output file name.

Developing Promoter/TFBS Databases

Promoters/TFBS databases must be formatted by the user. Each line should include only the start coordinate of the promoter or TFBS and the strand *orientation* of the promoter or TFBS (0 for coding strand, 1 for *noncoding strand*).

Putative Rho-Independent Terminators

1. Download the RNAMotif directory by going to http://www.scripps.edu/mb/ case/casegr-sh-3.5.html and following the rnamotif-3.0.4.tar.gz. The directory contains a README file that includes instructions for installing, testing, and running RNAMotif.
2. Before running RNAMotif, the path to the directory containing the energy data must be specified. To do this, enter "setcnv EFNDATA" then either manually enter the path to the *efndata* directory or drag and drop the directory into the terminal window.
3. To run RNAMotif, drag and drop the rnamotif executable icon (found in the src directory in the RNAMotif directory) into the terminal window. Enter "-descr," then drag and drop the descriptor file followed by the genome sequence file.
4. The output of the RNAMotif search is written to the UNIX terminal window. Copy/paste this output from the terminal window to a text file. This output file can be used without modification as input in sRNAPredict2 searches.

Rho-Independent Terminators Using TransTerm

sRNAPredict2 is designed to extract coordinates from TransTerm output files or, when available, from published TransTerm databases.

1. Go to http://www.genomics.jhu.edu/TransTerm/transterm.html. If a TransTerm database is available at for the species of interest, copy and paste the published database from Internet Explorer, and save it as a text file. The format of the database should conform to the format of the sample TransTerm database provided in the sRNAPredict2 directory.
2. If a TransTerm database is not available for the species of interest, TransTerm can be downloaded at http://www.tufts.edu/sackler/waldorlab/sRNAPredict.html and is accompanied by a README file that includes instructions for its installation.
3. To run TransTerm, first drag and drop the icon of the TransTerm executable (which is found in the src directory) into the terminal window.
4. Enter -s then drag and drop the genome sequence file.
5. Enter -c then drag and drop the formatted ORF database file.
6. Enter -o, the name to be assigned to the output file, then -g.
7. The TransTerm output file can be used without modification as input in sRNAPredict2 searches.

Predicted Conserved Secondary Structure Using QRNA

QRNA is a program that utilizes BLAST-generated sequence alignments to identify patterns of sequence *homology* that likely represent conservation of

RNA secondary structure. A putative sRNA identified by sRNAPredict2 is reported to correspond to a region of conserved secondary structure (denoted by a "Y" in the "QRNA?" column of the output file) if that sRNA overlaps any region predicted by QRNA to encode conserved secondary structure (reported as "winner = RNA" in the QRNA output file).

1. Download the QRNA directory by going to http://selab.wustl.edu/cgi-bin/selab.pl?mode=software#qrna. The directory contains a PDF user guide (in the documentation directory) that includes instructions for installing and running QRNA.
2. Assign a location for the QRNA libraries by entering "setenv QRNADB" then dragging and dropping the lib directory into the terminal window (or manually entering the full location of the lib directory).
3. Before running a QRNA search, the BLAST output file must be converted to a QRNA input file. To do this, drag and drop the "blastn2qrnadepth.pl" Perl file from the scripts directory followed by the BLAST output file into the terminal window. These Perl scripts will create three new files in' the scripts directory. The file with ".q" extension will serve as the input file for QRNA.
4. To run QRNA, drag and drop the icon of the QRNA Unix executable (located in the src directory) into the terminal window. Next, assign values for the window (-w) and slide (-x) parameters and a name for the output file (-o). Finally, drag and drop the QRNA input file located in the scripts directory. Be aware that, depending on the size of the input file and the window and slide values, QRNA analysis may take many hours to be completed.
5. Once the QRNA analysis is completed, the output file can be used without modification as input in sRNAPredict2 searches. A sample QRNA output file (Sample_QRNA.txt) is provided in the sRNAPredict2 directory.

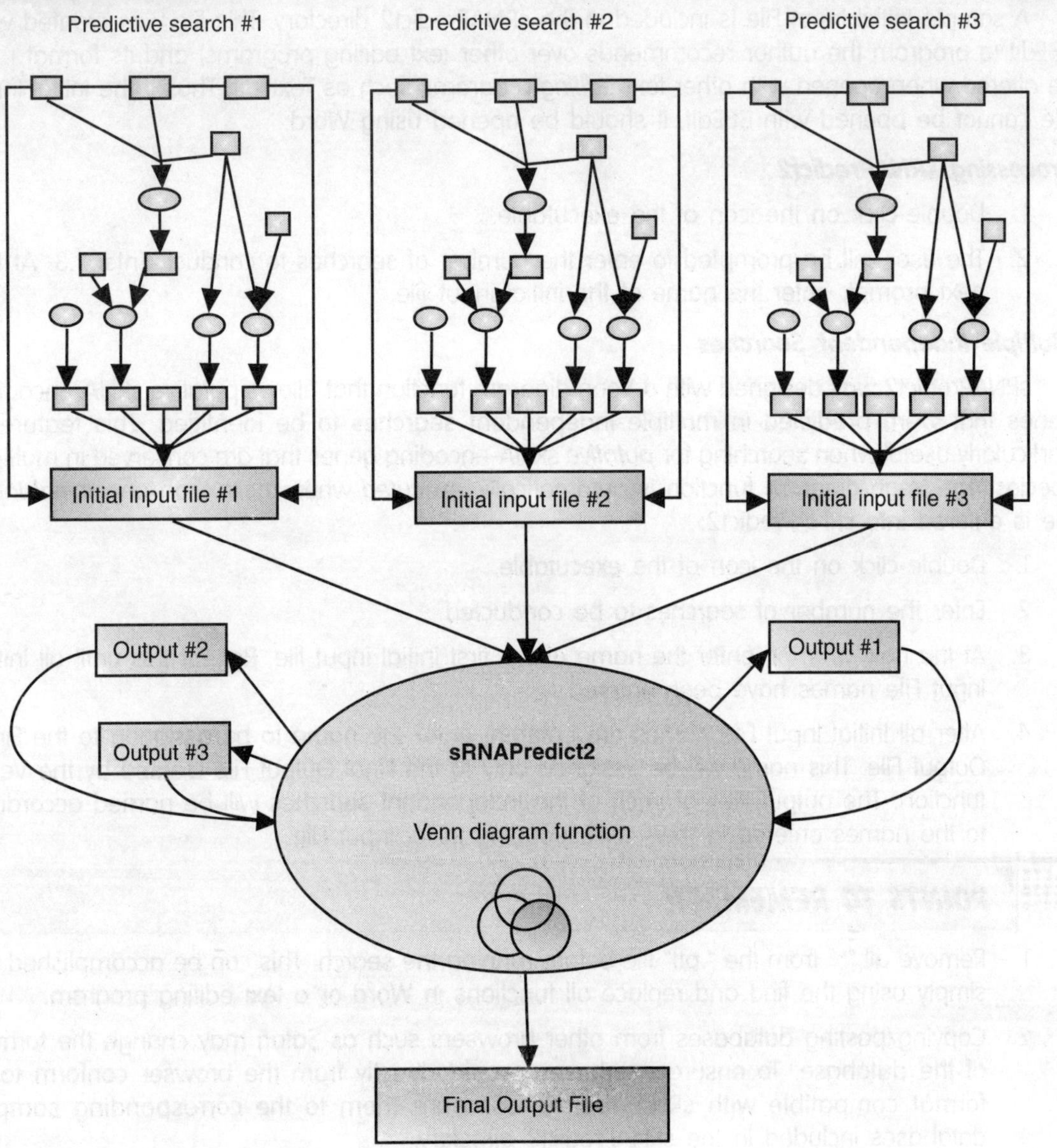

Figure 7.2: Schematic of an sRNAPredict2 search for putative sRNAs using multiple combinations of predictive features. The Venn diagram function is automatically executed following the completion of all individual searches to identify sRNAs predicted in multiple independent searches.

Stage IV

Developing the Initial Input File

The names of the primary data files, the desired name of the output file, and the values of various search variables are entered in sRNAPredict2 through an Initial Input File. If certain types of primary data files (such as a promoter database or a TIGR ORF database) are not to be included in the search, enter "none" in place of a file name.

A sample Initial Input File is included in the sRNAPredict2 directory. This file was created with BBEdit (a program the author recommends over other text editing programs) and its format may be altered when opened with other text *editing* programs such as TextEdit. Thus if the Initial Input File cannot be opened with BBEdit, it should be opened using Word.

Processing sRNAPredict2

1. Double-click on the icon of the executable.
2. The user will be prompted to enter the number of searches to conduct. Enter 1.3. At the next prompt, enter the name of the initial input file.

Multiple Independent Searches

sRNAPredict2 was designed with a Venn diagram function that allows putative sRNA-encoding genes that were predicted in multiple independent searches to be identified. This feature is particularly useful when searching for *putative* sRNA-encoding genes that are conserved in multiple species. The Venn diagram function is automatically *executed* when more than one *Initial Input* File is entered into sRNAPredict2.

1. Double-click on the icon of the executable.
2. Enter the number of searches to be conducted.
3. At the next prompt, enter the name of the first initial input file. Repeat this until all Initial Input File names have been entered.
4. After all Initial Input File names are entered, enter the name to be assigned to the Final Output File. This name will be assigned only to the Final Output File created by the Venn function. The output files of each of the independent searches will be named according to the names entered in their corresponding Initial Input File.

POINTS TO REMEMBER

1. Remove all "*" from the ".ptt" file before running the search. This can be accomplished by simply using the find and replace all functions in Word or a text editing program.
2. Copying/pasting databases from other browsers such as Safari may change the format of the database. To ensure databases copied directly from the browser conform to a format compatible with sRNAPredict2, compare them to the corresponding sample databases included in the sRNAPredict2 directory.
3. Add annotated sRNAs or riboswitches using the following format:
   "user gene name start coordinate end coordinate"

   For example:

   "user tmRNA 102333 102469"
4. The IGRExtract directory includes several sample files that can be used to test the program. When testing the program, enter "2160837" when asked to enter the size of the chromosome and "0" for the amount of sequence upstream/downstream of the IGR to be extracted. Compare the output file to "Test output.txt."

5. If input files are located in the same directory as the executable, entering only the names of the files is sufficient. If not, the entire path name must be entered (either manually or by dragging and dropping the files into the UNIX terminal window).

6. Comparison between two IGR databases will take significantly longer than a *comparison* between an IGR database and a whole genome but the former may be more sensitive for identifying short stretches of sRNA homology than the latter.

   If the IGR database of the partner species (the query sequence in the BLAST comparison) was created using IGRExtract, all "*" must be deleted or replaced from the FASTA names in this file before the BLAST comparisons are conducted (using the find and replace all functions in Word or a text editing program).

7. The start coordinate entered should correspond to the 3' boundary of the predicted promoter or TFBS. this is strand specific, i.e., if a promoter is predicted from positions 10 to 20 on the + strand, enter 20; if predicted from 10 to 20 on the - strand, enter 10.

8. The format of the output file of RNAMotif depends on the descriptor file used in the search. sRNAPredict2 was designed to extract coordinates from RNAMotif searches which use a specific descriptor file provided by D. Ecker. this file, "RNAMotif_descr.txt," is included in the sRNAPredict2 directory.

9. After compiling the source code using the "make" command, go into the transform file (in the src directory) and change the line "$path = `put a path here';" to a path pointing to the directory in which the file now resides, for example "$path = `/home/bob/favourite_programs/transform/src;'."

10. To ensure that the format of the QRNA output is *compatible* with sRNAPredict2, the BLAST comparison must be conducted as described in heading developing a trans term compatible input file. with the species of interest entered first (as the subject sequence) and the partner species entered second (as the query sequence) in the *command-line*.

11. The author has found that a window size of 100 and a slide position of 50 provided the best results for identifying experimentally confirmed *P. aeruginosa* sRNAs but different values might yield better results in other species. Be aware that increasing the window size can significantly increase the time it takes to complete a QRNA search.

12. sRNAPredict2 creates a number of intermediate output files that are passed from one function to another during the sRNAPredict2 search. these will be overwritten -every time the search is run unless removed from the directory in which sRNAPredict2 is located. Furthermore, if the name of the output file in the Initial Input File is not changed between searches, the previous output file is overwritten.

13. The Venn diagram function reports the total number of independent searches in which each predicted sRNA was identified. If each of the independent searches used conservation between the species of interest and a different BLAST partner as a predictor of sRNA-encoding genes, the name(s) of the BLAST partner(s) in which each sRNA was found to be conserved will be reported by the Venn diagram function in the Final Output File.

    The name assigned to each partner species in the Final Output File will be the name of the BLAST file OR the first word in the name of the BLAST file flanked by underscores. For example, if the name of the BLAST file is "BalGRBcO.txt" the BLAST partner name reported in the Final Output File will be "BalGRBc.txt." If the name of the BLAST file is "BalGR_Bc_0_.txt," the name in the Final Output File will be "Be."

Chapter 8

Protein-Protein Interaction

Proteomics as a science has depended on two major steps, namely, the separation and *identification* of individual components from a mixture or complex of proteins. *Historically*, the separation of proteins has been performed using two-dimensional gel electrophoresis (2-DE), combining *isoelectric focusing* of proteins in the first dimension with SDS-PAGE separation in the second.

While this technology has been available since 1975, tremendous advances have been made in the last 10 yr in the quality and *reproducibility* of protein separation, due to the advent of improved methodology such as immobilized pH gradients and "*narrow-range*" 2-DE, coupled with *simplification* and increased sensitivity of 2-D gel staining. It is not the intention of this chapter to focus on 2-DE methodology, because many excellent reviews are available elsewhere.

In more recent times, approaches to protein separation have included whole sample tryptic digestion followed by multiple rounds of liquid *chromatography* (LC) to separate peptides prior to their identification by mass spectrometry (MS). Such *technology* allows extremely rapid identification of individual proteins within a proteome.

As such, it is rapidly becoming the method of choice for answering the question, "is this gene expressed?" Initial reports suggest that well over 50% of a simple bacterial proteome, or 33% of a yeast *proteome*, can be identified in this large-scale manner. Recent reports suggest that the technology can also be used to characterize *post-translational* protein modifications.

For quantification or comparative analyses, peptides can be labeled with isotope tags (isotope-coded affinity tags, ICATs) that allow comparisons of peptide "*abundance*" for a number of peptides from within each protein on a *proteome*-wide scale. The second major step in proteomics is identification. Until recently, this required N-terminal Edman sequencing, an often slow and laborious process capable of identifying only a handful of proteins per day.

While this information is undoubtedly useful, proteomics requires the capacity for high throughput, high sensitivity, and high resolution. The advent of protein MS for identification rather than characterization purposes, revolutionized protein chemistry. The technology has advanced at such a rapid rate that many hundreds of proteins can now be identified in a single day and from femtomole (10^{-15}) amounts of *peptide* mixtures.

The combination of 2-DE or 2-DLC approaches with MS is now the accepted methodology for proteomics research. In this chapter, I will discuss how these technologies have been combined to yield new research discoveries in clinical and applied *microbiology.*

MAPPING BACTERIAL PROTEOME

2-DE provides a visual map of expressed proteins from a given bacterium. Many groups have begun to identify protein spots from these 2-D gel maps to create virtual reference libraries of visible proteins. These reference maps generally involve high-throughput mass *spectrometric* identification of abundant proteins from a range of 2-DE gels covering various pH ranges.

Such analyses can be broadened by taking a subcellular fractionation approach to map proteins specifically associated with bacterial membranes, or proteins secreted into the extracellular space. For whole cell proteomics, excellent reference maps have been established for a wide range of bacteria, including *Escherichia coli, Bacillus subtilis, Staphylococcus aureus, Mycobacterium tuberculosis, Mycoplasma pneumoniae,* and *Helicobacter pylori.*

Each of these maps contains the identifications of well over 100 protein spots, and they are useful for reconstructing physiological pathways under a chosen set of conditions. For example, an organism grown in rich. complex medium rarely expresses visible amounts of amino acid biosynthesis proteins, since *amino acids* are readily obtained from the medium.

On the other hand, under specific nutrient starvation conditions, or in the presence of different sole sugar sources, the organism will respond by expressing those proteins necessary to utilize the substrates present in that particular environment. Whole-cell reference maps also form the starting point for more complex studies, including strain and species comparisons, the *physiological* response to altered environmental conditions, and the regulation of gene expression.

Proteomics of Subcellular Products

Membrane and Extracellular Proteins

A great deal more useful information from bacterial proteomics can often be obtained by taking a *"pre-fractionation"* approach based on the separation of different subcellular fractions prior to 2-DE or peptide-LC separation. The stained 2-DE spot patterns are usually different, especially where the fractionation process leads to very pure fractions with little *contamination* from overlapping cellular compartments.

A second method is the "sequential" extraction process, where 2-DE is performed on solubility fractions, rather than sole cellular fractions. Subcellular fractionation has been used to examine bacterial membrane proteins from a wide variety of pathogenic species. The sodium carbonate enrichment-extraction method provides an excellent separation of Gram-negative outer membrane proteins (OMPs) and has been used for mapping and species comparisons in *Pseudomonas aeruginosa* and *Caulobacter crescentus.*

The method involves cell lysis by sonication or French Press and the removal of whole cells by centrifugation. Membranes are then precipitated in sodium carbonate, isolated by *ultra-*

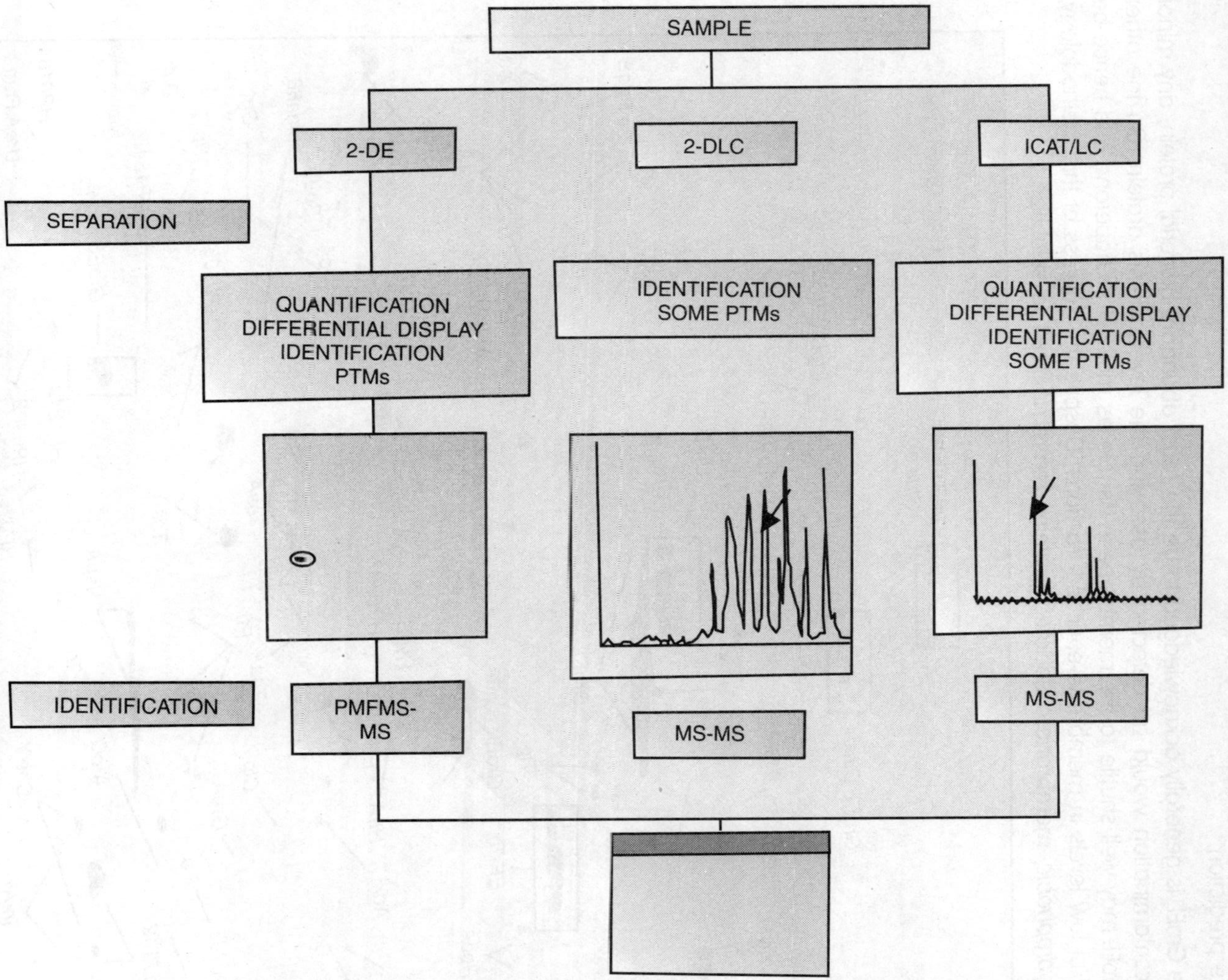

Figure 8.1: Schematic diagram of the proteomics process. Protein samples from whole cells, subcellular fractions, or affinity-purified complexes are separated by 2-DE, 2-DLC, or 1CAT prior to 2-DLC. Proteins and/or peptides are then identified by mass-spectrometric approaches including peptide mass fingerprinting (PMF) or tandem mass spectrometry (MS-MS).

centrifugation, and the proteins solubilized in 2-DE- or LC-compatible sample buffers. Separation of *P. aeruginosa OMPs* using 2-DE resulted in the identification of 189 proteins, where all except one (the ubiquitous chaperone GroEL) were predicted to maintain an outer membrane location by sequence prediction.

Since GroEL is generally acknowledged to be the most abundant bacterial protein, any minor cellular contamination would be detected by assaying the levels of this protein. On the other hand, GroEL may well shuttle to the membrane in its role as molecular chaperone and hence be detected at low levels in membrane-enriched fractions. Despite the success of this methodology, highly *hydrophobic*, membrane-associated proteins remain difficult to solubilize.

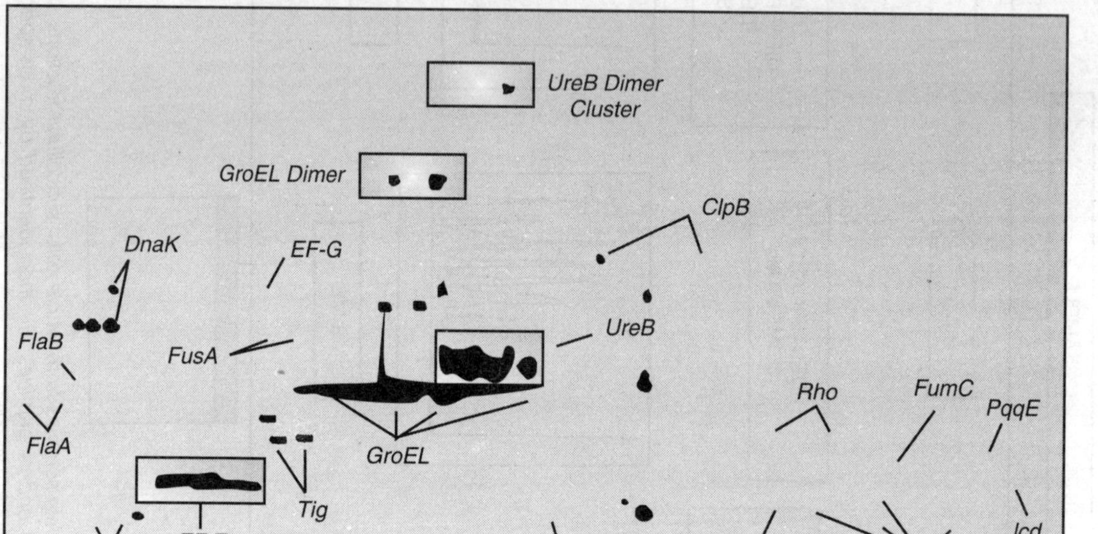

Figure 8.2: 2-DE reference map of H. pylori cytosolic proteins. Proteins were identified by PMF.

In *P. aeruginosa, only* 7 of the 189 identified proteins had positive theoretical hydrophobicity values, indicating that such proteins remain insoluble for electrophoretic purposes. When OMPs from *Leptospira interrogans* were separated using 2-DE following detergent extraction in Triton X-114, the major OMPs included several *lipoproteins* and proteins not previously characterized in the organism, including an abundant protein of 21 kDa.

A further enrichment step is to perform surface biotinylation and affinity selection prior to solubilization and 2-DE. In this case, provided the biotin label does not pass through surface pores and label *cytosolic proteins*, the affinity purification should select for surface-accessible proteins, and should also allow for the elucidation of surface-exposed peptide epitopes.

While Gram-negative membrane proteins are generally simple to enrich and separate, Gram-positive membrane proteins have been highly *problematic*. This appears to be a solubility issue, given that the thick peptidoglycan layer remains difficult to fractionate or remove entirely, and that the proteins themselves tend to be highly hydrophobic. In whole-cell analyses of the *S. aureus* proteome, envelope proteins were poorly represented, while subcellular fractionation of these organisms routinely resulted in fractions containing abundant cellular proteins.

For example, recent work by Hughes et al. characterized several *glycolytic* enzymes as being surface associated- in *Streptococcus agalactiae*. However, it remains unclear whether this is a true association or merely an artifact associated with poor enrichment of Gram-positive envelope proteins.

Investigations of bacterial surface proteins, particularly interstrain comparisons, provide one means of appraising potential vaccine and diagnostic candidates. A second fractionation can be used to examine proteins secreted by bacteria into culture supernatants. *Extracellular* proteins are often involved in pathogenic processes and include toxins, proteases, and other proteins involved in breaking down host defenses or the acquisition of *nutrients*.

Bacterial secretion is generally growth-phase dependent; therefore, the standard *methodology* is to grow cells in culture until the stationary phase is reached. The cells are then removed via *centrifugation* and filtration, and the free proteins in the culture *supernatant* are precipitated with ice-cold *trichloroacetic acid, acetone,* or *methanol*, alone or in combination.

With regard to the characterization of surface-associated proteins, fractions derived from culture supernatants are often a mixture of cytosolic proteins (present owing to cell *autolysis* and turnover during growth), membrane proteins (particularly those sheared off due to physical interactions in culture), and truly secreted proteins. The presence or absence of a signal sequence may aid in the identification of secreted proteins.

While this approach can be used for a variety of bacteria, it has been technically difficult to prepare *extracellular* fractions from organisms requiring protein additives such as serum in their growth medium, since these added proteins will also be detected using proteomics methodology. Recently, two groups have described studies of secreted proteins from *H. pylori,* an organism that is usually grown in the presence of serum.

Separation and identification of' culture supernatant proteins have also been performed for a variety of bacteria including *M. tuberculosis, B. subtilis, S. aureus,* and *P.aeruginosa* In *S. aureus,* culture supernatants were examined to detect virulence factors regulated by the SarA regulatory protein (a global regulator of S. *aureus* virulence factors), whereas in *P. aeruginosa* such comparisons were performed to examine differences between two strains with different modes of *pathogenicity*.

Development of Vaccine

The characterization of expressed proteins using proteomic technologies can be applied to the discovery of novel vaccine targets, antimicrobials, and diagnostics, and to determine what cellular effects newly synthesized drug compounds may induce. Affinity approaches, as outlined later in this chapter, are also useful for determining which proteins directly interact with a compound of interest.

Furthermore, proteomics is an excellent approach for monitoring the cellular response of bacteria to a given antimicrobial, first to provide a database of responses to known compounds, and then to begin screening the functions of new compounds in an attempt to characterize their mode of action.

As described above, particularly useful information about a microorganism can be gained by examining proteins associated with a chosen subcellular compartment. This becomes even more relevant when attempting to characterize immunogenic proteins from within a complex mixture. 2-DE combined with mass spectrometry and Western blotting using patient sera is a highly effective method of predicting potential vaccine targets.

However, this approach often detects highly abundant cytosolic proteins as immunogenic constituents, yet their role in pathogenicity remains controversial. It seems likely that proteins such as the chaperone GroEL and enzymes involved in glycolysis become immunogenic in the host, perhaps only following cell autolysis.

However, at least one study has shown convincing evidence that the "*housekeeping*" proteins *elongation* factor Tu, and pyruvate dehydrogenase beta subunit, are present as fibronectin-binding surface proteins in *M. pneumoniae.* This suggests that, at least in this *microorganism*, such proteins are capable of performing more than one function in vivo.

Prefractionation of surface-exposed and secreted proteins, combined with 2-DE and Western blotting, may provide much more focused information on potential vaccine targets. Furthermore, new methodology is also becoming available that utilizes surface display of genomic peptide libraries fused to *E. coli* OMPs, for screening against sera with high antibody titers.

This method also allows the identification of specific epitopes that interact with antibody. Interestingly, 60 *antigenic* proteins were characterized in *S. aureus,* and all were predicted to be surface or secreted proteins, suggesting that *immunoreactive* housekeeping proteins may well be an artifact when detected using the traditional Western *blotting* process.

APPLICATIONS OF PROTEOMICS

Proteomic "Signatures"

A proteomic signature is defined as "...the subset of proteins whose alteration in expression is characteristic of a response to a defined condition or genetic change". A signature may contain proteins involved in the same metabolic pathway, proteins with similar functions, and/or those with shared regulation. For example, proteins whose expression is up- or downregulated in response to oxidative stress include those responding directly to the environmental stimulus, as well as others (including general stress proteins and those with housekeeping functions) that are "*diagnostic*" or "*typical*" of the response, even though their functions are unrelated.

This approach can also be used to compare the responses of bacteria to antimicrobial compounds with both known and unknown modes of action.

The patterns of protein expression following exposure to the antimicrobials can be used to create a database of responses. Each signature is then compared with similar signatures, and used to predict the targets and mode of action of given compounds, both alone and by comparison to those with known targets.

For example, the proteomic signatures for the *S. aureus* response to cell wall-active antibiotics, show that this group of *antibiotics* induce or repress similar proteins, including peptide *methionine sulfoxide* reductase (which is indicative of oxidative stress) and GroES (which is indicative of protein synthesis stress), while antibiotics with different mechanisms cause wholly different protein expression profiles.

For large-scale screening, determination of the expression patterns produced by hundreds or thousands of compounds allows their *grouping* in terms of related mechanisms of action, based upon proteomic signatures.

Objectives

The widest use of microbial proteomics has been in the generation of maps of expressed proteins, and the application of these maps to the study of protein expression differences induced by changes in environment (e.g., heat/cold shock, *oxidative stress*) or *genetic background* (e.g., gene knockout and genetic overexpression). It is beyond the scope of this chapter to deal with every study that has used proteomics technologies in the study of bacterial physiology.

However, interested readers should consult the pioneering work of Neidhardt and VanBogelen in *E. coli,* and the work of Hecker in *B. subtilis.* These studies have elucidated the bacterial proteins involved in heat and cold shock, oxidative stress, the SOS response, nutrient starvation, the effect of antibiotics, and many other settings.

Proteomics can also be used to define *regulons,* or sets of genes and their corresponding proteins that are under the control or influence of a given regulatory system. The approach has been used to define proteins under the control of global regulators such as SigB in *B. subtilis* and *S. aureus,* the global virulence regulator SarA in *S. aureus,* Vfr in *P. aeruginosa,* and many others.

Proteomics technologies, especially in combination with "*transcriptomics,*" are therefore an indispensable resource for examining new regulatory systems in microbial genomes. In recent times, *proteomics* has been used as a tool for analyzing molecular differences between strains or species of bacteria with different phenotypes, e.g., virulent and *avirulent* strains of a given organism.

Excellent examples of this type of study include comparison of different clinical isolates of *H. pylori,* which was carried out in conjunction with Western blotting to detect immunogenic proteins, and comparisons of strains of *M. tuberculosis* and vaccine strains of *M. bovis BCG.* A similar study was performed to examine the protein differences between methicillin-susceptible and methicillin-resistant strains of *S. aureus*.

MODIFICATIONS AFTER SYNTHESIS

Posttranslational Changes

The major reason proteomics has established itself as an important technology in the postgenome era is its ability to focus on protein interactions and post-translational modifications (PTMs). While DNA microarray technology allows a comprehensive view of changes in gene expression induced by changes in environmental or genetic stimuli, by monitoring levels of RNA, it cannot monitor changes in protein modification induced by the same *stimulus*.

For example, although the levels of a particular mRNA may remain constant, a vast change in the phosphorylation state of the coded protein may occur. While it is true that there is a poor correlation between mRNA and protein abundance in higher organisms, and that mRNA is thus a poor predictor of protein levels in these organisms, this association has not yet been well tested in bacteria.

Microarray and proteomics studies of the *ctrA-cckA* two-component regulatory system in *C. crescentus* suggest that the correlation may be much better in prokaryotes. However, for a significant proportion of proteins changes in abundance under different biological conditions may be due to altered PTM state, rather than increased or decreased genetic expression.

This chapter concentrates on major PTMs such as phosphorylation, glycosylation, and fragmentation; however, analysis of PTMs in bacteria has not routinely been performed on a global scale.

Several methods exist to examine phosphorylation, including radiolabeling, antiphospho antibodies, and affinity enrichment prior to LC-MS, while the elucidation of specific phosphorylation sites relies heavily on comparative MS.

"Isoforms" of spots visualized on 2-DE gels rarely provide enough information to determine the PTM state of a protein species, unless combined with one of the above techniques. For 2-DE approaches, incorporating *radiolabel* into biologically active material provides one approach to analyze *phosphorylated* proteins globally.

However, the sensitivity of such labeling often means there is insufficient material remaining to allow identification and further characterization of the protein spots using MS, unless highly specialized sample concentration and desalting approaches are taken. Furthermore, the *sensitivity* of this approach also makes gel image comparisons with visible stains prone to errors.

A second method involves the use of overlapping antiphospho antibodies (including antiphosphoserine, phosphothreonine, and phosphotyrosine), but these approaches suffer from many of the same disadvantages as radiolabeling. Western blotting is not compatible with postseparation analyses such as MS, and triplicate 2-DE gels need to be performed prior to blotting and treatment with each of the available antibodies.

Furthermore, it is clear that the resolution and specificity of each antiphospho antibody is different, leading to some being more useful or reliable than others. Finally, no commercially

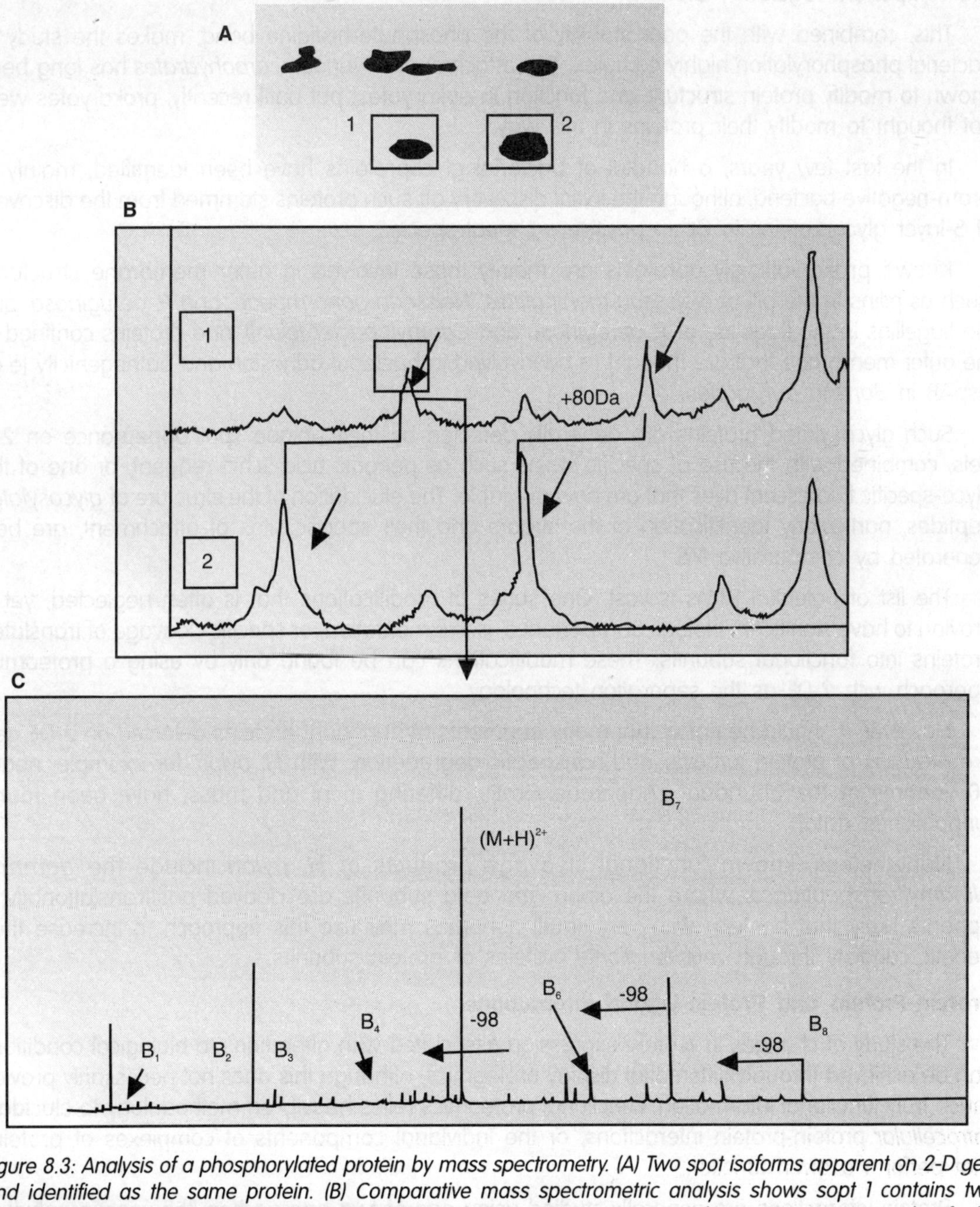

Figure 8.3: Analysis of a phosphorylated protein by mass spectrometry. (A) Two spot isoforms apparent on 2-D gels and identified as the same protein. (B) Comparative mass spectrometric analysis shows sopt 1 contains two phosphorylated peptides indicated by the addition of 80 Da. (C) Sequence analysis using MS-MS to identify the phosphorylated amino acid.

available antibodies exist for phosphohistidine, which is the crucial phosphorylation in bacterial two-component regulatory systems.

This, combined with the poor stability of the phosphate-histidine bond, makes the study of bacterial phosphorylation highly complex. The attachment of various *carbohydrates* has long been known to modify protein structure and function in eukaryotes, but until recently, prokaryotes were not thought to modify their proteins in this way.

In the last few years, a handful of bacterial glycoproteins have been identified, mainly in Gram-negative bacteria, although the initial discovery off such proteins stemmed from the discovery of S-layer glycoproteins in Gram-positive archaeal species.

Known prokaryotic glycoproteins are mainly those involved in outer membrane structures (such as pilins in the pili of *Neisseria meningitidis, Neisseria gonorrhoeae,* and *P. aeruginosa,* and the flagellins in the flagellae of *P. aeruginosa* and *Campylobacter jejuni),* and proteins confined to the outer membrane that are thought to be involved in bacterial adhesion and pathogenicity (e.g., OspAB in *Borrelia burgdorferi).*

Such glycosylated proteins are generally detected by their unique spot appearance on 2-D gels, combined with the use of specific stains such as periodic acid-Schiff reagent, or one of the glyco-specific fluorescent dyes that are now available. The elucidation of the structure of *glycosylated* peptides, particularly identification of the sugars and their specific sites of attachment, are best generated by comparative MS.

The list of potential PTMs is vast. One series of modifications that is often neglected, yet is proving to have significant biological importance, is fragmentation, or specific cleavage of translated proteins into functional subunits. These modifications can be found only by using a proteomics approach with 2-DE as the separation technology.

However, it should be noted that many fragments of abundant proteins detected on 2-DE gels are products of protein turnover and nonspecific degradation. With *H. pylori,* for example, nearly 20 variants of the abundant chaperone GroEL, differing in pI and mass, have been found (unpublished data).

Nonetheless, known functional cleavage products of *H. pylori* include the *gamma-glutamyltranspeptidase,* where the alpha and beta subunits are cleaved posttranslationally. It appears likely that bacteria with very small genomes may use this approach to increase their genetic capacity through *multifunctional* proteins or protein subunits.

Protein-Protein and Protein-Ligand Interactions

The study of changes in protein expression associated with alterations to biological conditions can be achieved through differential display proteomics, although this does not necessarily provide much truly functional information. Functional proteomics relies heavily on methodology to elucidate *intracellular* protein-protein interactions, or the individual components of complexes of proteins that perform a given function.

Protein interactions are generally studied using one of two approaches: the yeast two-hybrid system, or affinity capture combined with mass spectrometry of the individual components. As with global analyses of PTMs, little work has been performed to characterize protein complexes

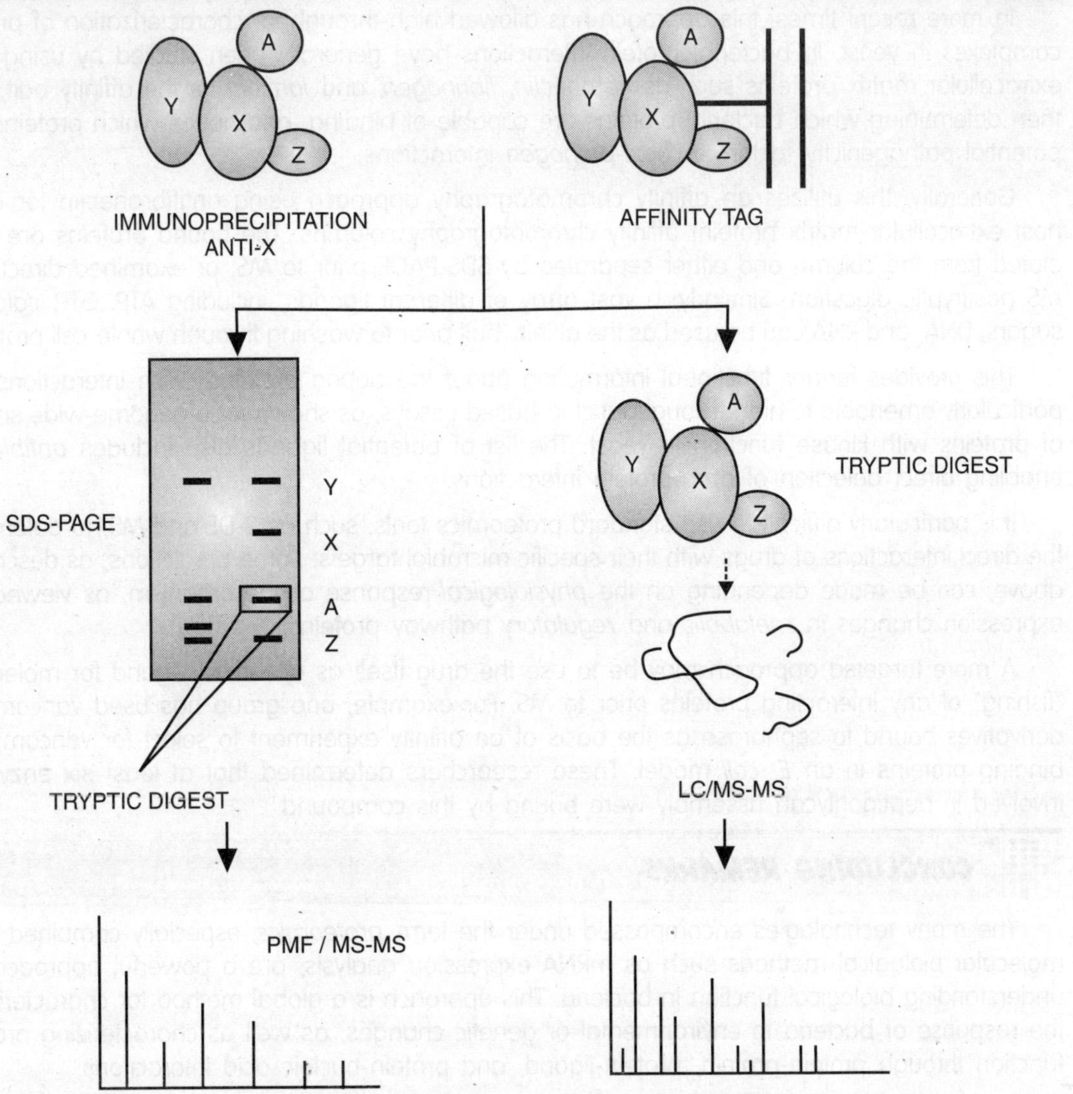

Figure 8.4: Identification of protein-protein interactions or components from a protein complex. Proteins are either immunoprecipitated with an antibody against a protein of interest, or purified using an affinity column. Complexes can be separated with SDS-PAGE or digested immediately with trypsin prior to 2-DLC/MS-MS.

or interactions in bacteria, although a genome-wide survey of interactions was performed using the yeast two-hybrid system in *H. pylori.*

The two approaches contrast in their utility for bacterial analyses. The yeast two-*hybrid system* has remarkably high-throughput and can assay an entire genome of "*baits*" and "*preys*" for potential interactions, but it remains highly predictive, and further assays must be carried out to verify whether the proposed interactions occur in vivo. The affinity-capture approach begins with a target "protein of interest" and uses antibodies to capture that target in vitro, and those proteins to which the target is also bound in the living cell.

In more recent times, this approach has allowed high-throughput characterization of protein complexes in yeast. In bacteria, protein interactions have generally been studied by using host extracellular matrix proteins such as *fibronectin, fibrinogen,* and *laminin* as the affinity bait, and then determining which bacterial proteins are capable of binding, and hence which proteins are potential pathogenicity factors in *host-pathogen* interactions.

Generally, this utilizes an affinity chromatography approach using antifibronectin (or other host extracellular matrix protein) affinity chromatography columns. The bound proteins are then eluted from the column and either separated by SDS-PAGE prior to MS, or examined directly by MS posttryptic digestion. Similarly, a vast array of different ligands, including ATP, GTP, calcium, sugars, DNA, and RNA can be used as the affinity bait prior to washing through whole-cell proteins.

This provides further functional information about the bound proteins. Such interactions are particularly amenable to high-throughput chip-based assays, as shown for a genome-wide screen of proteins with kinase function in yeast. The list of potential ligands also includes *antibiotics,* enabling direct detection of drug-protein interactions.

It is particularly difficult to use standard proteornics tools, such as 2-DE and MS, to determine the direct interactions of drugs with their specific microbial targets. Some predictions, as described above, can be made depending on the *physiological* response of the organism, as viewed via expression changes in *metabolic* and *regulatory* pathway proteins.

A more targeted approach may be to use the drug itself as an affinity ligand for molecular "fishing" of any interacting proteins prior to MS. For example, one group has used vancomycin derivatives bound to sepharose as the basis of an affinity experiment to select for vancomycin-binding proteins in an *E. coli* model. These researchers determined that at least six enzymes involved in peptidoglycan assembly were bound by this compound.

CONCLUDING REMARKS

The many technologies encompassed under the term *proteornics,* especially combined with molecular biological methods such as mRNA expression analysis, are a powerful approach for understanding biological function in bacteria. This approach is a global method for characterizing the response of bacteria to environmental or genetic changes, as well as characterizing protein function through protein-protein, protein-ligand, and protein-nucleic acid interactions.

Furthermore, proteins involved in host-pathogen interactions can be rapidly identified from within a complex mixture. Finally, proteomic studies are essential for *comprehending* posttranslational modifications of proteins that may be an influence in biological function, and that cannot be determined using a molecular biological approach. In the microbial *postgenome* era, proteomics has become, and will remain, an important tool for understanding the biology of microbial pathogens.

Chapter 9

Genetic Diversity in Bacteria

The traditional methods of typing bacteria are based on phenotypic characteristics such as *serotype* and phage *susceptibility*. Although these methods are widely used and remain important, they are gradually being replaced by methods that differentiate strains on the basis of genotypic differences, such as pulsed-field gel *electrophoresis* (PFGE), amplified fragment-length polymorphism (AFLP), and *mulilocus* sequence typing (MLST).

Both genotypic and *phenotypic* methods can be tailored to give information about strains that may be useful for disease management (e.g., drug resistance markers), but current *genotypic* methods generally rely on the analysis of anonymous markers that have no known phenotypic correlates. To illustrate this point, one of the features of MLST is that the products of the genes analyzed have essential *"housekeeping"* functions, and therefore it is highly unlikely that *polymorphisms* in the open reading frames would have gross phenotypic consequences.

A disadvantage of many of the early genotyping methods such as PFGE and restriction fragment-length polymorphisms (RFLP) was that laboratories found it difficult to standardize methods and compare data. Despite this, the genotypic methods are finding favour because they offer greater *flexibility* and *discriminatory* ability than the phenotypic methods.

Micro-arrays

The term *array* is applied to a range of different technological platforms. A common feature of all arrays is that they *comprise* a set of defined *nucleic acid* probes, each placed at specific X, Y coordinates on a surface. The *probes* on the array can be exposed to labeled target *nucleic acids,* and hybridization occurs if complementary sequences are present.

The term *microarray* is usually applied when probes are placed a very small distance apart (approx 200 μm). *Robotic* laboratory arrayers that exploit split-pin or ring-and-pin technology can produce microarrays comprised of > 20,000 *precisely* positioned probes on standard *low-fluorescence microscope* slides. The construction, handling, and potential applications of this type of microarray are described and discussed in this chapter.

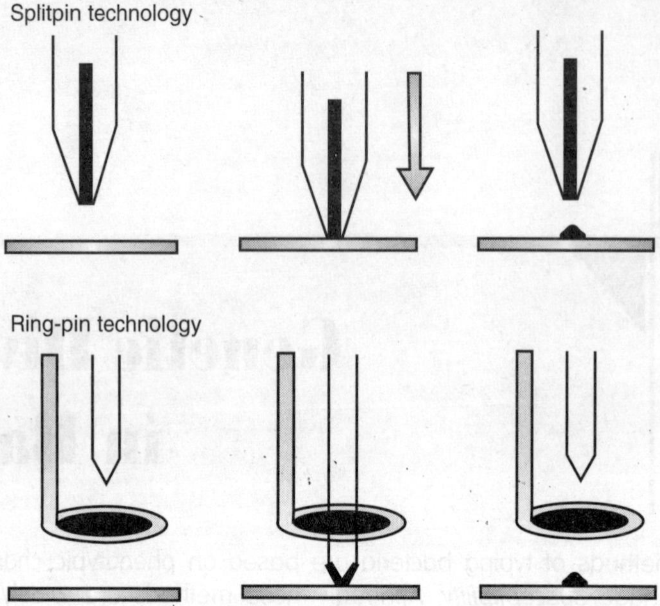

Figure 9.1: Schematic diagram of probe deposition onto the slides using split-pin technology (top) and ring-pin technology (bottom).

High-density microarrays are available *commercially* and comprise up to 100 to 200 times more probes per unit area than can be achieved using a laboratory arrayer. These arrays are produced using photolithography to build the DNA oligonucleotide probes *in situ (see,* for example, http://www.affymetrix.com)

Microarrays represent a powerful new method for analyzing bacterial *genotypes*. The main advantage of arrays is their *flexibility* in both the format of the array and the versatility of the probes to test sample material. The great promise of *microarray* technology lies in the fact that the probes can be used to detect DNA sequences linked to specific phenotypic characteristics. Future arrays might be used to generate typing data for *epidemiological* purposes linking disease presentation, *transmissibility*, and changes in the bacterial and host populations.

Bacterial Genomic Sequences

The number of complete bacterial genome sequences available is expanding rapidly. The technical and *logistical* problems presented by sequencing a few megabases of DNA appear to have been solved, principally owing to advances in robotics and the software for assembly of the data. Currently, the major *bottleneck* appears to be the full *annotation* of the available sequence data.

Many of the major bacterial pathogens that infect humans now have several fully sequenced strains, and precise companion of these genomes is now possible. This work illustrates the significance of lateral gene transfer and gene deletions as mechanisms in the evolution of bacterial species. Although genera vary in the degree of genomic plasticity evident, a common feature is the presence of a core of genes that are always present, together with a category of "*divergent*" genes that are present in only some strains. These genes are usually clustered, indicating that they are lost or acquired as a unit, and are often found in the vicinity of *chromosomal* elements that are associated with gene translocation (e.g., *transposons* and *bacteriophage* components). Comparison of complete *genome* sequences is likely to remain the most informative and *sensitive* method for typing bacterial strains.

Unfortunately, for the present, genome sequencing remains far from being a practical typing method for most purposes, because of its high cost. However, the rapid decrease in the cost per base of sequencing shows no current sign of coming to an end. Furthermore, technologies for resequencing (i.e., finding variants in a known sequence) in particular, are very promising in terms of low cost and high capacity.

For this reason it seems entirely plausible that in the not-too-distant future, comparison of complete genome sequences will be the primary method used for *epidemiological* typing. Bacterial gene microarrays use the newly acquired wealth of sequence information to provide a practical shortcut to this ultimate goal of *whole-genome* comparison.

Using arrays, genomes are compared gene by gene instead of base by base. This approach provides data comparable with that of sequencing in terms of the presence and absence of genes, but at a fraction of the current cost.

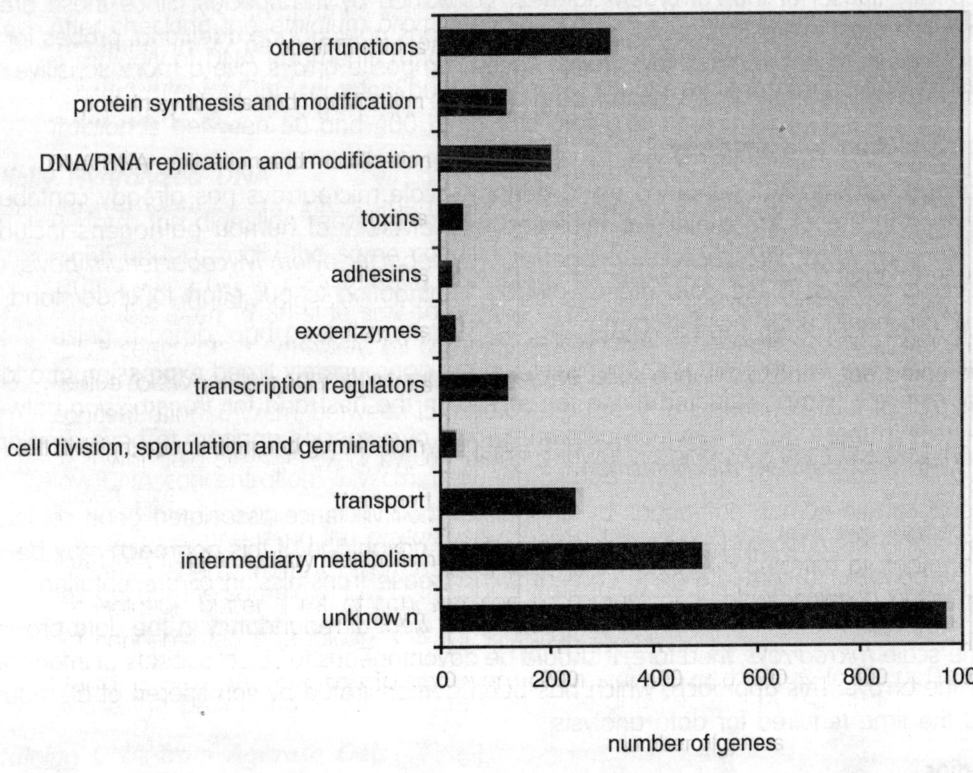

Figure 9.2: The functions of the 2595 ORFs of Staphylococcus aureus *strain N31.5 are shown. Approximately two-thirds of the ORFs encode proteins with function that are either known or suspected from homology.*

Genomotyping

Genomotyping is a new term applied to describe the analysis of bacteria by comparison of their genomes using microarrays. For example, Dorrell and

colleagues have described the use of arrays comprised of probes derived by amplification of clones produced for sequencing the genome of *Campylobacter jejuni*. Each probe was thus specific for a part of the genome of the strain used to generate the clones. When DNA from different strains was *hybridized* to the array, it was possible to identify sequences that were present only in the strain used to construct the array, from the pattern of probe *hybridization*.

The advantage of this approach is that it is relatively inexpensive, because a single primer pair (with *annealing* sites in the cloning vector) can be used to amplify all of the necessary probes. Unfortunately this method has the significant disadvantage that a single probe may include sequences derived from more than one gene and, *consequently*, the data may be *misleading*.

An alternative method of array construction, which avoids this drawback, is to design probes specific for each gene in a bacterial genome sequence. Probes for individual genes may be produced by amplification using specific PCR primers, or may be *oligonucleotides*. Computer software for the design of large numbers of PCR primer pairs and oligonucleotides is available.

An array of probes for every open reading frame (ORF) in a bacterial genome will include only a strain-specific subset of the "*divergent*" genes maintained by the species. Since these are, in some ways, the most interesting genes, most array designs now include additional probes for the "*divergent*" genes present in other key strains. These composite arrays give a more sensitive and complete genomotype than arrays based on the genes of a single strain.

Molecular Evolution and Diversity

The analysis of bacterial genomes using genome-scale microarrays has already contributed to our understanding of the molecular evolution and diversity of human pathogens including *C. jejuni, Vibrio cholerae, Streptococcus pyogenes, Helicobacter pylori, Mycobacterium bovis,* and *Staphylococcus aureus.* These data are a valuable contribution to our effort to understand the *pathogenic* mechanisms of these bacteria.

It is already clear that human pathogens rely on the appropriately timed expression of a large number of different genes, including those for survival in the host and for *transmission* between hosts. Diversity in these genes between different strains of a species appears to have developed in response to host defense *mechanisms*.

Typing of certain human pathogens by analysis of their virulence associated gene content is likely to be highly *discriminatory*, but the most significant advantage of this approach may be that it should be possible to correlate gene content with disease transmission or presentation.

From a typing viewpoint, there is likely to be a great deal of redundancy in the data provided by genome-scale *microarrays*; therefore, it should be advantageous to select subsets of informative genes for the arrays. This approach. which has been demonstrated by van Ijperen et al. reduces costs and the time required for data analysis.

New Studies

The current microarray format, in which each glass slide may carry several thousand probes, is well suited to *hybridizations* between target material and genome-scale probe arrays. However, although several arrays of less than 1000 probes can easily be accommodated on a single slide, it is not then a simple matter to hybridize each array to a different target.

Systems that allow hybridization of multiple targets to a single slide would be very convenient for typing applications. This equipment is likely to become available as *probe* sets for typing are developed further. *Genomotyping* has been the main focus of bacterial typing work using *microarrays.*

However, an alternative would be to develop arrays of probes each complementary to a different amplified fragment made by the AFLP method. Hybridization of the array to targets composed of AFLP fragments from an individual strain would result in a strain-specific pattern of reactions. The *discriminatory* ability of arrays of AFLPs would depend upon the choice of probes used, and could be very high.

In contrast to the *genomotyping* approach, the probes used would be selected on the basis of their contribution to the discriminatory ability of the typing method. In AFLP, the absence of a *fragment* in the amplified material shows only that one of the restriction sites is absent. This may indicate that a single base mutation has occurred in one of the sites or that the sequence is entirely absent.

REQUIREMENTS

Slides

Microarrays are usually printed onto glass slides coated with reactive groups. Slides with different coatings are available and are used in variations of the microarray method. There are two classes of slides in common use, and these are most suitable for spotting either PCR products or *oligonucleotides*. Slides for spotting PCR products are usually coated with polylysine or aminosilane, these surfaces bind unmodified DNA covalently via negatively charged phosphate groups.

The second class of slides, which are modified with aldehyde and epoxy groups, are usually recommended for printing oligonucleotides that have a free *aminolinker*. The amino group of the modified oligonucleotide binds to the active groups on the slide. This has the advantage that all of the *molecules* are bound in the same place and have the same *orientation* on the slide.

This is important for relatively short oligonucleotide sequences, since steric hindrance can significantly affect the specificity and stability of probe/target *hybridization*. Although *nonmodified* oligonucleotides can be covalently linked to *epoxy* slides, longer probes (>50-mer) are recommended, and care should be taken to obtain good crosslinking by UV treatment or baking.

Amplified Probes

In microarray technology, the PCR products or oligonucleotides spotted onto the slide are designated *probes.* Amplified probes are usually produced by PCR using gene-specific primers, with the genomic DNA of the organism of interest as substrate. Each PCR product represents a specified ORF and may either be an ORF-specific fragment or the complete ORF. Following amplification, the probes are checked for purity and identity.

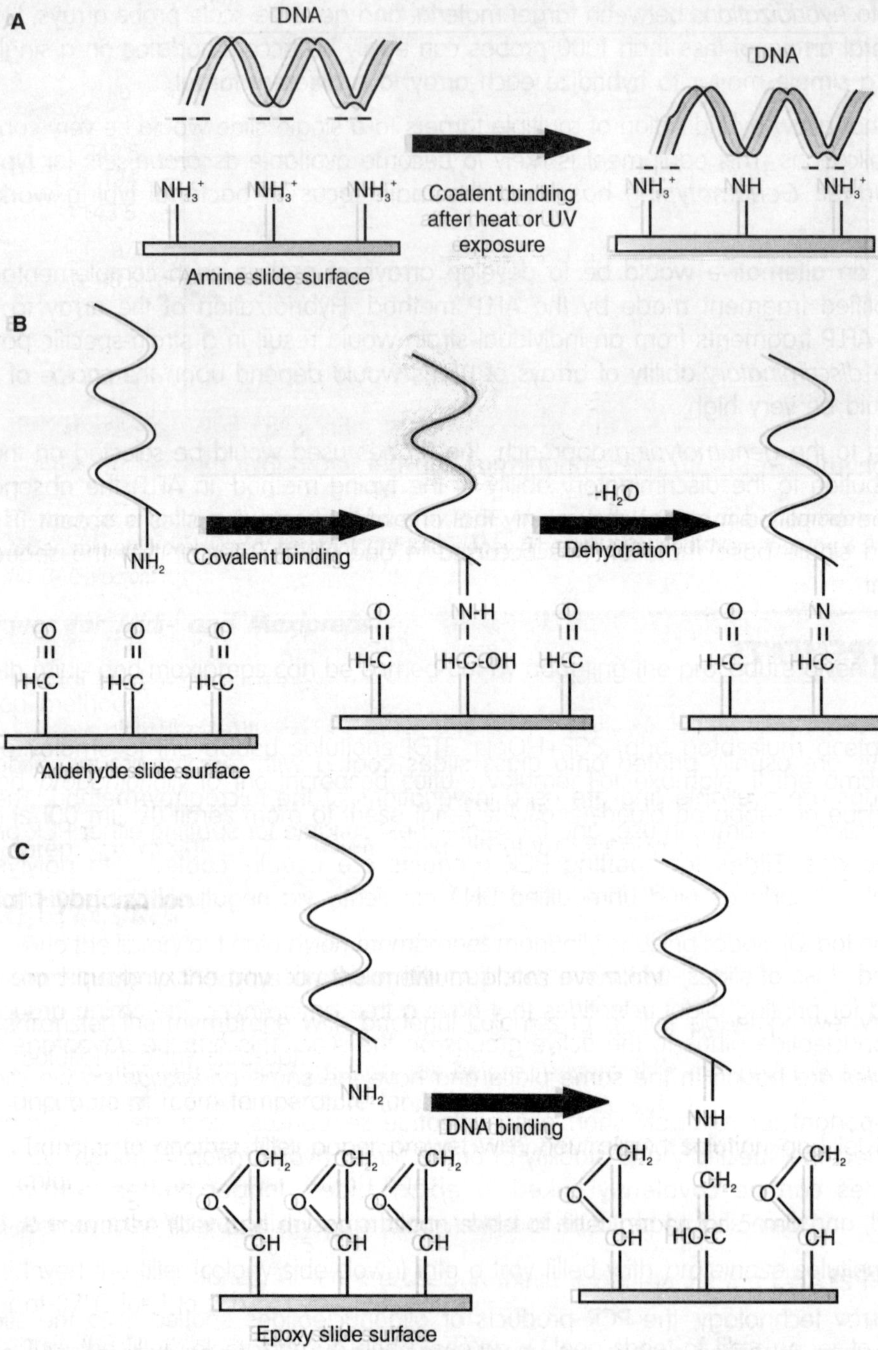

Figure 9.3: Binding of DNA onto different slide surfaces. (A) Binding of DNA to amine slide in which positive amine groups covalently bind negatively charged DNA phosphate backbone. Binding of amino-linked oligonucleotides to (B) aldehyde and (C) epoxy slide surfaces by an active binding of the free amino group.

The most common approach is to analyze the products of each PCR by agarose-gel *electrophoresis* to check for the presence of a single amplicon of the expected molecular size. A proportion of products that varies depending on the resources available, are then sequenced to provide a positive identity check prior to arraying. When oligonucleotides are used as probes, most of these steps are unnecessary.

Oligonucleotides ordered from commercial companies are purified and ready to print as soon as they arrive in the laboratory. This saves the time and resources required to run PCR reactions, analyze the products. and redesign any PCRs that do not give the expected product. However, short oligonucleotides are best bound to the glass surface via an aminolink group, and this increases the cost of synthesis.

Depending on the length, oligonucleotides have the advantage of giving more specific hybridization signals. The longer PCR-amplified products, however, might detect similar genes with partial *homology* to the gene of interest, which may be useful for the detection of gene homology.

The probes are dissolved in solutions that interact *physically* with the slide surface via the *depositing* device to give aliquots of consistent volume that dry as round spots of equal size and shape. The spotting solution must also be of suitable volatility to ensure rapid drying of the deposited spots while preventing excessive evaporation of the sample during the printing process. This is also influenced by the local temperature and *humidity*.

A common spotting solution is 50% *dimethyl sulfoxide* (DMSO), which is normally used on *nonactivated* surfaces. However, DMSO can interfere with some active groups on the slide surface, and it is less suitable for the printing oligonucleotides than alternatives such as saline-sodium citrate (SSC) solution. A wide range of cornmercial spotting buffers are also available.

Changes in the volumes of the probe solutions dispensed into microtiter plate wells for spotting *inevitably* occur over time, even when care is taken to avoid the evaporation of solvent. To preserve valuable probes, plates affected by evaporation may be dried down fully and the probes redissolved in the original volume of solvent. It is usually unnecessary to adjust the volume for probe used in a previous print run.

Objectives

In the usual nomenclature, the *"target"* is the nucleic acid that is hybridized to the array. A wide range of target formulations arc used. The well known application of microarrays to gene-expression studies uses labeled cDNA made from mRNA extracts by reverse transcription. Multiplex PCR products may be used, and genomic DNA may be most appropriate for typing or screening for the presence or absence of genes.

Target nucleic acids are labeled with fluorescent dyes by either direct or indirect *incorporation* methods. The most common dyes used are cyanine three (Cy3, green) and cyanine five (Cy5, red). These two dyes have different absorption and emission spectra and can therefore be used in the same hybridization experiment.

The control and a test sample are each labeled with a different dye, then mixed and hybridized to the array. The hybridization signal is detected with a high resolution scanner. The result at each probe is either a green, red, or yellow (mixed) signal. Green or red signals indicate that only one of the targets has hybridized to the probe, and yellow shows that both control and test sample have *hybridized*.

Direct incorporation of the Cy dyes by some polymerases is not always efficient due to steric hindrance by the large Cy groups. The degree of inhibition depends upon the structure of the Cy dye, and this may cause inaccuracies in experiments in which the ratio of two dyes is measured. Alternative indirect *labeling* methods have been developed using *amino-allyl-modified* dNTPs.

The small amino-allyl dNTPs are incorporated into the target nucleic acid at a rate similar to that of unmodified nucleotides, and Cy dye is then bound to the target DNA by ester bonding. This approach gives good yields of highly labeled target, and thus minimizes experimental variation. Furthermore, as all Cy dyes can be incorporated equally, the *normalization* of signal intensities is less problematic.

Prehybridization Buffer

During the three stages of a microarray experiment various buffers are used. The first step is to treat the slide with a blocking or prehybridization buffer before addition of the target, the purpose being to prevent nonspecific binding by blocking remaining sites on the glass surface that could interact with *nucleic acids*. Blocking buffers usually contain bovine *serum albumin* (BSA), sodium dodecyl sulphate (SDS), and SSC, but, depending on the application, competitor DNA may also be added.

For example, addition of *C_0t- I* DNA to the buffer reduces the hybridization of repetitive elements by binding to the labeled repetitive target sequences, while tRNA acts as a blocker of nonspecific hybridization. The next stage is the hybridization of target to the array. Various commercial hybridization buffers are available, but in-house buffers are also widely used. The use of formamide in the hybridization buffer has the advantage that the DNA remains *denatured* at lower temperatures.

Hybridization at a lower temperature results in reduced evaporation from the target solution during *incubation*; evaporation leads to high background signals in locations where target has dried. Following *hybridization*, several washes are carried out to remove nonspecifically bound target.

The stringency of the *post-hybridization* washes affects the strength of the hydrogen bonding between the probe and target. Washing stringency can be adjusted by changing the temperature and SSC concentration (*stringency* is raised at higher temperatures and lower salt concentrations). SDS is *autofluorescent* and should be omitted from the buffer used for the final post-hybridization wash.

Hybridization Studies

Water-tight chambers (Corning, Schiphol-Rijk, The Netherlands; and Genetix Limited, New Milton, UK) that fit the microarray slide exactly are used for hybridization. The chambers prevent the slide from drying out when used within a hybridization oven, but they can also be used in a water bath. Hybridization takes place under a coverslip so that a small amount of target is evenly spread over the array.

Automatic hybridization stations are also commercially available. In these systems, hybridization takes place in a low-volume chamber with access ports for the addition of buffers. An advantage of this equipment is that the target solution can mix more freely than under a coverslip so the target has a greater opportunity to interact with the probe array; this increases signal intensities and ensures that all positions on the array are equally exposed.

Scanners and Software

A variety of fluorescent scanners are now available for microarray work. The machines generally include a highresolution confocal microscope scanner (.10 to 4 μm), and have lasers with the appropriate wavelengths as light sources. The standard lasers excite Cy3- (532 nm) and Cy5- (635 nm) labeled targets, but up to six different lasers are now included with recent designs. Scanners generally come with user-friendly operation software and are therefore simple to use.

The Affymetrix 428™, the Axon Genepix 4000[1], and the PerkinElmer ScanArray™ are most commonly used. Similarly, a variety of software packages are available for *microarray* image and data analysis. Although there are differences, most programs have basic functions to allow measurement of signal intensities and to present the data in convenient formats (i.e., *histograms*, *scatterplots*, and so on)., The complexity of microarray data presents great challenges to the analysis software.

The available programs are under continuous development, and new upgrades contain more and better tools for *normalization*, *analysis*, and *statistical analysis*. Any microarray software listing will be rapidly outdated due to the frequent introduction of new programs. Software should be selected to fulfil the needs of the individual user. Bacterial comparative genomics is a relatively simple application, and consequently the major software consideration is likely to be ease of use.

PROCEDURES

Preparation and Printing of Probes

Probes often comprise specific parts of ORFs that are amplified by PCR and then purified before printing. In the case of oligonucleotide microarrays, the probes are obtained from custom synthesis houses and may be *aminolinkermodified* depending on the application and slide coating employed.

The probes spotted onto the slides must be very specific for the detection of the target gene or genes, since cross-hybridizations lead to array results that are very difficult to interpret. The steps required to prepare oligonucleotide or amplified probes are summarized in elsewhere in this chapter. Microarrays are usually spotted using *automated robotic arrayers*, although manual instruments are available that may be suitable for producing arrays comprised of relatively few probes.

Optimal environmental conditions for printing are between 40% and 50% humidity at around 20°C. These conditions prevent spots from drying too quickly. Handling of the slides after printing depends on the slide coating. Crosslinking by UV illumination is recommended for some slides, but quoted energy values vary and convincing optimization data are not available. Baking at 80°C for up to 2 h with occasional reintroduction of water by "fogging" the slide over warm water is recommended for some slides.

Some protocols call for both UV and heat treatments, while others recommend curing at room temperature. Currently, most authors and slide manufacturers take an empirical approach to probe attachment, and a wide variety of conditions are satisfactory.

Hybridization and Posthybridization Techniques

For genomotyping experiments where the aim is to show the presence or absence of genes within the genome of a particular bacterial strain, the target is comprised of labeled total DNA. Cy dyes can be incorporated directly into representative DNA by primer extension using a random priming kit (Invitrogen, Paisley, UK). Purification of the target prior to hybridization is very important.

For each hybridization on a 22 mm × 22 mm surface, 15 tL of probe is needed. The procedure is summarized in elsewhere in this chapter. Labeling of nucleic acid using direct methods is inefficient for some combinations of enzyme and modified *nucleotide*. Cy dyes in particular have a large 3D structure and are therefore relatively *inefficiently* incorporated into DNA.

Indirect labeling methods have been developed that use relatively small aminoallyl dNTPs as substrate for the enzymatic reaction. These reactive primary amines react with NHS ester-modified Cy dyes in a nonenzymatic second step.

Methods of Analysis

Currently, the most time-consuming part of microarray experiments is data analysis. The reason for this is that many factors must be considered before any conclusions can be drawn from the raw data. The first steps in data analysis involve normalization of the images. Standard methods of normalization calculate either the mean or median intensity for a microarray image.

This value is then used to correct the intensity of each spot. The mean or median of all corrected spot intensities is finally adjusted to center around one, and then converted to $\log_{10}$ form so that the points follow a normal distribution centered around zero. Using logarithmic transformed data helps to normalize the distribution of the data and improves the applicability of statistical tests.

Most statistical tests work with differences, although scientists are more interested in a fold difference. This problem is solved by working with a log ratio, as $\log a - \log b = \log a/b$. The next step in analysis is to detect outliers. If the intensity of a spot is influenced by the strong fluorescence of a dust particle on the slide, this spot should be excluded from further analysis.

In this way the data are reduced to a series of values that are in the form of ratios when different targets, each labeled with a distinct fluor, have been hybridized to the array. At this point the question to be considered is whether any differences observed are reproducible or significant. To answer this question, duplicate experiments are required, which in turn generate yet more data to analyze.

REALISTIC GOAL

Microarrays for typing is clearly not a holy grail, i.e., the objective of a quest that is unlikely to be realized without divine intervention. However, current microarray technology is not well-suited to high-throughput typing of many different strains. *Considerable* hands-on time is required for each sample, and the analysis can be *time-consuming*, especially while setting up a new application.

At present, genomotyping using microarrays should be viewed as a complementary method that can provide large quantities of gene-content data and is appropriate in that context. These comparative genomic data allow the construction of detailed overviews of the relationships between strains and are extremely valuable.

We anticipate rapid progress in the development of protocols and automation for microarray hybridization. These developments are likely to lead to the use of genomotyping as a routine tool for bacterial typing. This is certainly a realistic goal.

POINTS TO REMEMBER

1. Genomic DNA hybridization on microarrays. When genomic DNA is used for the preparation of target material, up to 2 μg can be labeled directly using Cy-labeled nucleotides and Klenow enzyme. Removing enzyme and unincorporated nucleotides is crucial for a low background signal after hybridization.

   The target is dried under vacuum and dissolved in 15 μL of *hybridization* buffer. After denaturing, the target can be applied to the slide, covered with a disposable coverslip, and incubated in a chamber at the appropriate temperature. After hybridization, slides are washed several times in 2X SSC buffer containing 0.1% SDS, and finally in 2X SSC buffer. Slides are dried by *centrifugation* in 50-mL tubes.

2. Probe preparation. Probe preparation by PCR amplification includes the following steps: First, specific oligonucleotide primers have to be designed using the appropriate software, followed by the PCR *amplification* of the product.

   After removing the enzyme and unincorporated nucleotides, the identity of the products can be verified by size and/or sequencing. Finally, the *concentrations* are adjusted where necessary and probes are diluted in the appropriate *spotting* solution. If probes are prepared by *oligonucleotide* synthesis, fewer steps are involved.

   Oligonucleotides are designed using *specialized* software and *hioinformatics* knowledge, and users decide on the basis of the length of the oligonucleotide whether an aminolink is necessary. Oligonucleotides are produced synthetically by commercial companies. After adjustment of the concentration, probes are ready for printing.

3. Array spotting technology. Various techniques for spotting small volumes of probe solutions have been described. The most common method of array printing currently is the use of capillary reservoir pins (split pins), which deposit picoliter volumes of probe onto the glass surface.

   The ring-pin system is similar in that pins are used to deposit the probe solution, but a loop filled with probe solution acts as the reservoir with the pin passing through the probe meniscus each time a spot is printed. Another method uses an *inkjet-type* spotter that fires the probe onto the surface without touching the surface.

Chapter 10

Genome Alignment

As genomes evolve, mutational forces and selective pressures introduce *rearrangements* via inversion, *transposition*, and *duplication*/loss processes. Lateral transfer can introduce novel gene content, or in the case of *homologous* recombination, introduce more subtle changes such as allelic substitution.

Through genome comparison we hope to first identify differences among organisms at the genome level and then infer the biological significance of differences and similarities among related organisms. Traditional sequence alignment methods were developed to *accurately* align individual gene sequences where rearrangement rarely occurs.

When aligning two or more genomes, shuffled regions of orthologous sequence may be interspersed with paralogous and novel sequence regions, forcing a *genome* alignment method to map segmental *homology* among genomes. Once a segmental homology map exists, the alignment task can be approached using traditional alignment methods based on dynamic *programming* and pair-hidden Markov model heuristics.

In genome alignment, it is important to go beyond simple classification of sequence regions as either homologous or unrelated. Local alignment programs such as Basic Local Alignment Search Tool provide such functionality. We define a global genome alignment to be a complete catalog of *orthologous*, *xenologous*, *paralogous*, and *unrelated* sites among a group of genome sequences.

All sites must be assigned to one of the given categories, and in the cases of *orthology*, *paralogy*, and *xenology*, the related sites must be identified. Furthermore, homologous sites should be grouped into regions of maximal *collinearity* such that the leftmost and rightmost site in any group defines a breakpoint of rearrangement. We refer to such maximal collinear sets of homologous sites as locally *collinear* blocks (LCBs) because they cover a "*block*" of sequence without any internal genome rearrangement.

Such a *categorization* of sites implicitly defines breakpoints of rearrangement, recombination, duplication, and insertion and deletion processes. A genome alignment lends itself to downstream evolutionary inferences such as rearrangement *history, phylogeny, ancestral* state prediction, and detection of selective pressure in coding sequence and in noncoding sequence.

Most current genome *alignment* systems, including Mauve, construct an incomplete form of genome alignment as defined above. Shuffle-LAGAN aligns both single-copy and repetitive regions in pairs of genomes, but does not classify them as either orthologous or paralogous. *Mauve* aligns *orthologous* and *xenologous* regions, but does not distinguish between the two cases.

Mauve also aligns orthologous repeats, but does not align paralogous repeats. Mulan and M-GCAT both construct alignments similar to Mauve. Recent versions of MUMmer identify and align orthologous and paralogous sequence, but among pairs of genomes only. Some earlier genome alignment systems have treated *segmental* homology mapping as a separate step and thus assume sequence collinearity.

Mauve performs five basic. steps when constructing a genome alignment:

1. Search for local multiple alignments (approximate multi-MUMs).
2. Calculate a phylogenetic guide tree using the local alignments.
3. Select a subset of the local alignments to use as anchors.
4. Conduct recursive anchoring to identify additional alignment anchors.
5. Perform a progressive alignment of each LCB.

The METHODS section describes each of these five steps in greater detail.

Before performing an analysis with Mauve, a researcher should first ask whether genome alignment is the right analysis. If the answer is "yes," the next question should be "Is Mauve the right tool for the job?" Mauve performs best when aligning a relatively small number of closely related genomes. It can align genomes that have undergone *rearrangement* and lateral transfer.

However, as mentioned previously, the current version of Mauve constructs only an incomplete form of genome alignment. Mauve genome alignments alone do not provide a suitable basis for inferences on *paralogous* gene families. Mauve is also limited in its ability to align rearranged and large (e.g., > 10 Kbp) collinear segments that exist in only a subset of the *genomes* under study.

And, in general, the level of nucleotide similarity among all taxa should be greater than 60% procedre. explains the reasons for these *limitations* in more detail and gives hints on choosing alignment parameters so as to mitigate any *potential* problems.

REQUIREMENTS

1. Windows, Mac OS X 10.3+, or Linux Operating System.
2. Mauve Multiple Genome Alignment software.
3. Four *Yersinia* genomes in GenBank format:
    a. *Yersinia pestis* KIM, accession number: AE009952.
    b. *Y. pestis* C092, accession number: AL590842.
    c. *Y pestis* 91001, accession number: AE017042.

d. *Yersinia pseudotuberculosis* IP32953, accession number: BX936398.

4. Two drosophila genomes:
    a. *Drosophila melanogaster* assembly 4.0 in GenBank format.
    b. *Drosophila yakuba* assembly 2.1 in FastA format.
5. Example output files located at http://gel.ahabs.wisc.edu/mauve/chapter/.

PROCEDURES

The following sections provide a deeper look into how the five main steps in Mauve's alignment algorithm contribute to the global genome *alignment* process.

Local Multiple Alignments

Mauve uses a seed-and-extend hashing method to *simultaneously* identify highly similar unique regions among all genomes. The method used in current Mauve releases remains similar to that described in but has been extended to *approximate matching* using spaced seeds.

In addition to finding matching regions that exist in all genomes, the algorithm identifies matches that exist only among a subset of the genomes being aligned. The local multiple alignment method is very efficient in practice and typically requires less than 1 min per bacterial-size genome to find local *alignments*, and around 3 h for each *mammalian* genome on a standard *workstation* computer.

Evolutionary Guide Tree

As part of the alignment process, Mauve calculates a guide tree via genome distance *phylogeny*. Rather than using pairwise BLAST hits to estimate distance, Mauve uses the local multiple alignments identified in the previous step. The average amount of novel sequence among each pair of genomes is calculated using the local multiple alignments. This value is used as a *pairwise* distance measure to construct a phylogenetic tree via *Neighbour* Joining.

Marking Anchors

In addition to local multiple alignments that are part of truly homologous regions, the set of approximate multi-MUMs may contain spurious matches arising as a result of *random* sequence similarity. This step attempts to filter out such *spurious* matches while determining the *boundaries* of LCBs.

Local alignments are clustered together into LCBs and each LCB is required to meet a minimum weight criteria, calculated as the sum of lengths of its *constituent* local alignments. Local alignments contained in LCBs that do not meet the weight criteria are deleted.

Gapped Alignment

The initial anchoring step may not be sensitive enough to detect the full region of homology within and surrounding the LCBs. Using the existing *anchors* as a guide, two types of recursive anchoring are performed *repeatedly*. First, regions outside of LCBs are searched to extend the *boundaries* of existing LCBs and identify new LCBs. Second, unanchored regions within LCBs are *searched* for additional alignment *anchors*.

Progressive Alignment

After gathering a complete set of alignment anchors among all *genomes*, Mauve then performs either a MUSCLE or a CLUSTAL W *progressive* alignment. CLUSTAL alignments use the previously calculated genome guide tree. The progressive alignment *algorithm* is executed once for each pair of adjacent anchors in every LCB, calculating a global alignment over each LCB.

Tandem repeats < 10 Kbp in total length are aligned during this phase. Regions > 10 Kbp without an anchor are ignored. For additional details and a more in-depth algorithmic analysis refer to ref.

Parameters of Alignment

To accurately align a set of genomes using the aforementioned steps, it can be helpful to carefully tailor the provided Mauve alignment parameters to better suit the characteristics of the genomes being compared. The following sections provide a detailed explanation of each *configurable* alignment parameter.

Match Seed Size

The seed size parameter sets the minimum length of local multiple alignments used during the first pass of anchoring the alignment. When aligning divergent genomes or aligning more genomes simultaneously, lower seed sizes may provide better *sensitivity*. However, because Mauve also requires the matching seeds to be *unique* in each genome, setting this value too low will reduce *sensitivity* (small k-mers are less likely to be unique).

Default Seed Size

Setting this option will allow Mauve to *automatically* select an initial match seed size that is appropriate for the length of sequences being aligned. The default seed size for 1 MB *genomes* is typically around 11, around 15 for 5 MB genomes, and continues to grow with the size of the genomes being aligned up to 21 for mammalian genomes. The defaults may be conservative (too large), especially when *aligning* more *divergent* genomes.

LCB Weight

The LCB weight sets the minimum number of matching nucleotides identified in a *collinear* region for that region to be considered true *homology* vs *random* similarity. Mauve uses a greedy breakpoint elimination algorithm to compute a set of LCBs that have the given minimum weight.

By default an LCB weight of three times the seed size will be used. For many genome comparisons the default LCB weight is too low, and a higher value will be desired. The ideal LCB weight can be determined interactively in the alignment display using the LCB weight slider.

Determine LCBs

If this option is disabled Mauve will identify local multiple alignments, but will not cluster them into LCBs. See the *description* of match *generation* in the command-line interface chapter.

Assume Collinear Genomes

Select this option if it is certain that there are no rearrangements among the genomes to be aligned. Using this option when aligning collinear genomes can result in improved alignment accuracy.

Island and Backbone Sizes

An island is a region of the alignment where one genome has a sequence element that one or more others lack. This parameter sets the alignment gap sizes used to calculate islands and backbone segments.

Full Alignment

Selecting the "Full alignment" option causes Mauve to perform a recursive anchor search and a full gapped alignment of the genome sequences using either the MUSCLE or the ClustalW progressive alignment method. Disabling this option will allow Mauve to rapidly generate a *homology* map without the time consuming *gapped-alignment* process.

Examples

Yersinia Genomes

The *Yersinia* genus is responsible for diseases as common as gastroenteritis and as infamous as the plague. Of the 11 known *Yersinia* species, *Y. pestis* became the most notorious when identified as the cause of the bubonic and pneumonic plague. *Y. pestis* can be further classified into three main biovars according to three historically recognized *pandemics*: *Antiqua* (~1000 AD), *Medievalis* (1300-1800), and *Orientalis* (1900-Present).

Additionally, it is believed that *Y. pestis* is a clone that has evolved from *Y. pseudotuberculosis* as recent as 1500 yr ago. The recent and rapid evolution of *Yersinia* species provides an excellent example for study with Mauve. In this example we will align and analyze the four currently finished genomes of *Yersinia: Y. pseudo-tuberculosis* IP32953, *Y. pestis* KIM, *Y. pestis* 910001, and *Y. pestis* C092.

Launching Mauve

Under Windows, Mauve can be launched directly from the Start Menu. On Mac OS X, Mauve is distributed as a stand-alone application and can be run from any location. On Linux and other Unix variants, simply run. /Mauve from within the Mauve directory to start the Mauve Java GUI.

Down Loading Data

To align these four *Yersinia* genomes, we first need to download the four input genome sequence files and load them into Mauve. Mauve accepts the following genome sequence file formats: FastA, Multi-FastA, GenBank flat file, and raw format. FastA and GenBank format files with the genome of your organism can usually be downloaded from NCBI at ftp://ftp.ncbi.nih.gov/genomes/. The fna files are in FastA format and the .gbk files are in GenBank format; for this example we will use the gbk files.

Once Mauve has started up, simply select "Align..." from the "File" menu to access the Alignment dialog box, as shown in Figure elsewhere in this chapter: The top entry area lists the sequence file(s) containing the genomes that will be aligned.

To add a sequence file, click the "Add sequence..." button and select the file to add. The Windows and Mac OS X versions of Mauve support drag-and-drop, allowing sequence files to be added by dragging them in from the Windows Explorer or Mac OS Finder.

Locating Mauve

The location where Mauve stores its alignment results can be set using the *"Output file:"* text entry field. By default Mauve creates output files in the system's temporary storage directory. Under Windows this is usually C : \Documents and Settings\ <username> \Local Settings\Temp\ or /tmp under Mac OS X and Unix.

Because the location of the system's temporary storage is rather obscure, we heartily encourage users to select a different output location. In this example, we specify C:\example\output. mln to store the Mauve output.

Mauve Configuration

By default, Mauve configures the alignment parameters so they are appropriate for aligning closely related genomes with *moderate* to high amounts of genome rearrangement. However, some alignment parameters can be modified to better suit alignments involving more distantly related genomes, or with a lower amount of rearrangement.

For example, when aligning more divergent genomes, the seed size can be reduced to find additional alignment *anchors* and achieve greater alignment coverage over the genomes. Another option disables the full alignment process, allowing Mauve to quickly generate a simple comparative picture of genome organization.

For this example, we start with the default parameters, and later use the interactive LCB weight slider to determine an LCB weight that excludes most spurious rearrangements (1500). We then recomputed the alignment with the LCB weight set to 1564.

Global Alignment

To start the genome alignment, click on the "Align..." button. Once Mauve finishes its global alignment of the four *Yersinia* genomes, we are ready to interactively inspect the results. Should the alignment fail, please check the console log. The Mauve Alignment viewer enables manual evaluation of both the proposed global homology and the nucleotide level alignment in the context of genome annotation

Alignment Display

The alignment display is organized into one horizontal "panel" per Input genome sequence. Each genome's panel contains the name of the genome sequence and a scale showing sequence coordinates for that genome.

Additionally, each genome panel contains one or more coloured block outlines that surround a region of the genome sequence that has been aligned to part of another genome.

Some blocks may be shifted downward relative to others; such blocks are in the reverse complement (inverse) orientation relative the reference genome. Regions outside blocks were too divergent in at least one genome to be aligned successfully. Inside each block Mauve draws a similarity profile of the genome sequence.

The height of the similarity profile corresponds to the average level of conservation in that region of the genome sequence. Areas that are completely white were not aligned, presumably because they contain lineage specific sequence. The height of the similarity profile is calculated to be inversely proportional to the average alignment column entropy over a region of the alignment. In Figure 11.1, coloured blocks in the first genome are connected by lines to similarly coloured blocks in the remaining genomes.

These lines indicate which regions in each genome are homologous. If many genomic rearrangements exist, these lines may become overwhelming. The boundaries of coloured blocks indicate the breakpoints of genome rearrangement.

Navigation

The alignment display is interactive, providing the ability to zoom in on particular regions and shift the display to the left and right. Navigating through the alignment visualization can be accomplished by using the control buttons on the toolbar immediately above the display. Alternatively, keyboard shortcuts allow rapid movement through the alignment display.

The keystrokes Ctrl+up arrow and Ctrl+down arrow zoom the display in and out, whereas Ctrl+left arrow and Ctrl+right arrow shift left and right, respectively. When moving the mouse over the alignment display. Mauve will highlight the aligned regions of each genome with a black vertical bar, and clicking the mouse will vertically align the display on the selected orthologous site.

Studying Sequences

If the aligned genome sequences were in GenBank files containing annotated features Mauve will display the annotated features immediately below the sequence similarity profiles. Annotated CDS features show up as white boxes, tRNAs are green, rRNAs are red, and misc RNA features are blue. Features annotated on the reverse strand are on a track immediately below the forward strand track.

Repeat region features are displayed in red on a third annotation track. Mauve displays the product qualifier when the mouse cursor is held over a feature. When a feature is clicked, Mauve shows a detailed listing of feature qualifiers in a popup window. For performance reasons, the annotated sequence features appear only when the display has been zoomed in to view less than 1 Mbp of genome sequence.

Discussion

Now that the viewer has been explained in detail, we can use it to analyze the results from the *Yersinia* genome alignment. These four *Yersinia* genomes have a rich and well-studied evolutionary history. In ref. elsewhere in this chapter it was reported that transmission by fleabite is a recent evolutionary adaptation that distinguishes *Y. pestis,* the agent of plague, from *Y. pseudotuberculosis* and all other enteric bacteria.

The high level of sequence similarity between *Y. pestis* and *Y. pseudotuberculosis* implies that only a few minor genetic changes were needed to induce fleaborne transmission. The question is thus; can we identify these changes using the Mauve genome alignment system?

From the global view presented in Figure 11.3 it is difficult to see individual nucleotide substitutions and deletions, so we need to exploit some of the advanced features of the interactive viewer, such as the zoom and gene annotation.

Figure 11.4 shows a zoomed in view of the one gene responsible for adhesion to the gut of host organisms. The variability in conservation of this gene could contribute to the differences in pathogenicity between *Y pestis* and *Y pseudotuberculosis* reported in ref. elsewhere in this chapter.

Additionally, using the Mauve alignment viewer we can see that the plasmid *pMTl*, which mediates infection of the plague flea vector, is present in all of the *Y. pestis* genomes, but not in *Y. pseudotuberculosis*.

A number of recent studies indicate that microbial genomes evolve and adapt by integrating novel genetic elements through lateral transfer. Such novel regions are commonly referred to as genomic islands (GIs) and appear as large gaps in the genome sequence alignment. Mauve identifies large alignment gaps that correspond to putative islands and saves their sequence coordinates in a ".islands" file.

Similarly, regions conserved among all genomes are frequently referred to as backbone, and appear as large regions of the alignment that contain only small gaps. Mauve saves the sequence coordinates of backbone segments in a ".backbone" file. Despite their recent speciation, *Y. pestis* and *Y. pseudotuberculosis* contain a number of GIs.

Genomes of D. melanogaster and D. yakuba

In contrast to the previous example, where each genome contained around 5 million nucleotides, the two *Drosophila* genomes involved in this comparison are each over 100 million nucleotides in size. Although Mauve can efficiently analyze smaller microbial genome comparisons in memory, genomes as large as Drosophila would require 2-3 GB of RAM to align in memory.

Fortunately, Mauve can utilize two or more scratch disks lo reduce memory requirements during generation of unique local alignments-the most memoryintensive part of the alignment process.

Thanks to this feature, aligning the two complete *Drosophila* genomes using the Mauve Aligner can be accomplished in four steps:

1. Create sorted mer lists (.sml) for each genome using the scratch-path parameter to identify two or more disks that can be used. The same command must be run once to create an SML file for each genome and once to generate the local alignments (MUMs).
   a. Command to generate the SML file for *D. melanogaster:*

      >mauveAligner -mums -scratch-path=/diskl -scratch-path=/disk2 -output =drosophila. mums dmel.gbk dmel.gbk.sml dyak.fas dyak.fas.sml
   b. Run again to generate the SML file for *D, yakuba:*
      >mauveAligner -mums -scratch-path=/disk l -scratch -path=/disk2 -output =drosophila. mums dmel.gbk dmel.gbk.sml dyak.fas dyak.fas.sml
   c. And one more time to generate the local alignments (MUMs): >mauveAligner -mums -scratch-path=/disk l -scratch-path=/disk2 -output=drosophila. mums dmel.gbk dmel.gbk.sml dyak.fas dyak.fas.sml
2. Generate initial alignment anchors without actually aligning:

>mauveAligner -match-input=drosophila- mums -weight=10000 -no-gapped-alignment -no-recursion -no-lcb-extension -output=drosophila_lcb.mums dmel.gbk dmel.gbk.sml dyak. f as dyak. fas . sml

3. Configure MUSCLE alignment parameters to reduce memory usage and speed up the gapped alignment process:
   >mauveAligner -match-input=drosophila-lcb- mums -lcb-match-input -muscle-args="-stable -maxiters 1 -diagsl -sv" -output-alignment=drosophila.xmfa dmel.gbk dmel. gbk. sml dyak. f as dyak.fas.sml
4. View the output files using the Mauve alignment display.
   >java Xmxl200m jar Mauve.jar drosophila.xmfa

Analyzing the results with the Mauve alignment display, we can see evidence of the two main lineages of gypA and gypB gypsy elements most likely resulting from multiple lateral transfer events reported in ref. elsewhere in this chapter.

See Note elsewhere in this chapter for converting the alignment output file into a signed gene-order permutation matrix for use with systems for rearrangement phylogeny such as the GRIMM/MGR server or BADGER. Additional command-line options are shown by running mauveAligner with no arguments, see Figure elsewhere in this chapter.

POINTS TO REMEMBER

1. If encountering a java.lang.OutOfMemoryError: Java heap space error, try increasing the heap space by running Mauve with the -Xmx command or closing any unused Mauve alignment windows.
2. The LCBs generated by Mauve in the alignment output file can be transformed into a signed gene-order permutation matrix by supplying the -permutation -matrix - output= <directory> commandline argument. The permutation matrix makes suitable input to systems for rearrangement phylogeny such as the GRIMM/MGR server or BADGER.
3. When aligning divergent genomes, the seed size parameter can be reduced to between 9 and 13.
4. If the console window is not present or has been closed, it can be reopened in the Help menu->show console.
5. In the Mauve alignment viewer the LCB connecting lines can be hidden (or made visible again) by typing Shift+L (pressing shift and L simultaneously).
6. Similar to all existing genome alignment systems, it is possible that Mauve may generate inaccurate alignments. Alignments can be manually edited using the Cinema-MX alignment editor that has been incorporated into the Mauve interface.

   To access this feature, right click on the suspect LCB and then select "Edit this LCB." Then, the Cinema-MX alignment editor window will appear and allow for dynamic alignment correction, and when finished the Mauve interface will update with the adjusted alignment.
7. If alignment is unacceptably slow or using too much memory, try using ClustalW instead of MUSCLE.

Chapter 11

Prokaryotic Artificial Chromosome

Bacterial artificial chromosomes (BACs) represent a very useful *cloning* system for large DNA fragments and utilize the *Escherichia coli* F factor as their backbone. *E. coli* F factor allows strict copy number control of the clones so that they aarc stably maintained at one to two copies per cell. The stability of the cloned DNA during propagation in *E. coli* host is substantially higher in lower copy number vectors than in *multi-copy counterparts.*

The BAC vector permits the cloning of DNA up to 350 kb in *E. coli.* Because of its clone stability and case of use, the BAC cloning system has emerged as the system of choice for the construction of large-insert genomic DNA libraries for *humans, animals,* and *plants.* However, BAC libraries have been difficult to construct with average insert sizes > 150 kb.

By using more reproducible *megabase-size* DNA preparations, and sizing and elutlng procedures, we have constructed BAC libraries from human DNA with an average insert size of 202 kb and from *Arabidopsis* DNA with an average insert size of 190 kb. In this chapter, we describe advanced *methodologies* for *generating* larger insert BAC libraries from animals, plants, and bacteria.

BAC Vectors

The pBAC108L vector is the first version of a BAC vector. After transformation, clones carrying the human DNA insert had to be selected by colony hybridization with *labeled* human DNA. The next version, pBeloBACII. allows lacZ-based positive color selection of the BAC *clones* that have insert DNA in the cloning sites. The nucleotide sequence of pBeloBACII is available from GenBank (accession no. U51113).

Additional BAC vectors have been derived from pBAC108L or pBeloBAC11. pECSBAC4 and plndigoBAC536 have a unique *EcoR*I cloning site and pBACwich has been designed for plant *transformation* using biolistic *bombardment.* BIBAC2 is designed for Agrobacterium-mediated plant transformation. pBACe3.6 allows positive selection for insert-containing BAC clones through inclusion of the *sacBII* gene.

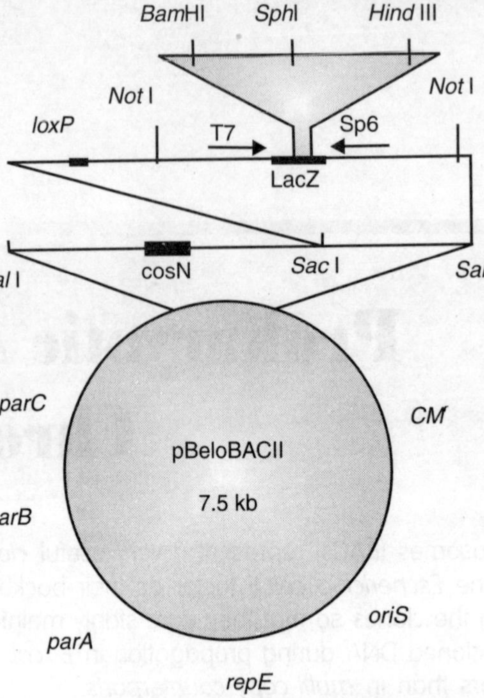

Figure 11.1: Diagram of pBeloBAC11 vector. CM^r, chloramphenicol resistance.

REQUIREMENTS

1. LB medium: 1 % tryptone, 0.5% yeast extract, 1 % NaCl.
2. QIAGEN Plasmid Maxi Kit (Qiagen).
3. Plasmid-Safe ATP-dependent DNase (Epicentre).
4. QIAyuick PCR Purification Kit (Qiagen).
5. Spermidine (Sigma).
6. Shrimp *alkaline phosphatase* (SAP) (USB).
7. TE: 10 mM Tris-HCl; 1 mM EDTA, pH 8.0.
8. InCert agarose (FMC).
9. Digestion buffer: 1 mg/ml, of proteinase K; 1 % *N-lauroylsarcosine*, 4 mM *dithiothreitol.* 0.5 M EDTA, pH 8.0.
10. Isolation buffer: 10 mM Tris-HCl, p11 9.5: 10 mM EDTA; 100 mM KC1: 4 mM spermidine: 1 mM spermine; 0.5 M sucrose. Add 0.1% β-mQrcaptoethanol before use.
11. Lysis buffer: 0.5 M EDTA, pH 8.0: 1 % N-lauroyl-sarcosine. Add 1 mg/mL of proteinase K before use.
12. Phenylmethylsulfonyl fluoride (PMSF) stock solution: 50 mM PMSF in isopropanol.

13. Cellulase buffer: 20 mM MES, pH 5.5: 1 mM MgCl,; 0.6 M sorbitol; 2% Onozuka cellulase RIO (Gallard Sehleeinger); 0.1% Maeerase pectnase (Calbiochem).
14. Wash buffer: 50 mM Tris, pH 8.0; 0.6 M sorbitol; 1 mM MgCl2, 100 mM b*-mercaptoethanol.
15. Glycerol, Ultra Pure (Gibco-BRL).
16. SeaKem LE agarose (FMC).
17. 0.5X TBE: 45 *mM* Tris: 45 mM boric acid; 1 mM EDTA, pH 8.3.
18. 1X TAE: 40 mM Tris-aeetate: 1 mM EDTA, phi 8.0.
19. Gene Pulser (Bio-Rad).
20. SOC: 2% Bacto-tryptone; 0.5% yeast extract; 10 mM NaCl, 2.5 mM KCl, 10 mM MgCl2, 10 mM $MgSO_4$, 20 mM glucose, pH 7.0.
21. X-Gal stock: 20 mg/mL in dimethylformamide.
22. Isopropyl-43-o-thiogalaetopyranoside (IPTG) stock: 200 mg/mL in ddH2O.
23. LB freezing medium: LB medium supplemented with 36 mM $K2HPO_4$, 13.2 mM $KH2PO_3$. 1.7 mM Na citrate, 0.4 mM $MgSO_4$, 6.8 mM $(NIt._a)2SO4$, 4.4%c (v/v) glycerol, 12.5 µg/mL of chloramphenicol.
24. GTE: *50 mM* glucose: 25 mM Tris, pH 8.0: 10 mM EDTA, pH 8.0, 0.1 mg/mL of RNase A.
25. 0.2 N NaOH+1% sodium dodecyl sulfate (SDS) solution.
26. Potassium acetate solution: 60 mL of 5 M potassium acetate: 28.5 mL of glacial acetic acid. 11.5 mL of ddH2O, pH 4.8-5.2.
27. AutoGen 740 DNA Isolation System (AutoGen, Japan).
28. KoAe/phenol/chloroform/ethanol (.AutoGen).
29. Q-hot (Genetix, UK).
30. Biomek2000 (Beckman).
31. Hybond N+ membranes (Amersham Pharmacia Biotech).
32. Denaturing solution: 1.5 M NaCl, 0.5 N NaOH.
33. Neutralizing solution: 1 M Tris, pH 7.5; 1.5 M NaCl.
34. Proteinase solution: 50 mM Tris. p1-I 8.0: 50 mM EDTA, pi{ 8.0: 100 mM NaCl: 1% N-lauroylsarcosine, 100 mg/L of proteinase K.

PROCEDURES

Making BAC Vector

Isolation and Purification

1. Inoculate a single *E. coli* strain DH10B colony containing the BAC vector into 5 mL of LB medium containing 30 µg/mL of *chloramphenicol* and grow at 37°C overnight.
2. Inoculate three culture flasks containing 1 L of LB plus *chloramphenicol* (30 µg/mL),

prewarmed to 37°C, with 1 mL/L of culture from step 1. Grow at 37°C with shaking to a cell density of approx 1 × 10⁹) cells/mL (A_{600} = 1.0–1.1).

3. Harvest the cells by centrifugation at 4°C: for 15 min at 6000g.
4. Isolate the BAC vector using the Qiagen Plasmid Maxi Kit (five Qiagen-tip 500 columns for a 3-L preparation).
5. Use Plasmid-Safe ATP-dependent DNase to selectively digest bacterial DNA while leaving plasmid DNA intact. The efficiency of the DNase can be improved by adding two restriction enzymes that do not cut the plasmid. For 10 µ.g of pBeloBAC11 vector, use 150 U of DNase, 50 U each of *ClaI and MIA (NEB),* and NEB buffer 2. Incubate for 2 h at 37°C.
6. Inactivate Plasmid-Safe ATP-dependent DNase by incubating at 65°C for 20 min.
7. Remove small nucleotides by using the QIAquick PCR Purification Kit.

Dephosphorylation of BAC Vector DNA

1. Digest the BAC vector (10 µg) to completion with 100 U of HindIII or *EcoRI* (Gibco) in the appropriate buffer supplemented with 4 mM spermidine°C for 2-4 *h.*
2. Clean up the digested DNA with the QIAquick PCR Purification Kit and resuspend in 1.00 µLof 10 mM Tris buffer (pH 8.0).
3. Dephosphorylate the vector DNA by adding SAP (1 U/)µg of DNA) and incubating at 37°C for 2 h.
4. Clean up the dephosphorylated vector with the QIAquick PCR Purification Kit, dissolve in 100 µL of TE, and store at 20°C.

High Molecular Weighs Genomic DNA

Initial Sources for BAC Cloning

From Animals

1. Incubate 8-10 mL of frozen semen at 37°C for 2-3 h.
2. Suspend the semen in 30 mL of TE and centrifuge at 7000g for 5 min at 4°C.
3. Resuspend the sperm in 30 mL of TE and harvest by centrifugation at 5000g for
5 min at 4°C. Repeat the procedure two more times.
4. Embed the resuspended sperm in agarose (0.5% agarose) at a concentration of 1.5 × 10^8/ mL using a hemocytometer for human, chimpanzee, or mouse.
5. Digest the agarose-embedded large DNA for 2 d with digestion buffer. Change the digestion buffer once during the digestion.
6. Rinse the sample with 0.5 M EDTA (pH 8.0) three times and store in 0.5 *M* EDTA at 4°C.

From Plants

There are two general methods for preparing megabase-size DNA from plants. The nuclei method is *universal* and works well for several *divergent* plants.

The DNA is more concentrated and contains lower amounts of chloroplast or mitochondrial DNA. The protoplast method yields highquality *megabase-size* DNA with minimal *breakage*, but each plant species requires a different set of conditions to *generate protoplasts.*

Nuclei Technology

1. Grind 30-50 g of the fresh tissue to a fine powder in liquid nitrogen with a mortar and pestle, and transfer the powder to an ice-cold 500-mL beaker containing 200 ml- of ice-cold isolation buffer.
2. Repeat step 1 several times in order to obtain enough nuclei to embed.
3. Filter the homogenate on ice *sequentially* through one layer each of 140- and 80-μm *nylon* net or filter mesh.
4. Add 1/40 vol of isolation buffer supplemented with 20% Triton X-100 to the *filtrate* and mix well.
5. Pellet the nuclei by *centrifugation* in a fixed-angle rotor at 2000g for 15 min at 4°C.
6. Discard the *supernatant* fluid and resuspend the pellet in 1 to 2 mL of ice-cold *isolation* buffer.
7. Count the nuclei with a *hemocytometer* and mix the *nuclei* with agarose in isolation buffer (without β-mercaptoethanol, 0.6% agarose) at a *concentration* of 2.2 × 108 nuclei/mL (for the diploid plants with haploid genomic sizes of 800-1000 Mb;). The *optimum* number of *nuclei* depends on the genomic size of the plant.

   Usually it is difficult to count the nuclei, and, thus, when preparing nuclei for the first time, we optimize the nuclei concentration *empirically*. We suggest optimizing the nuclei concentration by embedding two or three different concentrations of nuclei. We usually pipette a 0.5 vol of 1.8% InCert agarose solution into the *prewarmed* nuclei suspension and mix well (1 vol of agarose solution plus 2 vol of nuclei *suspension,* so that the final agarose *concentration* is 0.6%%).
8. Pour the mixture into plug molders on ice.
9. Incubate the plugs in *lysis* buffer for 24 h at 50°C with gentle *shaking* to *degrade* the proteins.
10. Replace this buffer with new lysis buffer for a further 24-h incubation at 50°C.
11. Wash the plugs three times (three times for a l-h incubation at 4°C) by *immersing* in 0.5 M EDTA, pH 8.0, containing 0.1 mM PMSF to inactivate the proteinase K.
12. Store the plugs in 0.5 *M* EDTA. pH 8.0, at 4°C until use.

Protoplast Preparing

This preparation is suitable for *maize* and *Arahidopsis.*

1. Cross-cut 5-10 g of the young leaves into 0.5-1-mm strips.
2. Infiltrate about 80 ml- of *cellulase* buffer through a 0.4-μm filter using a vacuum.
3. Transfer to four to six large *Petri dishes* and digest at room temperature for 3-5 h.

4. Remove the broken cells by sequential filtration through a 140-μm mesh and then an 80-tun mesh.
5. Pellet the protoplasts at 300g for 10 min and gently resuspend in 50 ml, of wash buffer.
6. Pellet the protoplasts at 300k for 10 min and gently resuspend in 1-2 ml, of wash buffer.
7. Count an aliquot under a microscope by using a hemocytometer and embed in isolation buffer (without b*mercaptoethanol). Adjust the final *concentration* to 8 × 10^7 protoplasts/ mL (for the diploid plants with *haploid* genomic sizes of 2500-3000 Mb;. We usually *empirically* optimize the protoplast concentration by embedding two or three different concentrations of protoplasts.
8. Make agarose plugs as described earlier.

From Bacterial Cells

1. Grow cells in 50 ml., of an appropriate medium (e.g., LB) to an OD_{600} of 0.8 to 1.0 (4-5 × 10^8 cells/mL).
2. Harvest cells by centrifugation at 4000g for 10 min at 4°C.
3. Resuspend the cell pellet in the same volume of 10% glycerol (Ultra Pure) and spin at 4000g for 10 min to pellet.
4. Resuspend the cell pellet in 2.5 mL of 10% glycerol.
5. Mix cells with agarose (InCert) in water to make a final agarose concentration of 0.5%. Optimize the cell concentration empirically by embedding two or three different concentrations of cells.
6. Incubate the DNA plugs in lysis buffer at 50°C with gentle shaking for 2 d and store at 4°C in 0.5 M EDTA (pH 8.0). Change the buffer once during the digestion.

Incomplete Breakdown

Once megabase-size DNA has been prepared. it must be fragmented and DNA in the desired size range isolated. In general, DNA fragmentation utilizes two general approaches: physical *shearing* and *partial* digestion with a restriction enzyme that cuts relatively *frequently* within the genome. Here partial digestion with restriction enzymes is used.

1. Equilibrate 50 μg of sliced plugs with a 150-μL mixture (total reaction volume of 200 AL): *corresponding* reaction buffer; 0. 1 mg/mL of acetylated bovine serum albumin: 4 mM *spermidine*: and diluted *HindIII* or *EcoRI* such as 0, 1, 2, 4. 8, 16, and 50 U/200 μL) at 4°C with appropriate shaking for 1 h.
2. Transfer the reaction mixture to a 37°C water bath and incubate for 1 h.
3. Add 1/10 vol of 0.5 M EDTA, pH 8.0. and place the tubes on ice.
4. Place the partially *digested* DNA plugs on an *agarose gel* molding plate using a comb and pour the warm (37-45°C) 1 %o agarose gel (SeaKem LE) in 0.5X TBE.
5. Perform pulsed-field gel electrophoresis (PFGE) using the following conditions: 6.0 V/cm, 90-s pulse. 0.5X TBE buffer. 14°C for 18 h.

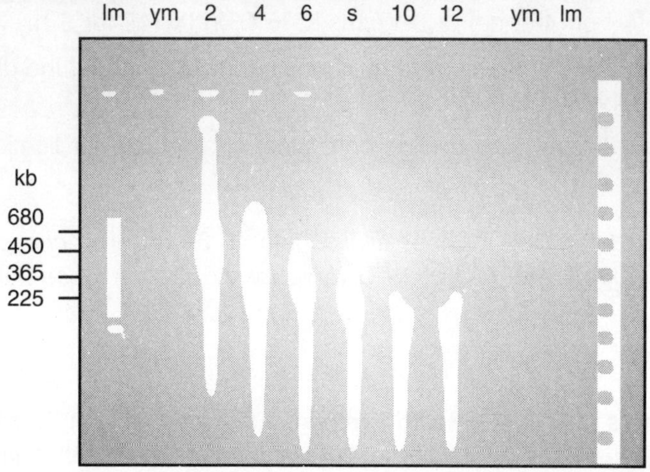

Figuer 11.2: Partially digested human megahase-size DNA. 1m, lambda ladder: ym. Sac-charonrvces cerevisiae (Bio-Rad). Lanes 3-8, human DNA partially digested with EcoRI: (2) 2 U, (4) 4 U, (6) 6 U, (8) 8 U, (10) 10 U, (12) 12 U, The genomic DNA was subjected to PFGE on a 1 % agarose gel in 0.5X TBE using a switch time of 90 s at 6 V/ cm and 14°C for 18 h.

6. After checking the *ethidium bromide-stained* gel, select the enzyme concentration giving a *majority* of DNA fragments ranging from about 300 to 800 kb. It depends on the DNA concentrations of the samples, but usually the actual size of the DNA fragments from this *fraction* is between 50 and 400 kb in size owing to DNA trapping.

Partially Hydrolyzed DNA

1. Perform the digestion reaction on a large scale by carrying out several reactions (five to eight) under exactly the same conditions as previously determined.
2. Place the partially digested DNA plugs on an agarose gel *molding* plate side by side using a comb, and pour the warm (37-45°C) 1 %% agarose gel (SeaKem LE) in 1 X TAE.
3. Perform PFGE using conditions adjusted to the DNA *concentration* of the *plugs* (high DNA concentration: 6 V/cm, between 15- and 39-s pulse, IX TAE buffer. 12°C for 16-18 h; moderate DNA concentration: 6 V/crn, between 40- and 50-s pulse, IX TAE buffer, 12°C for 16-18 h: low DNA concentration: 5 V/cm, between 5- and 7-s pulse, I X TAE buffer. 12°C for 14-16 lr: *see* Note elsewhere in this chapter).
4. Cut DNA fragments horizontally (about 5 mm each) ranging from about 200-600 kb from the gel (for the sample of low DNA concentration, excise the DNA band in the compression zone), and separately carry out a second size selection in 1 % agarose (SeaKem LE) using these conditions: 4 V/cm, 7-s pulse, 1X TAE buffer. 12°C for *8* h.
5. Excise the DNA hand in the compression zone and keep in 1X TAE at 4°C.

Obtaining DNA from Agarose Gels

1. Put the gel slice and an appropriate amount of I X TAE in dialysis tubing (0.75 in. diameter; Gibco).

2. Elute the DNA from the gel at 3 V/cm, 10-16°C, in IX TAE for 3 h.
3. Reverse the polarity of the current by changing the direction of the dialysis tubing and run at 3 V/cm in the same buffer for 30 s.
4. Remove the solution using a wide-bore pipet tip and use for ligation.

Ligation

Ligate the electroeluted DNA without *dilution* into 40-80 ng of *dephosphorylated* BAC vector in a total volume of 100 μL with 6 U of T4 DNA ligase (USB) plus *ligase* buffer at 15°C for 20 h.

Electroporation

Methods

1. Inoculate flasks of SOB (2% Bacto tryptone, 0.5% yeast extract, 10 mM NaCl. 2.5 *mM* KCl [pH 7.0] without Mg^{2+}) by *diluting* a fresh saturated (overnight) *culture* of DH10B (1:1000).
2. Grow with shaking at 37°C until the 0135% reaches 0.7 (no higher than 0.8). This should take approx 5 h.
3. *Harvest* cells by spinning at 4000g for 10 min at 4°C.
4. *Resuspend* the pellet in 1 vol of cold, sterile 10% *glycerol* equal to the *original* culture volume.
5. Shake it *vigorously* several times and spin at 4000g for 10 min at 4°C.
6. Carefully pour off the *supernatant* (the pellet will be quite loose) and *resuspend* the cells in cold, sterile 10% *glycerol* equal to the original culture volume.
7. Shake vigorously several times and spin at 4000g for 10 min at 4°C.
8. Pour off the *supernatant,* resuspend the cells in the volume of glycerol remaining in the centrifuge bottle. Pool the cells in one small centrifuge tube and add a few *milliliters* of cold, sterile 10% glycerol.
9. Spin at 4000g for 10 min at 4°C.
10. Pour off the supernatant and resuspend the cells in cold, sterile 10% glycerol using 1.5-2.0 mL/L of initial culture.
11. Aliquot into microfuge tubes (100-200 μL per tube) and freeze quickly in a dry *ice-ethanol* bath. Store cells at -80°C.

Process of Electroporation

1. Transform 1-2.5 μL of the *Ligation* material into 20-25 μL of *E. coli* DH 10B *compêtent* cells by using Gene Pulser and 0.2-cm cuvet under the conditions of 100 Ω, 12.5 kV/cm, and 25 μFa. This *usually* gives a time constant of approx 2.4 *s*.
2. Transfer the *electroporated* cells to 15-mL culture tubes with 0.5-1 mL of SOC and incubate at 37°C for 1 h with *shaking.*
3. Spread the SOC medium onto LB plates containing 12.5 μg/mL of *chloramphenicol,* 50μg/mL of X-Gal, and 25 μg/mL of IPTG, and incubate at 37°C for 24-36 h.

4. Pick white colonies into 384-well *microtiter* plates containing LB freezing medium.
5. Incubate the *microtiter* plates at 37°C overnight and store at -80°C.

Preparation of BAC DNA

Methodology

1. Inoculate a *colony* into a 15-ml- culture tube containing 5 mL of I .B plus 12.5 u,g/ml of *chloramphenicol,* and incubate overnight at 37°C with shaking.
2. Centrifuge the culture at 2000g at 4°C for 15 min using a *tabletop* centrifuge.
3. Pour off the supernatant and resuspend the cell pellet with the remaining solution in the tube.
4. Transfer the suspension to a 1.5-mL *microfuge* tube on ice, add ice-cold GTE, and mix by inversion several times.
5. Add 200 µL of freshly prepared 0.2 *N* NaOH+1% SDS solution, mix by gentle inversion several times, and *incubate* on ice for 5 min. At this stage the cells will lyre and the solution will grow clear and viscous.
6. Add 300 µL of the *potassium* acetate solution, gently invert the mixture, and return to ice. The addition of *potassium acetate* solution will cause the formation of a *flocculent precipitate.*
7. Centrifuge the mixture at full speed in a microfuge at room temperature for 15 min.
8. Transfer 750 µL of the supernatant fluid to a clean microfuge tube without disturbing the pellet.
9. Add 0.6 vol of cold *isopropanol* (450 µ.L) and centrifuge at full speed in a *microfuge* for 15 min to pellet the DNA.
10. Remove the *supernatant,* rinse the pellet with 1 mL of cold 70% ethanol, and dry upside down completely.
11. Add 40 µL of TE and digest 8-10 µL of this DNA solution with *Nod* to free the insert from the BAC vector.
12. Run 1 % agarose gels in 0.5X TBE for 15 h at PFGE conditions of 6 V/cm, initial switch time of 5 s. and final switch time of 15 s.

Technique of Miniprep

1. Inoculate the BAC clone into 3 mL of LB with 12.5 µg/mL of chloramphenicol within the tube specific for the AutoGen 740 DNA *Isolation System* and culture at 37°C with *shaking* overnight. *Transfer* the tubes into the AutoGen 740 DNA *Isolation* System.
2. Use the *alkaline* lysis method by setting up "System 1" in the panel of the AutoGen 740. Follow the user's manual provided by the *manufacturer* (Reagent 1: 10 mM Tris-HCl, pH 8.0, 1 mM EDTA, pH 8.0, 100 mg/l. of RNase A: Reagent 2: 0.2 N NaOH, 1 % N-lauroylsarcosine: Reagent 3: KoAc/phenol/chloroform/ ethanol [AutoGen]; Reagent 4: isopropanol: Reagent 5: 70% ethanol).

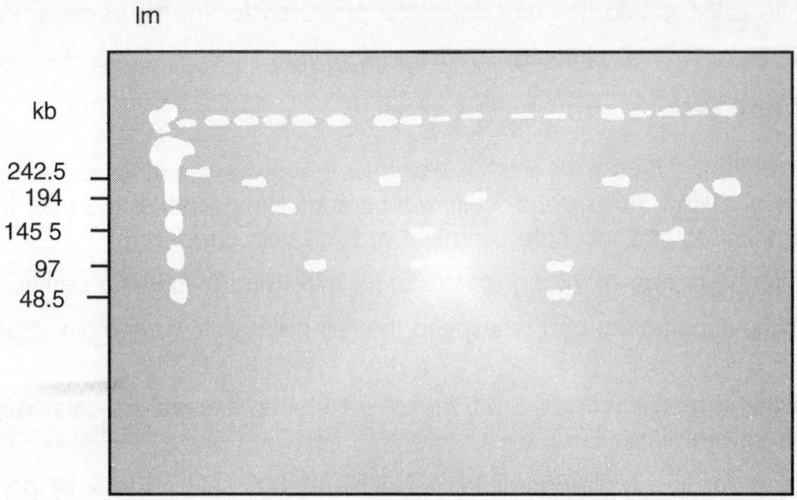

Figure 11.3: Analysis of random human BAC clones by PFGE. lm, lambda ladder. Lane 1, lambda concaterner; lanes 2-21, AutoGen miniprep products of recombinant BAC clones digested with Nod. *Note that the 6.9-kb hand in each lane is the BAC vector.*

Techniques for Midi- and Maxipreps

Both midi- and maxipreps can be carried out by adapting the procedure given for the manual miniprep method.

The volume of the added solutions (GTE, NaOH+SDS, and potassium acetate) should be adjusted *proportionally* to the increased culture volume. For example, if the amount of starting culture is 100 mL, 20 times more of these three *solutions* should be added in each step than for a 5-mL prep.

Colonial Hybridization

1. Grid the library out onto *nylon membranes* manually or using robots (Q-bot or Biomek2000), and incubate the trays on the medium plates overnight.
2. Transfer the membrane with bacterial colonies to a filter paper prewet with denaturing solution (colony side up).
3. Incubate at room temperature for 10 min.
4. Transfer to another filter paper prewet with *neutralizing* solution and let it stand for 10 min.
5. Remove the filter and dry on a clean sheet of filter paper for 5 min.
6. Invert the filter (colony side down) into a tray filled with *proteinase* solution and incubate at 37°C for 1 to 2 h.
7. Turn the filter colony side up and place on a clean sheet of filter paper to dry for 5 mm.
8. CJV-crosslink the filter.
9. Bake the filter at 80°C for 1 to 2 h and store in a dark place at room temperature.

POINTS TO REMEMBER

1. The plugs can be *sliced* into *fragments* that are about the same size as beads without any *appreciable* DNA breakage and then used for *partial* digestion.
2. Restriction enzyme *incubation* for an extended time may show *exonuclease* activity.
3. For other animals that have different genome sizes, you should adjust the *concentration* of cells.
4. For other plants that have different *genomic* sizes, you should adjust the concentration of *nuclei*, or *protoplasts*.
5. We suggest preparing the nuclei as described and in the final step *embedding* three different concentrations of nuclei (e.g., 0.5 mL of concentrated nuclei + 1.5 ml, of isolation buffer. I mL of concentrated nuclei + 1 mL of isolation buffer, and 2 mL of concentrated nuclei + 0 mL of isolation buffer).
6. The filters can be stored at room temperature for years, and each can be used for 10-20 *hybridizations*.
7. Using a Q-hot, 384-well microtiter plates containing BAC clones are spotted onto 22 × 22 cm Hybond N+ membranes. Bacteria from 72 plates are spotted twice onto one membrane. resulting in 27,648 unique clones on each membrane. Alternatively. smaller nylon membranes (12 × 8 cm) can be inoculated with a 384-prong High Density Replicating Tool from microtiter plates using the Biomek2000 robot.
8. The final yield is approx 20-40 μg of DNA from 3μL of media.
9. The amount of restriction enzyme and the digestion time depend on the number of restriction sites in the genome; usually 1-50 U for 5 min to 1 h.
10. When using highly *concentrated* DNA plugs for a partial digestion. apply a *shorter* pulse time (between 15-30 s) to spread out the DNA on the gel. This avoids *trapping* of smaller DNA segments. *Conversely*, for samples of low DNA *concentration*, use either a very short pulse (between 5 and 7 s) or a long pulse (90 s) to avoid too much *spreading* of the DNA on the gel.
11. A lower field strength (9-13 kV(cm) yields a lower number of *clones* but a higher *average* insert size.
12. We suggest preparing the *protoplasts* as described and in the final step embedding three different *concentrations* of protoplasts (e.g.. 0.5 mL of concentrated protoplasts + 1.5 mL of isolation buffer. 1 mL of concentrated protoplasts + 1 mL of isolation buffer, and 2 mL of concentrated protoplasts + 0 mL isolation buffer).
13. Making *protoplasts* is specific for every plant species and needs to be optimized.

Chapter 12

Genomic Study in *Toxoplasma gondii*

Rapid genotyping provides an important tool for epidemiological, population, and genetic mapping studies in parasitic organisms. The protozoan parasite *Toxoplasma gondii* has a relatively large (approx 65 Mb) genome that is *haploid* throughout most of its life cycle. *T. gondii* has a highly clonal population structure consisting of three closely related, yet *genetically* distinct groups defined as types 1, II, and III.

These clonal types are highly similar, with on average only 1-2% divergence at the *nucleotide* level. Remarkably, the *majority* of strains of *T. gondii* possess one of only two alleles at each locus.

This *biallelic* pattern combined with the high degree of similarity suggests that a recent *meiotic* cross gave rise to the majority of existing strains of *T. gondii.* Amplification of specific DNA segments using *polymerase* chain reaction (PCR), followed by *restriction* fragment length *polymorphism* (RFLP) analysis is the most commonly used method for *genotyping* of *T. gondii.*

However, this method is labor-intensive and typically requires that each locus be analyzed independently. Analysis of *microsatellites* has also been applied to *T. gondii,* although this approach is limited by the relative *infrequency* of short *dinucleotide* repeats in this organism. Despite the close similarity of strains, *single-nucleotide polymorphisms* (SNPs) are abundant in *T. gondii.*

Sequencing of a total of 33 separate loci from three *T. gondii* strains, chosen to represent the three clonal *lineages*, reveals that SNPs occur approx every 400 by . For reasons that are unclear, the frequency of *polymorphisms* is higher in type I strains than either type II or type III strains.

A biallelic pattern is evident at every one of these SNPs, such that whereas one strain contains a unique allele (different nucleotide at that position), the other two strains are *identical.* Combined with its *haploid* genome, this simplifies SNP analysis in *T. gondii* in that only two alleles need to be screened at any given *locus.*

However, up to 40% of DNA *polymorphisms* are not recognized by available restriction enzymes and, therefore, a more rapid and robust method of screening SNPs would be *advantageous.* SNPs are also abundant in the mouse and human genomes and have been used for mapping

Table 12.1: Frequencies of Polymorphisms Between Three Major Genotypes of T. gondii.

Strain type	Unique SNPs	No. of RFLPs	Frequency of SNPs
Type I (RH strain)	85	47	1 in 272
Type II (ME49 strain)	44	22	1 in 527
Type III (CTG strain)	59	30	1 in 393
Total	188	99	

Based on sequencing of 33 separate loci totaling 23,192 bp.

and genetic *linkage* studies. Various techniques exist for typing SNPs, including single-strand *conformational* gel electrophoresis, RFLP analysis, and various PCR-based approaches. To genotype *T. gondii* efficiently using SNPs, we developed a new method to analyze multiple loci by *allelespecific* primer extension followed by *microarray* hybridization.

To take advantage of the high-throughput screening *capabilities* of DNA microarrays, we have used a *complementary oligonucleotide* tag approach first described by Hirschhorn et al.. This method uses pairs of *complementary generic oligonucleotides* that are optimized for similar *thermodynamic* properties.

One of the members of the pair is synthesized in conjunction with an allele-specific primer to detect the SNP and the other is affixed to a glass slide using an oligo-dT tail. This combination allows the *simultaneous* analysis of multiple SNP-specific *oligonucleotides* using one set of standard *hybridiration* conditions. A second method that we incorporated involves the use of *singlebase mismatch* primers to amplify SNPs specifically.

Although this *protocol* originally used a primer *extension* reaction based on RNA samples, we have added a preamplification step to generate double-stranded DNA templates that flank the SNP. Allele-specific extension reactions are conducted on a singlestrand PCR reaction using primers that *specifically* prime only one of the two possible alleles.

Finally, we have used the more efficient labeling protocol of amino-allyl dUTP *incorporation* followed by dye *conjugation* atter the PCR reaction. These improvements have been combined to provide a rapid, high-throughput analysis of genotypes in *T. gondii.* This approach is also applicable to *genotyping* of various organisms, including other *pathogens*.

REQUIREMENTS

1. *Cell lysates* of parasite strains to be tested.
2. PCR reagents, including lox buffer, dNTPs, and AmpliTaq (Roche, Indianapolis, IN).
3. 10-mg/mL Proteinase K (Promega, Madison, WI).
4. PCR primers for preamplifying the region surrounding SNPs (called SNP primers).
5. Allele-specific extension primers (15-nt oligo-dT plus 17-25 by of allele-specific sequences that define the SNP).

6. Shrimp alkaline phosphatase (Roche, Indianapolis, IN).
7. Exonuclease I (Amersham, Piscataway, NJ).
8. Monofunctional N-hydroxysuccinamide-ester Cy5 (no. PA23001) and Cy3 (no. PA25001) dyes. (Amersham, Piscataway, NJ). Dissolve dye in 72 pL *dimethyl sulfoxide,* aliquot 16 ×4.5 μL, dry the *aliquots* in a vacuum *evaporator,* and store *aliquots* at 4°C in a dessicator before use.
9. A solution of 0.05% NP-40, 10% *sodium dodecyl sulfate* (SDS). 10 mM aminoallyl-dUTP (Sigma, St. Louis, MO).
10. *Sonicated* herring sperm DNA (Promega, Madison, WI).
11. ArrayHyb LowTemp *hybridization* buffer (Sigma, St. Louis. MO).
12. PCR purification and *Nucleotide Removal Kits* (Qiagen, Valencia, CA).
13. Speed-Vac vacuum evaporator (Savant, Farmingdale, NY). 14. 0.1 M carbonate buffer, pH 8.8.
15. 20X SSC, pH 7.0 (3 *M* sodium chloride, 0.3 *M* sodium citrate).
16. Array hybridization chamber.
17. Poly-L-lysine (Sigma, St. Louis, MO).
18. Phosphate-buffered saline (PBS).
19. Erie Scientific glass microscope slides and Lifter-slip (Erie Scientific Corporation, Portsmouth, NH).
20. *Affymetrix* model 417 arrayer and 428 scanner (Affymetrix. Santa Clara, CA).
21. ScanAlyze version 2.35 (Michael Eisen, Stanford University, CA).

PROCEDURES

The method described here provides rapid, *multiplex analysis* of SNPs as diagrammed in Figure elsewhere in this chapter. The *strategy* is first to amplify a short *genomic* region, consisting of approx 100 bp, using primers that flank the SNP and thus amplify both alleles. This first PCR step can be multiplexed (referring to the combining of several gene-specific PCR *amplifications* combined in a single reaction) to allow *simultaneous* amplification of different loci. In the second step, a primer extension reaction is conducted using two *separate* samples.

one seeded with primers for the A alleles and one seeded with primers for the B alleles. The primers differ by one nucleotide at the 3'-end, which defines the SNP. This reaction also incorporates a modified dUTP that is labeled after *amplification* with Cy3 (A alleles) or Cy5 (B alleles). The resulting products are cohybridized to *complementary oligonucleotides* spotted on glass slides.

The recognition of these *complementary* oligonucleotides is based on a unique generic tag attached to the 5'-end of each allele-specific extension primer. This generic oligonucleotide tag does not represent a sequence in the genome of the *organism* but instead is chosen to provide uniform hybridization to a complementary generic oligonucleotide tag that is cross-linked to a glass slide.

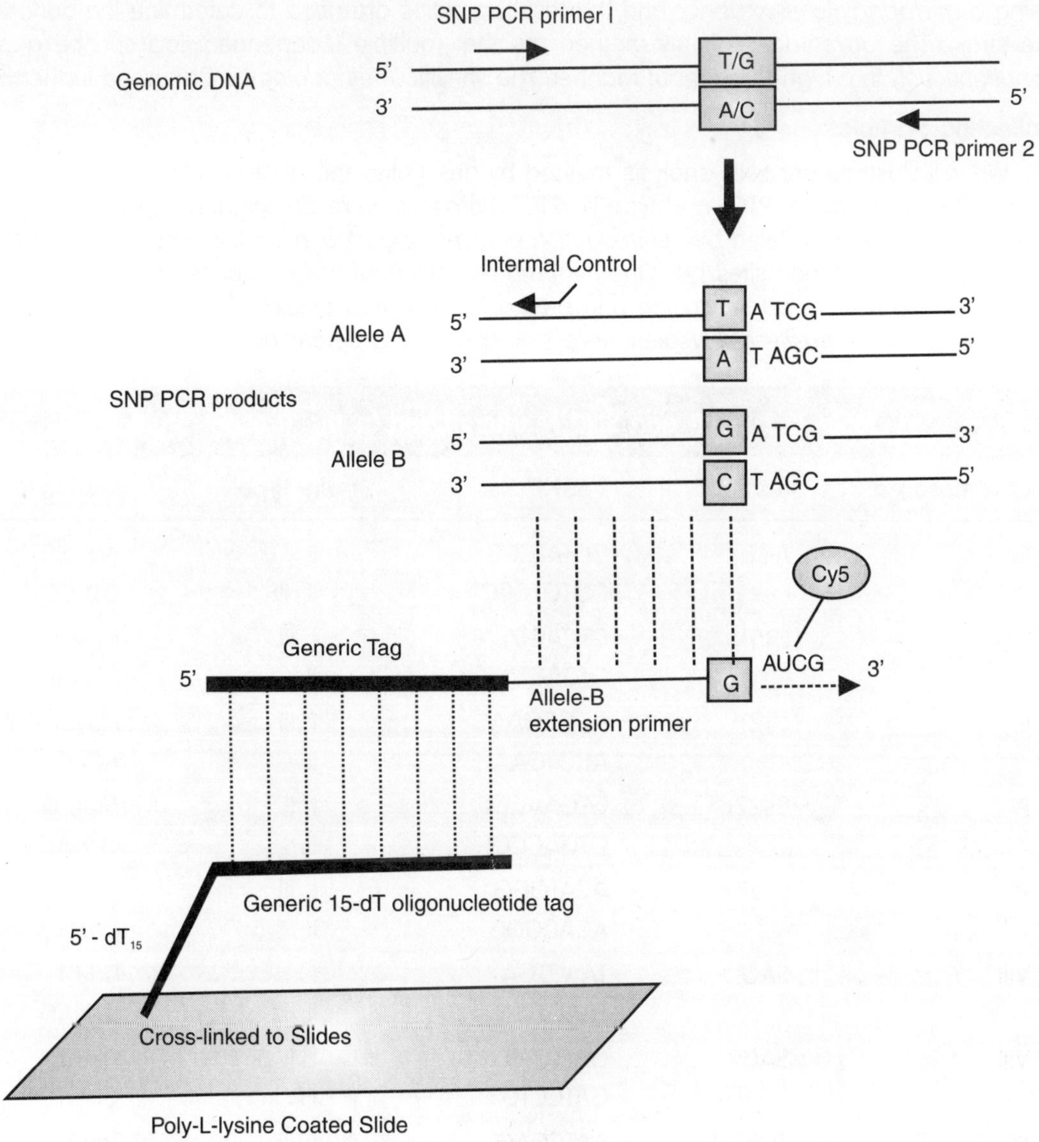

Figure 12.1: DNA fragments containing SNPs were amplified by multiplex PCR from genomic DNA. The SNP PCR products served as templates for the allele-specific primer extension reaction. Each extension primer of 17-25 bases contains a 5'-end generic tag complementary to a unique 15-dT oligonucleotide tag that was cross-linked to the slide. The 3'-end of the extension primers was an exact match to one of the two alleles (type A or B allele) for each SNP. The primer extension reaction was carried out in the presence of amino-allyl-dUTP that was later coupled with Cy3 or Cy5. The labeled multiplex primer extension products were pooled and hybridized to the oligonucleotide tag array. Two hybridization patterns were expected, a predominantly green color corresponds to type A alleles and a predominantly red color corresponds to type B alleles.

The relative hybridization of the A and B allele products to the *spotted oligos* are quantified using a *microarray* laser scanner and the resulting ratios are used to determine the genotype of the strain. The advantages of this method are that multiple independent loci can be analyzed *simultaneously* in a high-throughput manner. The specific *methodology* is discussed further below.

Collecting Samples

We will illustrate our SNP analysis method by describing the analysis of four *T. gondii* strains (type I, RH, GT 1; type II, PTG; and type III, CTG). Parasites were propagated in vitro as described previously and were pelleted by *centrifugation* and *resuspended* in 1 mL of PBS to a concentration of approximately 10^7 parasites/mL. One hundred microliters of 10X PCR buffer and 10 μL of 10-mg/mL *proteinase* K were added to the parasites and incubated at 37°C for 1 h, 50°C for 1 h, and 95°C for 15 min. *Parasite cell* lysates were stored at -20°C before use.

Table 12.2: RFLP and SNP Markers Used in the Present Study.

Chromosome	Marker	SNP	Strain type	RFLP site
Ib	ACT]	GGTATCCC	I	BsmF 1
		GGTGTCCC	II, III	GTCCC
II	L351	GTCACTA	I, III	Mae III
		GTCACTA	II	GTNAC
III	cB21	ACGCGAA	I, III	HpyCH4 IV
		ACGTGAA	II	ACGT
IV	Bs226	GTGAACCT	I, II	Mae III
		GTGACCCT	III	GTNAC
V	L358	ACATGGG	I, III	Nla III
		ACACGGG	II	CATG
VIII	SAG]	TAACTGA	I	Dde I
		TAAGTGA	II, III	CTNAG
VIII	SAG2	GATTCTG	I, II	Hinf I
		GATCCTG	III	GANTC
IX	TUB2	AATCGAG	I, III	Tay I
		AATGGAG	II	TCGA
X	GRA6	TTAAGCA	I, III	Mse I
		TTAGGCA	II	TTAA

SNPs are shown as bold and underlined letters for all markers. For each SNP, the top allele is denoted as A and the bottom as B. Strain types include I, 11, and III. The listed restriction enzymes are those that recognize restriction site polymorphisms at these locations, as used previously to distinguish different *T. gondii* strain types.

Independent Genetic Markers

We have selected nine independent genetic markers that were previously typed by PCR-RFLP and SNP analysis to illustrate the methods described here. These markers are *ACT1, L*351, cB21, BS226, L358, *SAG1,* SAG2, *TUB2,* and *GRA6,* which are dispersed on 8 of the 11 *T. gondii* chromosomes, including Ib, II, III, IV, V, VIII, IX, and X, respectively.

Generic Oligonucleotide Array

Generic tags are selected from a previously published source, from a listing provided at www.genome.wi.mit.edu/publications/SBE-TAGS. The full-length reverse complement sequences of the tags should be synthesized with a 15-dT *extension* at their 5'-end and spotted onto poly-$_L$-lysine coated slides to create the genericl5-dT oligonucleotide tag array. The l5-dT stretch is used to *facilitate* attachment to the slide.

In our study of *T. gondii,* the 15-dT oligonucleotides were named to indicate the specific SNP loci that they were designed to detect: ACTI, TTTTTTTTTTTTTTTAGTTCGTGCTTACCGCAGAATGCAG; L351,TTTTTTTTTTTTTTTTTTACAGGATGTGCTCAACAGACGTT;cB21, TTTTTTTTTTTTTTTACATCAATCTCTCTGACCGT TCCGC; BS226, TTTTTTTTTTTTTTTTTTATGGTGATCAGTCAACCACCAGG;L358, TTTTTTTTTTTTTTTAGACTG CGTGTTGGCTCTGTCACAG; SAG 1, TTTTTTTTTTTTTTTGCCTTACATACATCTGTCGGTTGTA; SAG2, TTTTTT TTTTTTTTTGAGACACCTTATGTTCTATACATGC; TUB2, TTTT TTTTTTTTTTTGCCACAGATAATATTCAC ATCGTGT; and GRA6, TTTTT TTTTTTTTTTCCTCATGTCAACGAAGAACAGAACC.

Regulatory Mechanism

The positive control 15 nucleotide oligo-dT tag, TTTTTTTTTTTTTTTTTACATATCACAACGTGC GTGGAGGC, is also spotted onto the glass slides and used to monitor the efficiency of array hybridization. The *complementary* sequence of this tag should be labeled with Cy3 at its 5'-end (Cy3-GCCT CCACGCACGTTGTGATATGTA) and spiked in the array hybridization reaction.

Two negative control oligonucleotides, referred to as negative control-1 TTTTTTTTTTTTTTTCGC AGGTATCGTATTAATTGATCTG and negative control-2 TTTTTTTTTTTTTTTTACACATACGATTCTGCG AACTTCAA, are used to monitor nonspecific array hybridization. Additionally, two internal control oligonucleotides are used to monitor the efficiency of the *primer extension* reaction as further discussed under see elsewhere in this chapter The oligonucleotide tags used for this step in our *T. gondii* study were LT4-GRA6, TTTTTTTTTTTTTTT TATTGAAGCCTGCCGTCGGAGACTAA, and LT 10-cB21, TTTTTTTTTTT TTTTCACAAGGAGGTCAGACCAGATTGAA, and these were spotted onto glass slides with the other complementary oligonucleotides.

Preparation of Slides

Generally, it is prudent to spot different concentrations of the oligonucleotides containing 0.005% NP40 onto poly-L-lysine coated microscope slides to assess the best level. After arraying, slides are processed as described by Eisen and Brown. In brief:

1. After spotting, UV-treat the oligonucleotides to cross-link them to the *amino groups* of the *lysines* on the slides.
2. Block the slides by *submersion* in succinic anhydride to convert the remaining free animo groups to amides.

3. Briefly heat the slides by submerging them in boiling water to denature and *eliminate* the secondary structure of the arrayed oligonucleotides.
4. Air-dry and store in the dark at room *temperature*. We have used slides for up to 3 mo but find that after this time period the quality *deteriorates*.

Methods of Preamplification

In this step, small fragments of DNA that flank the target SNPs are amplified by *multiplex* PCR and the products are purified and used as templates to generate *allele-specific* probes for the array hybridization steps described in the next section.

In the *T. gondii* study, PCR primers were designed using PRIMER3.0 (http:/ /www.genome.wi.mit.edu) and used in a multiplex reaction to amplify 701 10-bp *fragments* containing the nine target SNPs.

In the specific case illustrated here, these primers were ACT1-F, CAAGCGCATCGCGTACCG; ACT1-R, ATGATCTAGCTGAGGCGG; L351-F, TGCAAGTGGCAACCTTGAAA; L351-R, TGATTCCTATT AACAGACTA; cB21-F, ACTTGGTGGATCCAG CGA; cB21-R, CCTTCGACGCTGTAGACT; BS226-F, TACACATGCACCT GCGGCGA; BS226-R, CCTCTTGCATCCTTGTCCCC; L358-F, CCGCGAA GAATGGA ATGTTC; L358-R, AAGGTTTCGCTTTCTCCGCC; SAGI-F, GCTGACTGGTTGCAACGAGA; SAGI-R, CTTATCACTCGAAGCGTTAC CCTGC; SAG2-F, TCTGCGAAGAAAACGAGCGC; SAG2-R, ACTGCAAC CCGTGAAACAAC; BTUB-F, TCTGCGTGACAGCTTCGCCA; BTUB-R, TCTCCGCGATTCTCCACAGG; GRA6-F, CAGCAGACAGCGGTGGTG TT; and GRA6-R, TTGCTGTCCACCGCTCGAAC.

Before being used in a multiplex PCR reaction, each SNP PCR primer set is tested individually to ensure that only one DNA fragment of the correct size is amplified from genomic DNA. For the multiplex PCR reaction, the 50-µL mixture should contain 2 µL of *T. gondii* lysate, 200 µM each of the dNTPs, 1.5 mM of $MgCl_2$, 150 nM each of the nine sets of *forward* and reverse primers, and 1.25 U of AmpliTaq.

The reaction is carried out for 30 cycles at 95°C for 30 s, 55°C for 30 s, and 72°C for 1 min. After amplification, the multiplex PCR products are treated with 4.5 U of SAP (Roche) and 25 U of exonuclease 1 at 37°C for 30 min and 96°C for 15 min. Products are purified by Qiagen PCR Purification Kit following the *manufacturer's* instruction and eluted in 50 µL of H_2O. The purified PCR products are concentrated to 20 µL by *vacuum evaporation*.

Allele-Specific Primer

In this step, the purified SNP-containing PCR products created in elsewhere in this chapter are used to generate *allele-specific* sequences in a primer *extension* reaction in the presence of amino-allyl-dUTP. Then, the *incorporated aminoallyl-dUTPs* are coupled to *monofunctional* NHS ester Cy3 or Cy5 dye to label the products prior to use for array hybridization.

Each pair of extension primers consists of 17-25 by that are complementary to one of the two alleles at that locus. Both members of the pair are coupled to the same *generic* oligonucleotide tag at the 5'-end. The 3'-end of the *extension* primers is an exact match to one of the two alleles for each SNP.

A complementary generic oligonucleotide tag containing 15 nucleotide oligo-dT is cross-linked to the slide. In the *T. gondii* study, 12 generic oligonucleotide tags were selected from a pool of previously described oligonucleotide tag sequences and combined with paired extension primers. The *extension* primers were designated as allele-A (type I strains) or -B primers (non-type I strains).

For the following *extension* primers, the first 25 nucleotides are complementary to a generic tag spotted onto the array slides, the underlined 17-25 *nucleotides* are specific to *T gondii* sequences, and the two SNP alleles are indicated in parentheses. The extension primers used here were ACT 1-(A/B), CTGCATTC TGCGGTAAGCACGAACT AATGATCTAGCTGAGGCG GGA(T/C); L351(A/B), AACGTCTGTTGAGCACATCCTGTAAAG CTC TTC ACG TGT TGTC (G/A); cB21-(A/B), GCGGAACGGTCAGAGAGATTGATGTTACCGGCCC ATCCATATTC(G/A); BS226-(A/B), CCTGGTGG TTGACTGATCACCATA AACGGCGAAACTCTAGTGA(A/C); L358-(A/B), CTGTGACAGAGCCAA CACGCA GTCT GTTTCGCTTTCTCCGCCC(A/G); SAG1-(A/B), TACAAC CGACAGATGTATGTAAGGCTCAA AGATATTTTGCCAAAATTAA(C/G); SAG2-(A/B), GCATGTATAGAACATAAGGTGTCTCACCCGTGA AACAAC ACACAG(A/G); TUB2-(A/B), ACACGATGTGAATATTATCTGTGGCAG CTTCGCCAGTGTAAAT(C/G); and GRA6-(A/B), GGTTCTGTTCTTCGTT GACATGAGGTTCCGAAGGGGTCTGC(T/C).

Locating Regulatory Mechanisms

Two internal controls LT4-GRA6 (internal control-1) and LT10-cB21 (internal control-2) are included to monitor the efficiency of the primer *extension* reaction. These controls generate probes equally well from the SNP PCR products containing either one of the two alleles. The control oligonucleotide LT4-GRA6, TTAGTCTCC GACGGCAGGCTTCAAT CGAGCGGTGGACAGCAA, is designed to generate a product using the GRA6 SNP PCR product as template.

The control oligonucleotide LT10-cB21, TTCAATCTGGTCTGACCTCCTTGTGAGAG AAGCGAATAG TCTACAGC, uses the cB21 SNP PCR product as a template to generate a product. For both controls, the oligonucleotides are designed so that the primer extension reactions proceed in the opposite direction from the *allele-specific* primer extensions, and, therefore, they do not interfere with the *genotyping* of cB21 or GRA6. The products generated by these reactions are detected by hybridization to *complementary* oligonucleotide tags affixed to glass slides.

Applications of Primer

All extension primers should be tested in the absence of SNP PCR products to check for *nonspecific* priming. In our study, the primer pair L358-(A/B) showed strong signal from both alleles, and was therefore eliminated from further *experiments.* Two parallel extension reactions (denoted as Reaction-A and Reaction-B) of 50 µL are prepared, both containing 5 µL of purified primary PCR products, 50 µL dATP, dCTP, and dGTP, 16.7 *µM* dTTP, 33.3 µM of *aminoallyl-dUTP* (Sigma), and 1.25 U of AmpliTaq. In Reaction-A, each of the nine Allele-A primers and the two internal controls are added to a final *concentration* of 150 nM.

In Reaction-B, each of the nine Allele-B primers and the two *internal* controls are added to a final concentration of 1.50 nM. The extension reactions consists of 35 cycles of 94°C for *30* s, 55°C for *30* s, and 72°C for l min. After primer extension, the products are purified by Qiagen Nucleotide Removal Kit following the manufacturer's instruction, eluted in *50* µL of H_2O, dried using vacuum evaporation, and *resuspended* in 10 µL of 0.1 M carbonate buffer (pH 8.8).

Amplified PCR Products

Amplified PCR products are labeled by direct chemical reaction with derivatized *Cy3* or Cy5 to form a covalent adduct with the amino-allyl-dUTP that has been incorporated during PCR, as described at http://cmgm.stanford.edu/ pbrown/protocols/aadUTPCoupl.i.ngProcedure.htm.

1. Mix 4.5 μL of monofunctional N-hydroxysuccinamide-ester Cy3 or Cy5 with 4.5 μL Reaction-A or Reaction-B products, respectively.
2. Incubated at room temperature for 1 h in the dark.
3. Quench the dyes by adding 4.5 μL 4 M *hydroxylamine* and *incubate* for 15 min at room temperature in the dark.
4. Add 35 μL 0.1 M NaOAc, pH 5.2, to each reaction to *neutralize* the pH.
5. Combine the products of the Cy3 and Cy5 reactions.
6. Purify the probes using the Qiagen Nucleotide Removal Kit, elute in 50 μL of H_2O, dry down by vacuum evaporation, and resuspend in 4 μL of H_2O.

Array of Complementary Genes

Labeled PCR products are cohybridized to the array of *complementary* generic oligonucleotides spotted onto glass slides. The procedure is analogous to *hybridizations* used in many microarray studies.

1. Denature the 4 μL products obtained above at 96°C for 5 min, snap cool on ice.
2. Add 1 μL of 1 nM positive oligonucleotide control (Cy3-GCCTCCACGCACGT TGTGATATGTA), 1 μL of 5-mg/mL herring sperm DNA, and 12 μL of ArrayHyb LowTemp hybridization buffer.

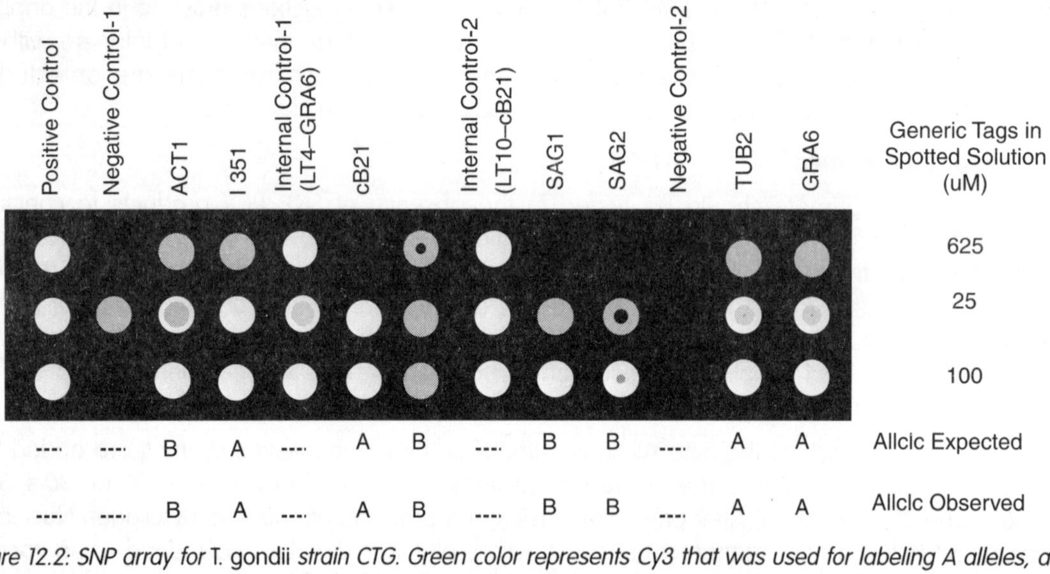

Figure 12.2: SNP array for T. gondii *strain CTG. Green color represents Cy3 that was used for labeling A alleles, and red represents Cy5 that was used for labeling the B allele. Each row was spotted with a different concentration of 15 nucleotide oligo-dT tags, from 6.25 to 100 μM, top to bottom. The positive control labeled with Cy3 shows the expected hybridization to its complementary oligonucleotide. The negative controls show little to no signal and the two internal controls show a combination of Cy3 and Cy5 signals. The SNP alleles were correctly identified for all alleles except ACTI that did not perform well for all strains.*

Table 12.3: Summary of Cy3/Cy5 Ratios for Array SNP Analysis

Marker	Toxoplasma strains			
	RH (Type I)	GT1 (Type I)	PTG (Type II)	CTG (Type III)
ACT1	A (2.4)	A (1.2)	B (2.5)	B (0.9)
L351	A (5.9)	A (5.1)	B (0.5)	A (3.4)
cB21	A (3.8)	A (3.4)	B (0.8)	A (6.1)
Bs226	A (2.7)	A (2.2)	A (1.7)	B (0.04)
SAG1	A (16)	A (14)	B (0.7)	B (0.02)
SAG2	A (4.6)	A (2.9)	A (3.2)	B (0.1)
TUB2	A (10.8)	A (8.8)	B (0.7)	A (8.2)
GRA6	A (4.6)	A (3.9)	B (1.0)	A (4.4)

A and B represent the allele types. Numbers in parentheses are Cy3/Cy5 ratios. The A allele was labeled with Cy3, therefore ratios ≥ 1.0 indicate the A allele and ≤ 1.0 indicate the B allele. All of the markers except *ACT1* provide clear results that correctly discriminate between the alleles.

3. Cover array slide with Erie Scientific Lifter-slip.
4. Load PCR products to slide from the side of the Lifter-slip, allowing it to spread across the slide by capillary action.
5. Incubate slides in a hybridization chamber at 45°C for 16 h.
6. Wash slides by immersion at room temperature in 2X SSC/0.1 % SDS for 2 min, 1X SSC for 2 min, and 0.2X SSC for 5 min.
7. Centrifuge slides at 100g for 5 min to quickly remove residual wash solution.
8. Scan slides using an Affymetrix model 428 laser scanner and store the results as a .tif data file.

Studying Polymorphism

An example of a SNP array detecting polymorphisms in the strain CTG is shown in Figure 12.2. *Analysis* of this array was performed using ScanAlyze soft ware. Signal intensities for Cy3 and Cy5 *channels* were adjusted by subtracting background hybridization based on several *randomly* chosen. areas outside the spots.

Additionally, signals were normalized by setting the ratio of Cy3:Cy5 = 1, based on the signal intensities of the two internal controls. The summary of Cy3:Cy5 ratio for SNP typing is shown in table 12.3. Eight of the nine SNP markers (the exception being ACT I) correctly identified the alleles for the four *T. gondii* strains tested here.

The range of differences in hybridization of the correct allele-A vs correct allele-B ranged from 4-fold (in the case of *GRA6)* to almost 50-fold (in the case of Bs226).

These results demonstrate that allele-specific primer extension tag arrays provide a powerful method for SNP genotyping in *T. gondii.* The technique described here will be useful in the genetic analysis of multiple strains of many species of pathogenic organisms.

POINTS TO REMEMBER

1. In our study of *T. gondii* polymorphisms, seven of nine SNPs were correctly genotyped by microarray. One SNP marker (L358) showed nonspecific signal in the absence of SNP PCR products and was *eliminated.* Another marker (ACT 1) could not distinguish the two *alleles* among all strain types, a result that was likely because of *nonspecific* priming.

   Redesign of the *allele-specific* primers could potentially solve these *problems.* However, given the abundance of SNPs, it may be easier simply to pick another nearby *polymorphism* to analyze.
2. It is desirable to establish a cutoff level for interpreting the array hybridization results before *genotyping* unknown samples. In the *experiments* presented here, the range of differences in hybridization of the correct allele-A vs correct allele-B was fourfold or higher for all the SNP markers except ACT I. In practice, a threefold cutoff might be *reasonable.* For those loci that cannot be distinguished based on array data, another *genotyping* method such as RFLP is preferred.
3. Direct labeling of allele-specific primer extension products by incorporating Cy3dCTP and Cy5-dCTP also provides good detection of SNPs using the *microarray* approach. The main disadvantage is that the background is higher for Cy3-dCTP. For these reasons, we prefer the *amino- ally* l-dUTP incorporation followed by dye labeling after PCR.
4. The design of allele-specific extension primers is critical to generate specific probes for SNP detection. Because the last nucleotide at the 3'-end of the primer has to be specific and fixed at the SNP site, the options for primer design are limited.

   The primers with 16-25 bases were selected to hybridize to either the sense strand or the complementary strand of DNA templates. In our *experiments,* candidate allele-specific extension primers generated from both *strands* were checked for melting temperature (T_m), selfpriming, or dimerization using Oligo Analyzer (http://www.idtdna.com/program/main/home. asp). A T_m of 55-60°C was achieved by adjusting the length of the primers, because their location is determined by the position of the SNP.

   Primers that were predicted to self-prime or dimerize were eliminated and those that passed this screening were paired with a list of generic tags. *Chimeric* primers that were predicted to self-prime or *dimerize* were eliminated.
5. The advantage of SNP microarray is the capacity to genotype strains on a highthroughput scale compared to, for example, RFLP analysis. Currently, the genetic linkage map of *T. gondii* consists of over 50 RFLP-PCR markers that provide *reasonably* good coverage over the 11 separate *chromosomes.*

   Application of microarray SNP genotyping to these markers would enable rapid *genetic linkage* analysis of complex biological traits. SNP microarray typing will also allow more rapid analysis of the large numbers of new isolates required for *epidemiology* and *population* biology studies.

6. Array slides should be used within 2 mo after preparation. Older slides gave low signal intensity and high *background*, making the results more variable. In our *experience, array slides* older than 3 mo were not useable.

Chapter 13

Clonal Complex

The realization over 30 yr ago that particular serovars of *Escherichia coli* were strongly associated with disease led to the *bacterial* clone concept. In subsequent decades it has become well established that all isolates of *pathogenic* species are not equal, and that in a typical pathogenic species there are a small number of genotypes (clones) that are greatly over-represented among those recovered from disease or from a particular type of disease.

The recognition that particular clones are much more strongly associated with disease than others was first established using *serological* methods, but most of our current knowledge of the genetic structure of bacterial populations, and of the association of particular genotypes with disease, is due to the introduction of improved methods for distinguishing the *diversity* of *genotypes* that are present within most bacterial species.

In particular, it was the introduction of *multilocus enzyme* electrophoresis (MLEE) into microbiology by Milk-man and Selander and Levin that provided the key experimental tool that led to a greatly increased understanding of bacterial populations. MLEE indirectly assesses genetic *variation* at multiple *housekeeping* loci by examining the *electrophoretic* mobility of their gene products on starch gels.

By numbering the different *mobility variants* at each locus, and by *equating* these with different alleles at the underlying genetic locus, a strain can be characterized by a string of integers that correspond to the alleles at each of the multiple housekeeping loci. MLEE was *introduced* into microbiology by *eukaryotic* population geneticists who had an evolutionary *perspective* that had been lacking from earlier studies of bacterial populations.

They stressed the limitations of serological methods, which take advantage of extensive variation in bacterial components that interact with the host *immune system*, and the importance of using neutral variation for strain characterization and for understanding the *evolutionary* relationships among isolates of bacterial species.

MLEE is highly discriminatory, and the identification in the early studies of isolates in *E. coli* populations that were indistinguishable in genotype, although recovered several decades apart, was a surprise, and was taken to imply that the alleles at different *genetic loci* in the population are not being shuffled by recombination.

Thus, the sexual processes of *conjugation* and *transduction* that had been studied so extensively in the laboratory were considered to be relatively *ineffective* at *promoting* genetic exchange and genetic diversification among *E. coli* strains in nature.

From these *pioneering* studies of *E. coil* emerged the view that bacterial pathogens have clonal population structures, which arise as a consequence of low rates of genetic exchange in nature. This view has changed *significantly* over the last decade, and there is now a general recognition that the extent of *genetic* recombination among bacterial species varies, such that there is a *spectrum* of population structures, from highly clonal through to essentially *nonclonal.*

Recently, MLEE has increasingly been replaced by multilocus sequence typing (MLST), and this chapter focuses on the use of the latter technique for *precise* strain characterization, and on some of the insights that both MLEE and MLST can provide about the nature and evolution of bacterial populations, and the concept of clonality. For more detailed accounts of some of the topics covered in this chapter there are a number of recent review articles.

BACTERIAL RECOMBINATION

The sequences of the genomes of bacterial pathogens have provided strong evidence for a major impact of recombination over evolutionary *timescales.* Many *genes* that are common to all isolates of a species appear to have been acquired from distantly *related* species, and there is evidence for more recent gene acquisition (and gene loss), because different isolates of the same species differ *substantially* in gene content.

Acquisition of novel genes may be of huge significance in the evolution of pathogens, since gene acquisition may, occasionally, produce a strain within a commensal species that has an improved mode of transmission to new hosts—e.g., production of a cough, vomiting, or diarrhea-*which we observe as an ability to cause disease.

However, the type of recombination that occurs at a frequency that can impact on population structure, and the rate of diversification of clones, which is the *focus* of this chapter, involves extensive sequence similarity (and thus occurs typically between strains of the same or very closely related species) and is mediated by *homologous* recombination.

Recombination in bacteria is, of course, very different from that in typical *eukaryotes,* where reassortment of the alleles from each parent occurs in every generation. In bacteria, *recombination* is not linked to reproduction and involves the replacement of a small region of the genome (typically, from a few hundred to a few thousand base pairs) of a recipient *bacterium* by the corresponding region from another isolate of the same or a closely related species.

To produce genetic change there must be both a sexual mechanism and opportunities for donor and recipient bacteria to meet. Transformation (in species that are *naturally* transformable) and *phage-mediated* transduction are probably the main mechanisms in nature. The rate of recombination is a key parameter in bacterial population genetics.

In the total absence of recombination, all genetic variation within house-keeping genes arises by mutational events, and *genetic novelties* (e.g., *adaptive mutations*) occurring in one strain are not able to spread horizontally into other strains. The absence of recombination has another important *consequence*; fitter *variants* that arise at intervals will increase in frequency in the population and, in the absence of *recombination*, may rise to high frequency in the population, leading to a loss of other *genotypes* (*periodic selection*;).

The continual *purging* of genotypes by repeated rounds of periodic selection leads to a population where there are fewer *genotypes*, and clones are much more apparent, than in populations where recombination is frequent. If rates of recombination are very high, those fitter variants that arise will increase in frequency in the population but, while doing so, will diversify rapidly as a result of the accumulation of frequent recombinational replacements.

The rapid diversification of any emerging fitter *genotypes* can result in an absence of *observable* clones within the population, and a nonclonal population, such as that of *Helicobacter pylori*, in which all isolates from *epidemiologically* unlinked sources may be different in genotype, as assessed using MLEE or MLST.

Populations in which recombination is either absent or extremely rare, or very frequent, probably represent the *minority* of situations, and most bacterial species are likely to have intermediate levels of *recombination*, and population structures that are between these highly clonal and *nonclonal* extremes.

NEUTRAL VARIATION IN POPULATIONS

As emphasized by Selander and colleagues, population structures and the relatedness between isolates of a bacterial species should be assessed by *examining* genes for which the genetic variation within the population is likely to be selectively *neutral*, or nearly so.

Genes that encode enzymes involved in essential metabolic processes (*housekeeping genes*) are therefore used in both MLEE and MLST, rather than, for example, genes that encode cell-surface components where strong selection for variation may be imposed by the host immune response. Thus, while genes involved in *pathogenicity* and *virulence* are of crucial importance, they cannot be used for defining *population* structures.

Rather, the evolutionary behavior of these genes has to be mapped onto the population structure and *relatedness* between isolates, as inferred from the analysis of *genetic* variation at *neutral loci*. MLEE has a number of limitations and has largely been replaced by MLST, which is based on the same principles, but assigns alleles at multiple house-keeping loci directly by nucleotide *sequencing* of an internal fragment of each locus.

MLEE identifies only that minority of the nucleotide sequence variation that leads to an alteration in the electrophoretic mobility of the gene product, whereas MLST identifies all genetic variation at the locus. Consequently, high levels of *discrimination* in MLST can be *achieved* using far fewer loci than are used in MLEE. Each different sequence at a locus is assigned as a distinct *allele* and, as in MLEE, the alleles at each *locus* provide an allelic profile.

Each allelic profile is described as a sequence type (ST), whichprovides a convenient *descriptor* for a strain and is *equivalent* to the *electrophoretic* type (ET) used in MLEE. Using multiple housekeeping loci is crucial for a number of reasons. Firstly, neutral variation accumulates slowly, compared with the more poorly characterized variation that is indexed by methods such as pulsed-

A

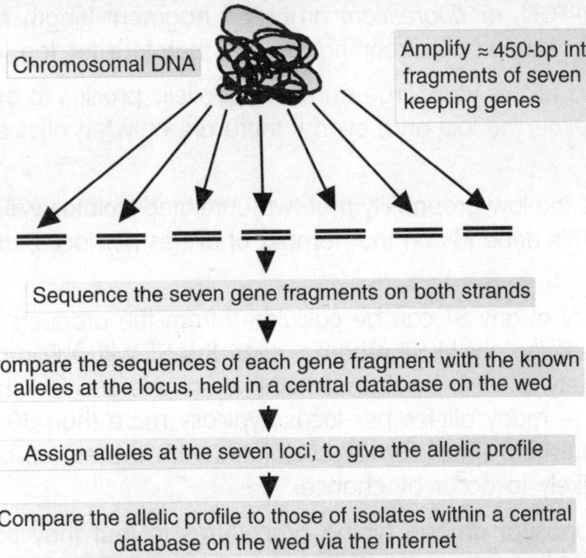

B

	Allelic Profile							ST
Strain 1	1	5	3	7	2	8	4	7
Strain 2	11	15	30	12	22	11	20	3
Strain 3	6	5	9	1	2	13	17	5
Strain 4	11	15	30	12	22	11	20	3
Strain 5	9	10	1	19	12	18	14	2
Strain 6	1	5	3	7	2	8	4	7
Strain 7	20	11	29	9	21	13	11	4
Strain 8	9	15	13	27	22	18	14	1
Strain 9	20	11	29	9	21	13	11	4
Strain 10	14	15	30	12	22	11	20	3
Strain 11	1	5	3	5	2	8	4	6
Strain 12	11	15	30	12	22	11	20	3

C

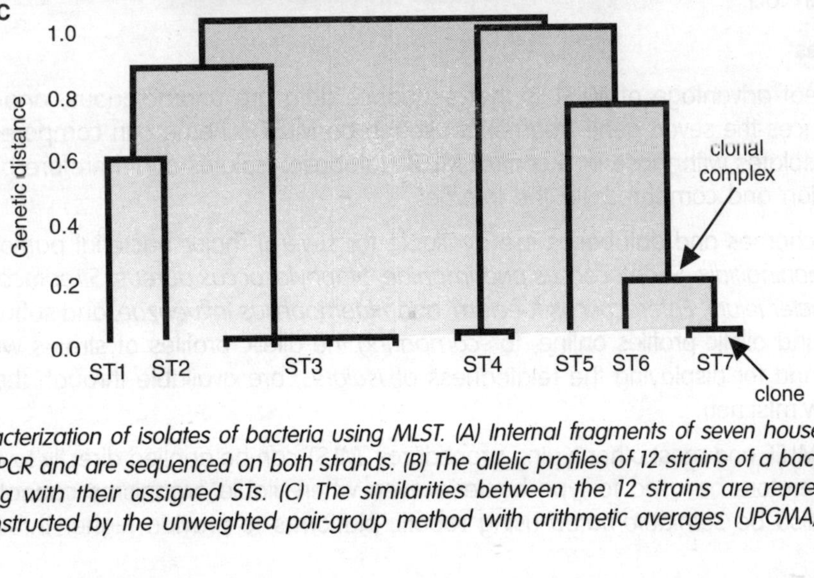

Figure 13.1: Characterization of isolates of bacteria using MLST. (A) Internal fragments of seven housekeeping loci are amplified by PCR and are sequenced on both strands. (B) The allelic profiles of 12 strains of a bacterial species are shown, along with their assigned STs. (C) The similarities between the 12 strains are represented as a dendrogram, constructed by the unweighted pair-group method with arithmetic averages (UPGMA).

field gel electrophoresis (PFGE), or *fluorescent* amplified fragment length *polymorphism* (fAFLP). The number of *alleles* at a single housekeeping locus is therefore far too low for typing.

The use of multiple loci allows very large numbers of allelic profiles to be distinguished. MLST schemes use seven *housekeeping* loci and, even if there are only ten alleles per locus, 10 million STs can be distinguished.

Of more importance is the low probability that two unrelated isolates will, by chance, have the same or very similar STs; this depends on the number of alleles per locus, and the frequencies of the *alleles* at each locus.

The expected frequency of any ST can be calculated from the products of the frequencies of each allele and, in a *discriminatory* MLST scheme, even the ST with the most common allele at each locus should be unlikely to occur by chance within a sample of a few thousand *isolates.* The current MLST schemes have many alleles per locus (typically more than 30, allowing >20 billion different STs to be distinguished), and *unrelated* isolates with the same, or very similar, allelic profiles are *extremely* unlikely to occur by chance.

MLST and MLEE are unusual among typing procedures in that they provide high levels of *discrimination*, but use neutral *genetic* variation that accumulates slowly. Most other typing *methods* obtain high levels of discrimination by searching out variation in the genome that accumulates very rapidly.

Those typing systems based on a single locus (typically, and inappropriately, these are based on a highly polymorphic locus subject to diversifying selection from the host immune system) are also *unsatisfactory* because they are sensitive to recombinational events.

Two identical isolates will look completely unrelated if a recombinational event in one isolate changes the allele at the single locus that is used for typing.

MLST provides *buffering* against this effect, as a recombinational replacement at one locus will not obscure the similarity of the two strains, since they will still have identical sequences at six of the seven loci.

Advantages

The great advantage of MLST is that sequence data are unambiguous, and any laboratory that sequences the seven gene fragments used in an MLST scheme can compare the sequences from their isolates with those in a central MLST database. Isolates can therefore be characterized with precision and compared via the Internet.

MLST schemes and databases are available for several major bacterial pathogens, including *Neisseria meningitidis, Streptococcus pneumoniae, Staphylococcus aureus, Streptococcus pyogenes, Campylobacter jejuni, Enterococcus faecium,* and *Haemophilus influenzae,* and software for assigning alleles and allelic profiles online, for *comparing* the allelic profiles of strains with those in the database, and for displaying the relatedness of *isolates,* are available through the MLST website (http://www.mlst.net).

Unlike MLEE and most other typing procedures, MLST can be applied directly to clinical samples and can therefore be used to type bacteria even when viable organisms cannot be recovered. MLST can also be *semiautomated* using robotic platforms to provide higher throughput.

PROKARYOTIC CLONE

Molecular typing procedures are often employed to establish the *relatedness* of isolates from some epidemiological setting. Those that are indistinguishable are typically assigned as members of the same clone. The concept of a clone is derived from studies of asexual *eukaryotes*, where the absence of sex results in *progeny* that are *genetically* identical.

In bacteria the situation is more *dynamic*, as their genomes are more plastic, and they are subject to genome re-arrangements (deletions, movement of insertion sequences, and so on) and, to a variable extent, to localized *recombinational* replacements. Thus, bacterial isolates assigned to the same clone may not be identical, as the recent descendents of the same common ancestor may differ somewhat in *genotype*.

This can be seen clearly using PFGE, where, at least in some species, even isolates from an outbreak of disease may differ slightly in their fragment patterns. In *microbiology*, the strict definition of a clone therefore tends to be loosened slightly, and clones are defined *pragmatically* as isolates that are indistinguishable, or highly similar, using a particular *molecular typing procedure*.

This loose definition of a clone is *unsatisfactory*, but unfortunately several terms that are difficult to define precisely are used in microbiology to describe groups of identical, similar, or related isolates. Thus, besides clones, we have *species*, *genomovars*, *lineages*, *clonal complexes*, and *strains*, to name but some of them. Part of the problem in defining these terms is the very different level of genetic diversity within named species of bacteria, and the differences in their population biology.

Molecular typing procedures that are highly resolving in most bacterial species (e.g., MLEE or MLST) would indicate that almost all isolates of some species, such as *Mycobacterium tuberculosis*, *Bordetella pertussis*, *Yersinia pestis*, *Bacillus anthracis*, and *Burkholderia mallei*, have the same genotype (i.e., are members of a single clone), while in others, such as *H. pylori*, all isolates from *epidemiologically* unlinked individuals typically would have different *genotypes*.

If we define a clone as a group of isolates that are indistinguishable by the typing *procedure* that is employed, using MLEE or MLST, the former named species would each be a single clone, or a few closely related clones, whereas the latter species would consist of an *immeasurable* number of clones.

The utility of the clone concept is that it attempts to recognize groups of isolates that have all descended from a recent common ancestor and that therefore may have *broadly* similar biological properties (*pathogenicity*, *host adaptation*, and so on). The problem is that *bacterial* clones are not real genetic entities, since the genotype of any fitter variant that arises will continually diversify as it becomes more prevalent in the *population*, to produce an increasingly variable cloud of *descendent* genotypes.

These variant genotypes may fall into many clones using a procedure (e.g., fAFLP) that *sensitively* detects minor variation across the whole genome, whereas they may all be seen as a single clone by MLEE or MLST, which look at *neutral variation* within a very small proportion of the genome. The challenge is to provide a sensible *nomenclature* that is *appropriate* for the biology of the organism.

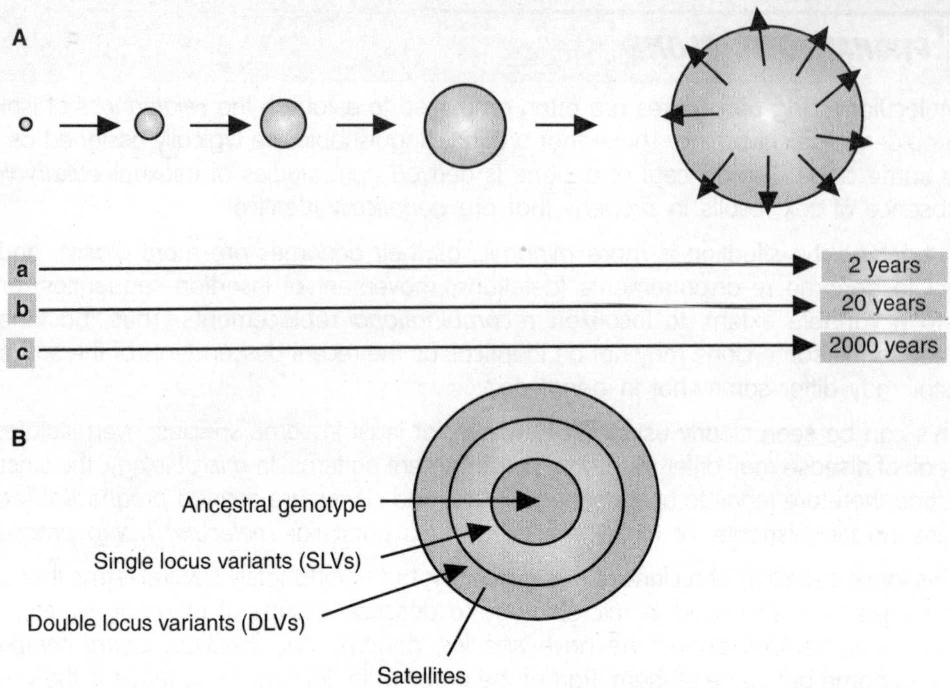

Figure 13.2: Diversification of a bacterial clone to produce a clonal complex. (A) The rise in frequency of a fitter genotype is represented by the increasing size of the sphere. (c). The rate of diversification probably depends mainly on the extent of recombination in the population. (B) MLST indexes variation in only about 0. 1 % of the genome, and identifies those isolates where no variation has yet occurred within any of the sequenced fragments of the seven housekeeping loci (ancestral genotypes), and those where the variation has involved one of these loci (SLVs), or two of these loci (DLVs). These three classes of genotypes are represented by the BURST diagram; isolates that are assigned to a clonal complex by BURST, but which are not the ancestral genotype, or SLVs or DLVs, are called satellites and are linked to the ST that is most similar in allelic profile.

Species that are *essentially monoclonal* using MLEE or MLST typically appear, on population genetic grounds, to be single clones within another named species, that were given species status because of the distinctive diseases that they caused. If we are to understand the epidemiology of diseases caused by *monoclonal* species, we need to use methods that discriminate the variant genotypes within these slowly diversifying clones.

For example, isolates from typhoid fever have, *historically*, been given the species name *Salmonella typhi*. However, population genetic analysis of the *Salmonellae* has shown that there is no justification for defining isolates from typhoid fever as a distinct species, since using MLEE we find that they are members of two closely related clones within the assemblage of clones that defines the species *S. enterica*.

Similarly, the nomenclature of some other "species" could be rationalized on population genetic grounds. *Y. pestis* could be re-assigned as the clone of *Yersinia pseudotuberculosis* that causes plague, and *B. mallei* as the clone of *B. pseudomallei* that causes glanders.

Using the terms *clone* or *clonal complex* to define the clusters of very similar genotypes that cause typhoid fever or plague or *glanders* is attractive, and has *biological plausibility*. For typhoid fever there is enough genetic *variation* between isolates to *discriminate* adequately using established typing procedures (e.g., PFGE or fAFLP) that are more sensitive *detectors* of minor genetic variation than MLEE or MLST.

This use of the genetic magnifying glass can go even further, as in the recent study of deliberate anthrax releases in the United States, where genome sequencing of *B. anthracis* isolates was used to identify some of the small number of *single-nucleotide polymorphisms* (SNPs) within this extremely uniform species.

Similarly, comparison of the genome sequences of multiple isolates of *M. tuberculosis* have revealed a number of SNPs that are being used to produce an unambiguous digital typing system that can resolve the minor variants of this uniform species. What we should call these very fine *subdivisions* within a monoclonal species is unclear, but if they are unlikely to have much *biological* meaning, it does not seem sensible to define them as clones.

Thus, *clone* would be an inappropriate term for the very fine (and presumably biologically unimportant) subdivisions within *B. anthracis* made for forensic purposes following a deliberate release, but it would be appropriate for defining the subdivisions within *M. tuberculosis* if they corresponded to *epidemiologically* important entities that varied in their *virulence, transmissibility,* or drug resistance profile.

Producing Clonal Complexes

Using MLST, we define isolates that are identical in allelic profile as a clone, and clusters of isolates with closely related allelic profiles as clonal complexes. The identification of clones and clonal complexes within samples of a *bacterial* population is typically achieved by using the genetic variation indexed by a molecular typing procedure to construct a *dendrogram*. This procedure is simple and useful but has a number of *problems*.

Firstly, *dendrograms* are *cumbersome* when very large numbers of isolates are analyzed, as occurs in MLST since results from many different laboratories may be submitted to a single central database. They also do not accurately represent the processes by which bacterial clones *diversify* to produce clonal complexes, and they fail to extract information about *clonal ancestry* and patterns of *evolutionary* descent that may be present in MLST data.

An algorithm, BURST, has been developed by Ed Feil to identify the isolates within a MLST dataset that are likely to be descended from the same recent ancestor. This program starts by identifying groups of isolates within a MLST dataset that have some *predefined* level of similarity (the default value is that all isolates are related to at least one other member of the group at five of the seven loci), and then predicts the ancestral genotype of each of these groups (clonal complexes), and the most likely patterns of evolutionary descent of all isolates within each clonal *complex* from the predicted *ancestral* genotype.

The program therefore gets around the problem of trying to display the relatedness of all of the isolates, and focuses on those that are similar in genotype, without being concerned about the relationships between these groups of *similar isolates*, which in many species will be impossible to *discern*, owing to the impact of *recombination*.

BURST predicts ancestral genotypes by assuming a simple but realistic model of bacterial

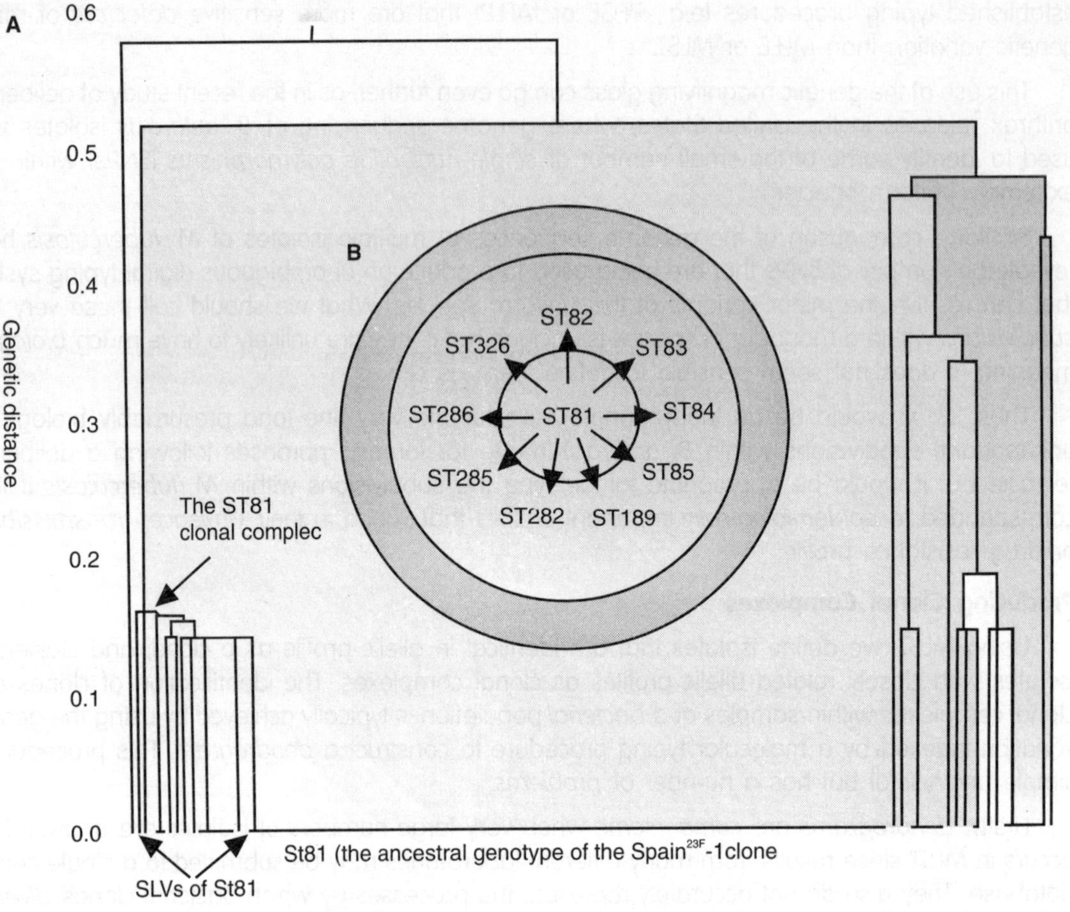

Figure 13.3: Identification and diversification of the Spain^{23F}- 1 clone of S. pneumoniae. *All isolates with allelic profiles that are similar (identity at a 3 loci by MLST) to that of the reference isolate of the Spain^{23F}–1 clone were extracted from the pneumococcal MLST database (http://spneumoniae.mlst.net). The relatedness of the resulting 119 strains was displayed as a dendrogram, constructed from the matrix of pairwise differences in their allelic profiles, by the UPGMA method (A). Isolates of the Spain ^{23F}- 1 (ST8 1) clonal complex are easy to recognize, since the cluster of closely related genotypes is well resolved from all other isolates in the MLST database, presumably because resistance did not emerge in a genotype that was prevalent within the pneumococcal population. BURST also identifies the ST81 clonal complex and predicts that ST81 is the ancestral genotype (B). ST81 is therefore placed in the central circle of the BURST diagram. The other nine STs in the clonal complex are all shown to differ from the ancestral genotype at only one of the seven MLST loci (i.e., they are all SLVs of ST81), and are displayed in the middle circle. The arrows indicate that they have arisen from the diversification of the ancestral genotype. No isolates differ from ST81 at two loci, presumably because there has not yet been sufficient time for DLVs to accumulate. The relationships of the isolates within the clonal complex are far better displayed using BURST than a dendrogram, as the latter gives no indication that all the variant STs in the complex are the diversification products of the ancestral genotype, ST81.*

evolution in which a genotype rises in frequency within the population (typically because it has some fitness advantage, such as increased transmission to new hosts, or antibiotic resistance) and, while doing so, *accumulates* point mutations and *recombinational replacements.*

The ancestral *genotype* therefore *diversifies,* firstly to produce variants that differ by MLST at just one of the seven loci (single-locus variants [SLVs]) and subsequently those that differ at two loci (double-*locus variants* [DLVs]). As *ancestral* genotypes initially diversify to produce a set of variants that differ at only one of the loci, the *ancestral* genotype is assumed to be the ST within the clonal complex defined by BURST that has the *largest* number of SLVs.

The ST of the predicted ancestral genotype, and the SLVs and DLVs, and those isolates that are within the complex, but are neither SLVs nor DLVs (satellites), are represented as a BURST diagram. The *representation* of the genotypes of a group of similar isolates of *pneumococci* and *meningococci* are displayed in Figures 13.3 and 13.4 using both a *dendrogram* and a BURST diagram.

Some caution is required in using MLST for purposes other than for identifying clones and clonal complexes and predicting ancestral genotypes of clonal complexes. MLST identifies isolates that are identical at all seven loci and those that have accumulated variation at one or more of these loci, but this represents only approx 0.1% of a bacterial genome.

For example, if we consider 1000 meningococcal isolates with an absolutely identical genotype and, to represent the genetic variation that has accumulated in contemporary descendents of this genotype, we were to pepper the genome with 200 small random recombinational replacements, some *isolates* would have no changes at any of the seven loci used in MLST (and would appear to have retained the ancestral genotype), whereas others would have *alterations* at one of the loci and would be recognized as SLVs.

However, if we chose a different set of loci for the MLST scheme, the same distribution of replacements around the genome would result in some of the SLVs in the first MLST scheme *appearing* to be ancestral genotypes using the second MLST scheme, and some that were *ancestral* genotypes will be SLVs.

In a real population, contemporary isolates with the ancestral genotype, and those that are SLVs, are derived from the same ancestor and, on average, will have received similar numbers of mutations or replacements, few of which will have altered a locus used in the MLST scheme. Isolates with the *ancestral genotypes* are, therefore, not necessarily a more *uniform, natural,* or fit group of isolates than the apparently diverse *collection* of SLVs.

BURST performs very well with species where rates of recombination are moderate. For example, it has proved excellent at resolving the clonal complexes and likely patterns of evolutionary descent within *methicillin*-resistant and *methicillin*-susceptible strains of *S. aureus* and, more challengingly, is capable of objectively assigning isolates of meningococci to a clonal complex.

However, in the latter species, the rapid diversification of clones results in very complex dendrograms, and a complex BURST diagram. BURST would probably not be effective at analyzing the even more diffuse clonal *complexes* within populations of pathogens that have higher rates of recombination than the meningococcus. However, other approaches are also likely to be unsuccessful.

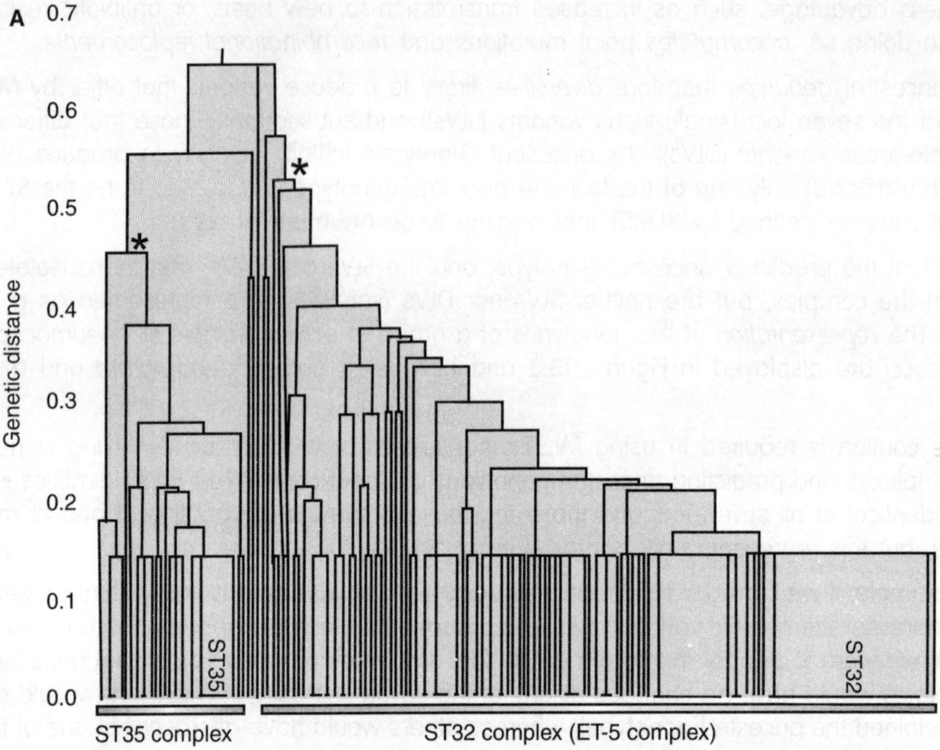

Figure 13.4: Identification and diversification of the ST32 (ET-5) clonal complex of N. meningitidis. *All isolates with allelic profiles that were similar to that of* N. meningitidis *strain 44/76 (identity at ≥ 3 loci by MLST), the reference isolate of ET-5, were extracted from the meningococcal MLST database (http://neisseria.org/nm/typing/mlst/). The relatedness of the resulting 163 isolates was assessed using a dendrogram (A). The dendrogram is complex, and provides no objective way of distinguishing those isolates that are predicted to be the diversified descendents of the ancestral genotype of the ET-5 complex, from those that are similar in genotype, but are not derived from this ancestral genotype. It also provides an inadequate representation of the way in which an ancestral genotype diversifies to form a clonal complex. BURST defines two clonal complexes among the 163 isolates, and there are only four isolates that do not belong to one of these clonal complexes. These isolates differ from each other, and from all isolates in the two clonal complexes, at ≥ 3 of the 7 MLST loci. The ancestral genotypes of the two complexes were predicted by BURST to be ST32 and ST35. The node that includes all isolates assigned to the ST32 and ST35 clonal complex by BURST is shown by an asterisk on the dendrogram.*
The two clonal complexes defined by BURST are separated on the dendrogram by the isolates that are not assigned by BURST to either clonal complex. The inferred relationships between the isolates within the ST32 (ET-5) clonal complex are shown as a BURST diagram (B). ST32 is strongly predicted to be the ancestral genotype, and is surrounded by a large number of SLVs (middle circle; there are so many STs that the ST numbers cannot be read), and by three DLVs (outer circle). This is a highly diverse clonal complex and the BURST diagram is complex compared to that in Figure elsewhere in this chapter. BURST has separated some SLVs into distinct subgroups, as these STs are themselves associated with a number of SLVs, and in the case of the ST33.subgroup, by three DLVs. BURST assigns isolates objectively into the ST32 (ET-5) clonal complex and shows the diversification of its predicted ancestral genotype far more clearly than a dendrogram. A greatly improved version of BURST (eBURST) is available at http://eburst.mlst.net. ***(Figure 13.4 Contd.)***

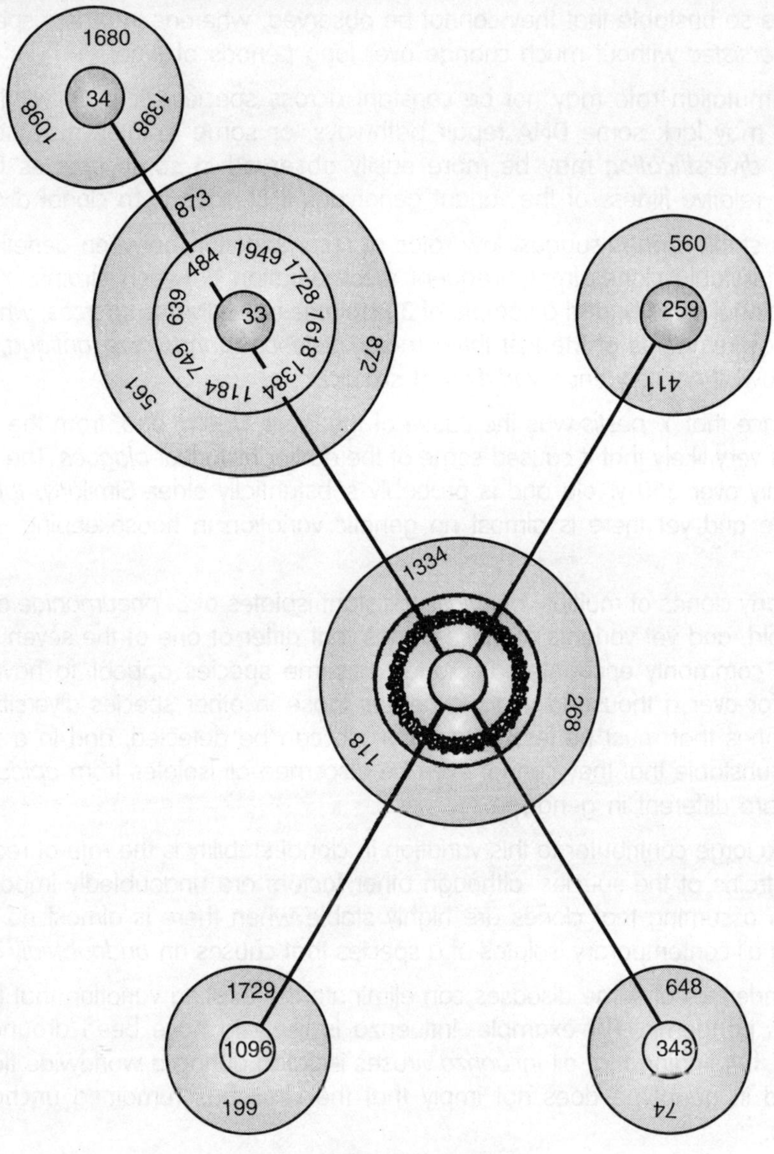

(Figure 13.4 Contd.)

MOLECULAR DIVERSITY

The relative impact of recombination and point mutation is a major determinant of the stability of bacterial clones. If, as a first *approximation*, we assume the mutation rate is constant, the stability of clones will be determined by the extent of recombination and the *recombinational* replacement size.

There appears to be a large variation in the stability of clones in different pathogens. In some species, clones are so unstable that they cannot be observed, whereas in others species, clones appear to have *persisted* without much change over long periods of time.

In reality, the mutation rate may not be constant across species, or even within a species (some *pathogens* may lack some DNA repair pathways, or some isolates may have mutator phenotypes), and *diversification* may be more easily observed in some species than others, depending on the relative fitness of the variant genotypes that arise from clonal diversification.

However, very stable clones suggest low rates of recombination between genetically distinct strains, and very unstable clones imply frequent recombination between strains. *Y. pestis* is a good example of clonal stability, and a sample of 36 isolates from diverse sources, which included *isolates* assigned as remnants of the last three major *historical pandemics* (*antiqua, medievalis,* and *orientalis*), showed no sequence *variation* at six loci.

There is evidence that *Y. pestis* was the cause of the Black Death, and, from the descriptions of the disease, it is very likely that it caused some of the earlier historical *plagues*. The pestis clone is therefore certainly over 650 yr old and is probably substantially older. Similarly, *tuberculosis* is an ancient disease and yet there is almost no genetic variation in housekeeping genes of *M. tuberculosis*.

In contrast, many clones of multiply *antibiotic*-resistant isolates of S. *pneumoniae* are probably only about 20 yr old, and yet variants of these clones that differ at one of the seven loci used in MLST are already commonly encountered. Clones in some species appear to have remained uniform by MLST for over a thousand years, whereas those in other species diversify so rapidly that variants of clones that must be less than 50 yr old can be detected, and in a few species clones may be so unstable that they cannot even be discerned-all isolates from *epidemiologically* unlinked sources are different in genotype.

It is likely that a large contributor to this variation in clonal stability is the rate of recombination between distinct strains of the species, although other factors are undoubtedly important. Some care is required in assuming that clones are highly stable when there is almost no variation at neutral loci among all contemporary isolates of a species that causes an *undoubtedly* old disease.

Worldwide pandemics of some diseases can eliminate the existing variation that has built up since the previous pandemic. For example, influenza is likely to have been around for many hundreds of years, but finding that all *influenza* viruses isolated during a worldwide flu pandemic are closely related in genotype does not imply that the virus has remained unchanged over hundreds of years.

Influenza has unusual features and, for bacterial pathogens, the availability of isolates from diverse *geographic* sources, and of old isolates stored in culture collections, or of sequences obtained from archaeological material, can provide supporting evidence for assigning clonal *stability* over substantial periods of time.

The rate of *diversification* of clones has implications for the way molecular *epidemiology* is carried out. It is convenient to distinguish two types of epidemiological investigation. Short-term epidemiology is concerned with recent events, such as whether a cluster of cases of disease in a community or hospital are unrelated and caused by different strains, or are an outbreak caused by the *transmission* of a single strain.

This type of *epidemiology* is usually *straightforward* as, for an outbreak, all of the isolates are derived from a very recent common ancestor (the isolate from the index case), and no *variation* is likely to have *accumulated* among the isolates within a period of a few weeks. Provided the typing method is sufficiently *discriminatory*, so that isolates from *epidemiologically* unlinked individuals within the community are usually *distinguishable*, the identity of the genotypes from the cases of disease is sufficient to establish the epidemiological link.

In many cases, MLST or MLEE may be unsuitable for short-term epidemiology, as a significant *percentage* of isolates circulating in the community may have the same *allelic* profile as those from the outbreak, and an additional more *discriminatory* method is needed to establish it is an outbreak rather than independent acquisitions of a prevalent clone.

Epidemiological investigations may also be concerned with longer-term events, such as the worldwide spread of important clones (global epidemiology), where the *relatedness* of isolates recovered from different countries in different decades may need to be assessed.

In this situation, the common *ancestor* of all of these isolates may have *arisen* several decades ago, and considerable genetic variation may have *accumulated* in this time, such that it is not always easy to establish whether or not *contemporary* isolates from different countries are closely related and share the same common *ancestor*.

This problem becomes greater as the rate of diversification of clones increases, and at sufficiently high recombination rates it will be impossible to establish that isolates descended from the same common ancestor that existed several decades ago are related.

For global *epidemiology* of pathogens where rates of recombination are high and *clones* diversify relatively rapidly, it is essential to use *genetic* variation that accumulates slowly. MLST is ideal for this purpose as it provides the required combination of an ability to discriminate between billions of genotypes and the use of genetic *variation* that *accumulates* slowly.

Methods that are more sensitive at detecting minor variation are much less appropriate for global *epidemiology*, as the accumulation of far too much variation may *obscure* the underlying *relatedness* of strains derived from a recent common ancestor.

SPECTRUM OF GENETIC VARIATION

There is a large spectrum of genetic variation at neutral loci among different species of pathogens, and this can impact upon the ability of MLST to produce a highly discriminatory typing scheme. For example, *M. tuberculosis* is so genetically uniform that a typing scheme would resolve the whole species into far too few STs to make the scheme useful for epidemiological purposes, and other typing methods need to be employed.

Similar uniformity at housekeeping loci occurs in *Y. pestis, B. anthracis, B. pertussis,* and, to a slightly lesser extent, in *Neisseria gonorrhoeae*. Even where there is adequate variation at housekeeping loci, MLST may produce a scheme that is of limited utility. For example, in a highly clonal species, where clones are stable over *hundreds* of years, MLST may readily identify these clones, but may provide little advantage over other established methods that achieve this aim.

For example, MLST would undoubtedly provide an excellent procedure for identifying the clone within *S. enterica* that causes the great majority of typhoid fever, but as clonal diversification at neutral loci appears to be very slow in this species, it would not provide the further resolution within the *typhi* clone that is needed to investigate the spread of *typhoid* fever.

There are also *limitations* at the other end of the spectrum of population structures. Very frequent recombination results in an absence of clones. Thus, if a fitter strain arises within the population, and begins to increase in frequency in the population, frequent *recombinational replacements* will lead to a rapid diversification of its genotype, to produce an increasingly diverse array of *genotypes* that soon cannot be recognized as having been derived from the same recent ancestor.

These populations therefore consist of a vast number of unique genotypes; all isolates from *epidemiologically* unlinked individuals will be different by MLST, and only very recent transmission events may be discerned. The absence of clones, or of clusters of related clones, therefore makes it impossible to address longer-term epidemiological questions, and we cannot expect to be able to observe the spread of *particular* clones within countries or worldwide.

H. pylori provides an example of such a nonclonal population; there is ample genetic variation for a highly *discriminatory* MLST scheme, but extremely frequent recombination precludes long-term epidemiology, and only very recent epidemiological events can be explored.

Thus, recent transmission of a strain within a family may be detected by the presence in family members of very similar genotypes, but the relationships between these strains and those in the *wider community* cannot be established. Of course, this is an inherent problem with *nonclonal* species, rather than with MLST as a typing procedure, and other methods would also fail to provide useful epidemiological information over longer timescales.

However, the majority of bacterial pathogens have levels of genetic diversity at housekeeping loci that are adequate for a discriminatory MLST scheme, and have rates of recombination that allow the identification of clones and of clusters of related clones (clonal complexes). For example, the MLST procedure was developed using a challenging species for global epidemiology, *N. meningitidis*, as this species (or at least serogroup B and C isolates) was known from previous studies to have a high rate of recombination, leading to clones which diversify very rapidly.

The fact that MLST works extremely well for *meningococci*, and that clonal complexes can be discerned using BURST, suggests that it will be useful for most bacterial species where there is adequate genetic diversity. MLST may also have applications for typing eukaryotic *microbes*. These *organisms* can raise additional problems, as some may be diploid, or they may be difficult to culture *routinely*.

An MLST scheme using six loci has been published for the diploid species *Candida albicans*. In this scheme, heterozygous nucleotide sites can be identified as there are two peaks of approximately equal amplitude on the sequencer trace, and these can be assigned the standard *ambiguity* codes. A more difficult problem arises if the organism cannot be cultured.

For example, in malaria, an MLST scheme might be possible, providing there was sufficient diversity at neutral loci, but it would probably have to be applied directly to blood samples, as purification to single clones is not realistic for large numbers of samples. Allelic profiles could then be assigned only where there was a single or a *predominant* parasite strain present in the blood.

This would preclude the use of the scheme for the most interesting epidemiological situations, where there are high rates of disease transmission and blood samples often contain multiple strains of the parasite. *Multilocus* typing schemes have recently been reported which mix *antigen-encoding* and housekeeping loci, or include rRNA loci.

The mixing of genes that are subject to very different evolutionary constraints is unwise, as it complicates the ways in which the data can be analyzed, and is likely to provide less reliable indications of the true relatedness between isolates than schemes based on seven housekeeping loci. These schemes should not be referred to as MLST.

BACTERIAL POPULATION STRUCTURES

Bacterial population structures are often described as clonal or nonclonal, to indicate that clones can, or cannot, be observed within the population. In reality there is a continuous *gradation* in population structures, from those in which recombination is very rare and clones are stable; through populations where recombination is a major contributor to clonal *diversification*, but unstable clones can be observed; to populations where recombination is so frequent that clones are not observed.

How do we decide whether bacterial populations are highly clonal, weakly clonal, or nonclonal, or perhaps more meaningfully, how do we measure the extent and impact of recombination in different bacterial species?

Periodic Selection

The extent of recombination within bacterial populations has usually been addressed by examining the level of linkage *disequilibrium* between the alleles at different loci within the *population*. If rates of recombination are very high, the alleles at different loci are randomly associated (linkage *equilibrium*), and the population consists of a vast array of unique *genotypes*.

Conversely, if recombination is very rare, those fitter genotypes that arise periodically within the population will increase in frequency and lead to the loss of other genotypes from the population (*periodic selection*). These fitter genotypes may rise to high frequency, and will diversify only slowly in the *absence* of recombination, resulting in a number of clusters of isolates with identical or very closely related genotypes.

The alleles at different loci in the population are therefore not randomly associated; i.e., there is significant linkage disequilibrium between alleles. Bacteria divide by binary fission, and a high frequency of *recombinational* replacements is required to break down the *inevitable* associations between alleles at different loci that occur as a consequence of this mode of *reproduction*.

A statistically significant departure from linkage equilibrium occurs in most populations, since complete randomization of alleles is rarely achieved. Recombination can therefore be considered to be very frequent if linkage *equilibrium* is observed, but the presence of significant *disequilibrium* establishes only that rates of recombination are not very high.

Significant departures from linkage equilibrium can occur because recombination cannot completely break down linkage associations, but it may also occur as a consequence of poor sampling. Artifacts introduced by sampling are likely to be particularly evident in those species where carriage is very common but rarely proceeds to disease (e.g., *meningococci, pneumococci, S. aureus*).

If in these species only 3 in 100,000 children get invasive disease, although 1 in 10 children carry the organism in the *nasopharynx* or *nose*, the vast majority of isolates in a truly representative sample of the population would be from carriers. In practice, most isolates of these species that have been studied are from disease, and are therefore a highly *biased* sample of the real population.

Sampling problems may also occur in situations where some isolates of a species gain particular medical importance and thus are greatly oversampled. For example, methicillin-resistant *S. aureus* and multiply antibiotic-resistant pneumococci have been studied extensively, with much less attention being paid to antibiotic-susceptible isolates of these species, which in almost all countries still cause the majority of disease; and even less attention has been paid to the carried population.

These types of sampling bias will tend to make populations appear to be more highly clonal than they really are, and will inflate the level of linkage disequilibrium, as often a restricted set of genotypes is associated with most cases of *invasive* disease, and with resistance to *antibiotics*.

Occurence of Clones

One of the original criteria of a clonal population was the presence of indistinguishable isolates in temporally and geographically separated samples of the population. The presence of clones is still often used to justify the *description* of a population or species as clonal. However, clones may be observed in most populations, including those in which rates of recombination are high, and the term *clonal* lumps many rather different population structures together.

For example, rates of recombination are considered to be high in serogroup B and C meningococci (the impact of recombination appears to be much smaller in *serogroup* A isolates), but clones are very obvious. In some studies this may be due to the analysis of invasive isolates, which are highly *enriched* for the few genotypes that have a substantially increased ability to cause *meningococcal* disease, but clones are also observed among the diverse collection of genotypes obtained from *asymptomatic nasopharyngeal* carriage.

In a recombining species of this type, recombination over evolutionary timescales may *randomize* the alleles at different loci in the population, but, within human timescales, *diversification* may still be slow enough for some genotypes to increase in frequency in the carried population as a consequence of a transmission advantage (increased fitness) or *stochastics* (drift).

Multiple isolates with the same allelic profile by MLST may therefore be observed within a small *representative* sample of the total meningococcal population. These transient clones will diversify over human timescales to produce increasingly variable clonal complexes, and the presence of the decaying *remnants* from the emergence and rapid breakdown of fitter genotypes will introduce significant linkage disequilibrium into all but the most highly recombinogenic populations.

A population in which recombination is the dominant mode of evolution may therefore still show the presence of clones, and may exhibit a significant departure from linkage equilibrium. Only at very high rates will these two features be absent from a bacterial population.

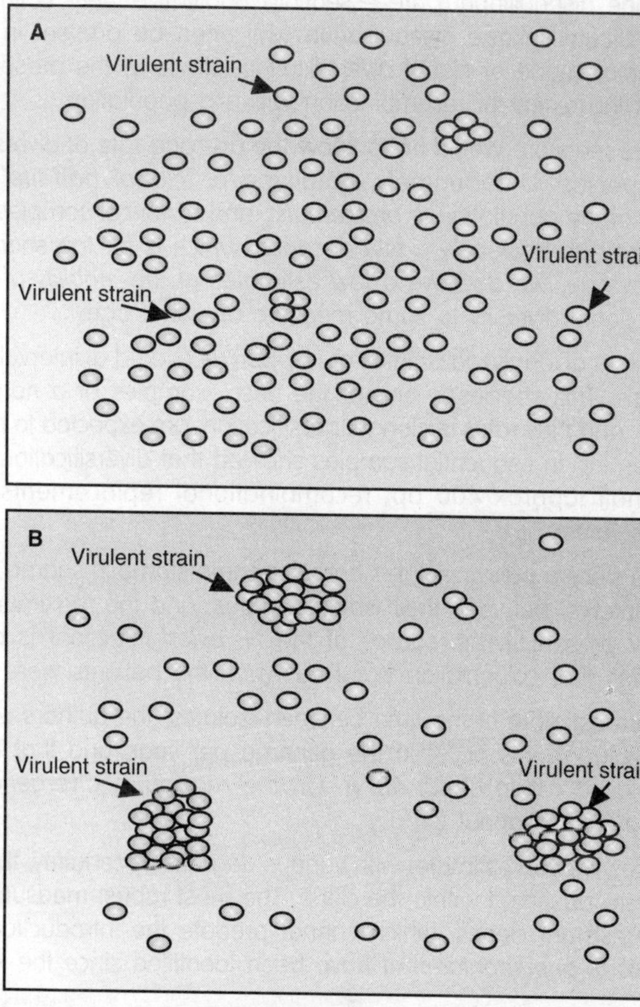

Figure 13.5: Representative sampling of a bacterial population. (A) shows the isolates present in a representative sample of the population of a species in which disease is rare compared to carriage. Since disease is very rare, these isolates are all from carriage. The similarities in the genotypes of the isolates are reflected by their distance apart; the great majority of isolates are spatially separated and therefore have unique genotypes. Two small clusters of isolates are present, which represent clones that are overrepresented in the carried population, as they have a slight fitness advantage. The great majority of genotypes in this diverse carried population are unlikely to cause disease. However, three genotypes, which are relatively rare in the carried population, have a greatly increased ability to cause disease (virulent strains). (B) shows a sample of the isolates of this species obtained exclusively from cases of disease. Most disease is caused by the three virulent strains circulating in this community, and these genotypes are therefore greatly overrepresented in the sample of isolates taken from individuals with disease. Inappropriate sampling of the real population can make a diverse weakly clonal population appear to be much more highly clonal than it really is, and can introduce highly significant linkage disequilibrium.

Clonal Diversification

Measures of linkage disequilibrium are essentially qualitative-there is linkage equilibrium or there is not; and significant linkage disequilibrium will often be present in populations where recombination is a major mode of clonal diversification. Similarly, the presence of clones is an *insensitive* indicator of the extent of recombination within a population.

A more quantitative measure would be to know the average rate of diversification of bacterial clones in different species. Unfortunately, obtaining a "*clonal half-life*" from analyzing a contemporary sample of the population is *problematic*, and historical samples of most *pathogens* provide little help as they go back only a few decades, which is far too short for measuring this type of parameter. However, we do have a few estimates of the *stability* of clones as, in some cases, we can relate clonal diversity to some measure of clonal age.

For example, Falush et al. analyzed isolates of *H. pylori* recovered at intervals from the stomachs of the same individuals. This species is one of the best examples of a nonclonal species, and rates of recombination, and thus rates of clonal diversification, are expected to be high. Examination of the sequences of ten loci in sequential samples showed that diversification occurs very quickly, predominantly by small (approx 400 bp) recombinational replacements (presumably from coinfecting *H. pylori* isolates).

The minimum time since a pair of isolates from the same stomach shared a common ancestor can be taken as the interval between their isolation dates, and the maximum time (unless both variants were already present in the source of the *H. pylori* infection) is based on the more conservative *assumption* that colonization occurred when the patients were children.

Using the minimum estimate of the time between isolates, the authors calculate that approx 60 recombinational replacements occur in the genome per year, and that half of the *H. pylori* genome may be replaced within about 40 yr. On the maximum *divergence* time, half of the genome would be replaced in about 2200 yr.

In a few other cases we can estimate, with various degrees of certainty, the age of clones and relate this to the diversity observed within the clone. The most robust measure of age is obtained for multiply antibiotic-resistant clones, which cannot predate the introduction of *antibiotics*. The multiresistant clones of *S. pneumoniae* that have been identified since the late 1970s provide a good example.

Figure elsewhere in this chapter shows the relatedness of the allelic profiles of the 97 isolates in the *pneumococcal* MLST database assigned to the Spain ^{23}F-1 clonal complex. Although 83 of these isolates have an identical *allelic profile* (the ancestral genotype, ST81), there are nine different variant allelic profiles (which arise predominantly by recombination) that differ at one of the seven loci used in MLST (SLVs), but none that differ at two loci (DLVs). Thus, within about 25 yr, variation is readily observed at one of the seven neutral loci indexed by MLST.

In other cases we have a date before which isolates of a particular clone have not been identified, as in the case of the ST32 clonal complex (ET-5 complex) of *N. meningitidis*, the first isolate of which was recovered in the 1970s, although of course it may have arisen earlier but remained undetected. In the meningococcal MLST database there are 31 isolates with the ancestral genotype of this clonal complex (ST32), and a large number of SLVs and DLVs.

Diversification of this clone into a highly variable clonal complex appears to have occurred very rapidly, which agrees with other data indicating a high rate of *recombination* in this species. In other species, clones are far more stable and appear to remain virtually unchanged in allelic profile over centuries.

RECOMBINATION FREQUENCY

Another approach to understanding the impact of recombination is to measure recombination rates in different species. This is not a trivial task, but MLST data provide a way of measuring the relative contribution of recombination, compared with point mutation, in clonal diversification. Clonal complexes within the population are first identified using BURST, and the ancestral allelic profiles of each are predicted.

The sequence changes that have occurred as the ancestral allelic profiles diversify to produce SLVs can then be examined, and can be assigned as the result of a point mutation or a recombinational replacement. *Summing* the number of mutational and recombinational events during these initial stages of clonal diversification provides a recombination/mutation ratio for the species.

These ratios can be expressed per allele (how often an allele defined by MLST changes to a variant allele by recombination or mutation), or per individual nucleotide site. The per allele ratios vary from about 10: 1 in favor of recombination in meningococci (52) and pneumococci to about 15:1 in favor of point mutation in *S. aureus* (*E. J. Feil* and *B. G. Spratt*, unpublished data). Presumably these values would be substantially lower than the *S. aureus* value in those species where clones appear to be stable over many centuries.

RELATIVE CONTRIBUTION OF RECOMBINATION

The relative contribution of recombination to clonal diversification varies greatly among bacterial species. What is the impact of recombination on the long-term evolution of different bacterial populations and our ability to extract information about the phylogenetic history of a species? If recombination is absent, the genetic variation at neutral loci can be used to reconstruct the phylogeny of the species, and variation at other loci (e.g., the presence or absence of *virulence* genes) can be mapped onto the resulting *phylogenetic* tree, to understand the origins of *pathogenic* lineages, host adaptation, and so on.

However, frequent recombination will preclude any attempt to construct a meaningful phylogeny of the species, and only relatively recent evolutionary events can be examined. In order to understand how much phylogenetic information can be extracted from MLST data, some idea of the impact of recombination within the species is required.

Statistical tests of congruence between the trees obtained for each of the seven loci used in MLST have been used to assess the impact of recombination. If *recombination* is rare, the tree obtained using sequence data from one locus should be similar (congruent) to that derived from each of the other loci.

However, if recombination is very frequent, the trees obtained from each locus will be different (noncongruent), and none will reflect the true phylogenetic relationships between the isolates. For this type of analysis, isolates that are very different in allelic profile should be selected to minimize any congruence that would be introduced by including multiple isolates of the same clonal complex. The approach determines whether the maximum likelihood tree obtained for each of the MLST loci is a better fit to the sequence data from the other six loci than it is to 200 trees of random topologies.

In some species (e.g., *N. meningitidis* and *S. pneunioniae)* the similarities between the trees are, in all cases, no better than random, indicating a total lack of congruence between loci; the elimination of all *phylogenetic* information, such that patterns of evolutionary descent cannot be discerned, is a long-term consequence of frequent recombination.

In other cases there are a few tree comparisons which show a similarity that is significantly greater than random *(H. influenzae),* and in other species most tree comparisons show significant congruence (*E. coli).*

The amount of phylogenetic information present within a dendrogram therefore differs among bacterial species; in some cases only the clusters of isolates at the tips of a *dendrogram* may have any validity, whereas in others even the relatedness between the deepest branches of the tree may approximate the evolutionary history of the species.

For example, in the meningococcus, the relatedness of very similar STs is likely to be meaningful, since these are the result of the recent diversification of a clone into a clonal complex, but the phylogenetic relationships between the different clonal complexes that are implied by a dendrogram are likely to be *meaningless.*

CONCLUDING REMARKS

Bacterial species vary considerably in the amount of genetic variation they contain and in the extent and impact of recombination. In many cases highly uniform bacterial species are misnomers, where a distinctive stable clone within a more appropriately *constituted* species has been raised to species status because the disease it causes is so distinctive.

Recombination in some species is the major contributor to the diversification of clones and, in a few *species,* recombination may be so frequent, compared with point mutation, that it prevents the emergence of observable clones. In other species, recombination may play a minor role, either because *colonization* with multiple genotypes rarely occurs, or because sexual *exchanges* are rare.

Recombination appears rarely to be so frequent that it prevents the appearance of clones; in this sense, most bacterial species can be called clonal.

However, rather than classifying species as clonal or nonclonal, it is more appropriate to consider the relative impact of recombination in each species, or in different subpopulations of the species.

Low rates of recombination produce mostable clones, and allow infra-species phylogeny to be reconstructed which, in an era of genome *sequencing* and *microarrays,* holds out the hope of understanding the changes in the genome along the different branches of an evolutionary tree-changes that have influenced important *biological properties,* such as virulence or host- or nicheadaptation. High rates of recombination result in unstable clones, a fact that needs to be considered when molecular epidemiological studies are undertaken. Unstable clones are likely to invalidate any attempts to produce an *intra-species phylogeny*, and may make it far more difficult to understand the pathways and processes by which virulent strains emerge within populations.

Chapter 14

Phylogenomics

In the decade since the completion of the first bacterial genome-sequencing project, an ever-increasing amount of genomic information has challenged our definition of a bacterial species and shed light on the dynamic nature of bacterial genomes. Initial results from comparative genomic (CG) analysis of whole-genome sequences from multiple strains of *Helicobacter pylori* and *Escherichia coli* revealed that although some genetic variation is in the form of differences in shared conserved gene sequences, the most important source of genetic variation within a bacterial species is at the level of gene content.

In the most comprehensive bacterial intraspecies CG study to date, the complete genomes of eight strains of *Streptococcus agalactiae* were recently compared. This study revealed that the *S. agalactiae* species can be described in terms of a "pan-genome," which consists of a "core genome" of genes shared by all isolates and accounting for approx 80% of any single genome, plus a "dispensable genome" of partially shared and strain-specific genes.

This confirms results from earlier studies and further affirms that bacterial genomes are in a constant state of flux. An increasing number of bacterial species have complete genome sequences available for more than one strain however, as of January 2006, of the 623 bacterial species whose genomes are either fully sequenced or in progress, there exist only 16 CG datasets comprising five or more strains.

Of note, these 16 species, and nearly two-thirds of all bacterial species with multiple strain coverage, represent plant or animal pathogens. Thus, although the study of intraspecies CG is important from the standpoint of understanding genome dynamics and evolution, it also has important practical implications. Each species can be thought of being comprised of a spectrum of genetic variants with potential differences in important phenotypes such as host specificity and pathogenicity.

The results from CG studies performed thus far have led to an increasing realization that obtaining the complete genome sequence of a single bacterial strain not only fails to address the genetic diversity comprised by the species but also seriously limits the extent to which these

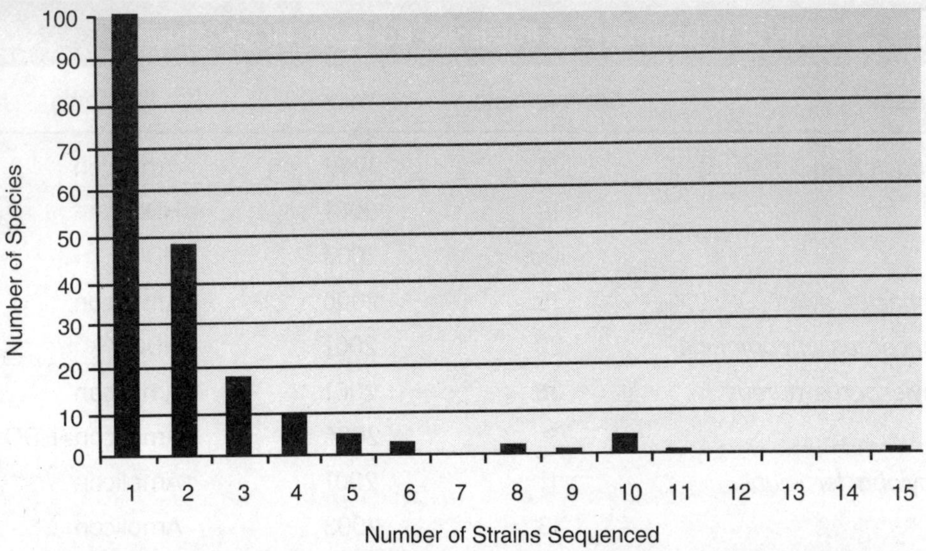

*Figure 14.1: Multistrain coverage in bacterial sequencing projects. As of January 2006, 847 bacterial genomes (representing 623 different species) have been or are currently being sequenced. Although multiple genome coverage is available for 91 species, only 16 have five or more genomes characterized (*532 species have single genome coverage).*

genomic data can be exploited for practical applications. For example, in exploiting bacterial genome sequence information toward the development of "broad spectrum" antimicrobial drugs or vaccines, the suitability of a gene target rests in large part on whether the gene is conserved in a large proportion of clinically relevant strains. A proof of concept for this multigenome CG-based approach to vaccine research has recently been described through the development of a universal Group B *Streptococcus* vaccine.

Although obtaining the complete genome sequence of multiple strains per bacterial species is highly desirable, these types of datasets are likely to remain limited in number in the near future. In their absence, the field of CG has benefited greatly from the development and rapid adoption of DNA microarray technology.

The development of DNA microarrays comprising every open reading frame (ORF) in a genome strain has become an integral part of nearly every bacterial genomics project, whereas an increasing number of laboratories have acquired the necessary infrastructure to carry out microarray experiments.

Although microarrays were originally applied in gene expression studies (i.e., transcript profiling), studies describing microarray-based comparative genomic hybridization (MCGH) soon followed. In the years since these initial efforts, MCGH has been used to study the CG of a large number of different bacterial species and has shown enormous potential as a cost effective and reasonably high-throughput approach to generating large CG datasets when whole-genome sequence data from multiple strains is lacking.

This chapter will provide an overview of currently published methods for DNA microarray-based bacterial CG, placing a special emphasis on data analysis and visualization issues. It is

Table 14.1: Summary of Published Bacterial MCGH Studies

Organism	Sample size	Year	Platform
Mycobacterium tuberculosis	14	1999	Amplicon
	19	2001	HDO
	100	2004	HDO
Helicobacter pylori	15	2000	Amplicon
Streptococcus pneumoniae	20	2001	HDO
Staphylococcus aureus	36	2001	Amplicon
	9	2005	Amplicon+HDO
Campylobacter jejuni	11	2001	Amplicon
	18	2003	Amplicon
	26	2004	Amplicon
	97	2004	Amplicon
	111	2005	Amplicon
Vibrio cholerae	10	2002	Amplicon
Streptococcus (Group A)	36	2002	Amplicon
Salmonella	22	2002	Amplicon
	26	2003	Amplicon
	79	2004	Amplicon
	40	2005	Amplicon
Listeria monocytogenes	50	2003	Shotgun/Amplicon
	52	2004	Shotgun/Amplicon
	113	2004	Shotgun/Amplicon
Escherichia coli	26	2003	Amplicon
	29	2003	Amplicon
	22	2004	Amplicon
	10	*2005*	LDO
Bacillus thuringiensis, B. cereus and B. anthracis	19	*2003*	Amplicon
Francisella tularensis	27	2003	Shotgun/Amplicon
	17	2004	Shotgun/ Amplicon

(Table Contd.)

Organism	Sample size	Year	Platform
Pseudomonas aeruginosa	18	2003	HDO
Yersinia	32	2003	Amplicon
	18	2004	Amplicon
Chlamydia trachomatis	14	2004	Amplicon
Bordetella	42	2004	Amplicon
Neisseria	38	2005	Shotgun/Amplicon
Brucella	6	2004	HDO
Mycobacterium avium	16	2005	Amplicon
Staphylococcus epidermidis	42	2005	LDO
Thermotoga maritima	11	2005	Amplicon

important to note, however, that this is a newly emerging field and that protocols and analytical methods are in an active state of development.

Gene Specific

Microarray

Gene-specific probes for a microarray can be obtained either through synthesis of oligonucleotides to be used directly as probes, or through the PCR amplification of gene-specific DNA fragments.

PCR Amplicon-Based Microarrays

The PCR amplicon-based microarray (henceforth referred to as "ampliconarray") platform, pioneered by the Brown laboratory at Stanford uses DNA fragments amplified by PCR (i.e., amplicons) from cDNA library clones as microarray probes, which are robotically deposited (printed) by "contact-spotting" as a high-density array onto a derivatized glass microscope slide.

This class of microarray, which can be used both for gene expression analysis and for MCGH, has been the most popular platform in bacterial microarray studies. The first description of bacterial MCGH was by Behr et al. in a study in which they examined the genome content of various Mycobacterium bovis strains used as vaccines against tuberculosis using an amplicon-array derived from the M. tuberculosis H37Rv genome strain.

To generate their array, they designed and synthesized PCR primer pairs for every gene to be included in the microarray ("ORFmer sets"), PCR-amplified each probe, and spotted the genespecific probes onto glass slides. Their basic method for MCGH used differentially labeled targets generated from the genomic DNA of tester and reference strains, which were then cohybridized to the microarray.

Once control- and reference-derived signals were compared, genes with differential hybridization signals were identified as likely to show copy number changes and/or sequence divergence between control and tester strains.

Using this approach, Behr et al. uncovered a number of genomic deletions in the vaccine strains with respect to the H37Rv genome strain.

Oligonucleotide-Based Microarrays

The original Gene Chip technology developed by Affymetrix relied on in situ synthesis of short 25-mer oligonucleotides (oligos) directly on the array surface using a proprietary photolithographic method. Affymetrix arrays have been used in a small number of studies.

A number of additional methods have been developed for in situ oligo-synthesis but they have yet to be widely adopted for MCGH. Although microarrays made by direct spotting of short oligos perform poorly compared to amplicon arrays, advances in high-throughput synthesis of "long oligos" (i.e., >50 bases), has led to the development of "long-oligo arrays" that perform on par with spotted amplicon arrays with the additional attractive feature that they can be directly printed without the need for a PCR amplification step.

Applications

Even though some commercial sources exist for microarrays (Affymetrix, MGW Biotech), ORFmer sets (Sigma Genosys), or long oligos for spotting (Operon), these are only available for a very select number of species. Thus, most groups will need to design custom-made arrays for their organism of interest.

At the present time, spotted arrays have been the platform of choice in bacterial MCGH in large part because the method allows research teams complete autonomy and flexibility in microarray design and production. In addition, the use of nonproprietary fabrication technology can substantially decrease operating costs in the long run.

Although most studies have used amplicon-arrays, oligo-arrays may eventually supersede them as the platform of choice because they are easier to implement because oligo-arrays do not require the additional amplification of probes, amplicon assessment, and amplicon purification steps, which are not a trivial endeavor.

An additional attractive feature of oligo-arrays is their increased hybridization specificity as a result of the increased sensitivity of oligos toward probe/target mismatches. Oligonucleotides can be designed to specifically minimize cross-hybridization to homologous (i.e., similar but nonidentical) sequences, which makes discrimination between alleles possible.

One caveat is that the increased specificity of oligo-arrays can come at the cost of decreased sensitivity in gene detection because unknown alleles of a gene may yield little to no hybridization signal, which can lead to erroneous gene presence/absence calls. Because additional or unknown alleles are generally absent from a microarray, the increased tolerance of amplicon-based arrays toward mismatches may allow for their detection, thus minimizing the likelihood of false gene-absence calls.

Although *in situ* synthesized oligo-arrays have not seen wide use in bacterial MCGH, they merit mentioning because the platforms are likely to play an important role in future CG research. The ultra-high density at which oligos can be arrayed with *in situ* synthesis-based platforms (such that multiple probes can be used per gene) and the increased hybridization specificity of oligos can be exploited for high-resolution MCGH, allele determination, and genomic resequencing.

Shotgun Clone

"Shotgun" clones are derived from genomic libraries that have been generated by cloning randomly sheared genomic DNA fragments (0.5-5 Kb) into a plasmid vector. The approaches described next have used plasmids in microarray construction either directly as probes for spotting or indirectly as templates for amplicon generation.

Low-Cast Microarrays

Because shotgun library construction and the subsequent plasmid isolation from thousands of individual clones is generally the first step in a typical bacterial genome-sequencing project, low-cost microarrays can be constructed that use these plasmid clones as templates for amplicon probes. The first MCGH application of this approach was described for a *Campylobacter jejuni* microarray made using clones derived from the genome-sequencing project.

Although Dorrell et al. first amplified plasmid inserts by PCR to generate their printing stocks, this approach for generating microarrays was relatively inexpensive because a common primer pair was used to generate all of the amplicons, obviating the need for the synthesis of gene-specific primers on a genome-scale.

Because of the concurrent characterization of the clones during the sequencing project, the authors were able to select a panel of clones representing the majority of genes in the genome strain. Nevertheless, the main disadvantage of shotgun clone microarrays is that it is generally impossible to completely avoid clones that contain the sequences from adjacent genes, confounding subsequent analysis.

Shotgun Screening

Recently, several groups have used the microarray platform to prescreen shotgun-libraries for strain-specific genes (i.e., absent or highly divergent in the genome strain) by directly spotting plasmid DNA from shotgun clones to produce microarrays. When hybridized against the genome strain, clones bearing sequences conserved in both strains would be expected to yield significant signal whereas probes derived from clones with "unique" sequences would be expected to yield reduced or no signal.

By performing a microarray-based prescreening step, the authors were able to narrow down on the clones bearing unique sequences, enabling them to focus sequencing resources on such clones thereby reducing sequencing costs and speeding up the discovery of novel "strain-specific" genes. This method is likely to play an instrumental role in the development of "pan-genomic" microarrays (i.e., containing nonredundant sets of conserved and strain-specific genes) when additional genome sequence information is lacking.

Microarray Platform

In conventional microarray platforms, the complete genome of a bacterial strain is represented on a microarray as a collection of probes derived from all of the individual genes contained in its genome. The presence or absence of these genes in a strain of interest is then assayed based on the hybridization pattern on the microarray.

Zhang et al. have described a novel application of microarray technology in which each probe on the array corresponds to whole genomic DNA from a bacterial strain. The array can thus contain thousands of whole genomic DNA samples from different strains arrayed at high density. These arrays can be used to assay for the presence or absence of a specific gene and/or allele in thousands of strains by hybridization to a gene-specific labeled target.

Although this method does not allow for whole-genome CG analysis *per se* (only one gene can be assayed at a time), it is well worth mentioning because it enables researchers to assay for the prevalence of a gene of interest (e.g., antibiotic resistance gene, virulence factor, and so on) in very large collections numbering thousands of strains. The method also represents a very efficient way for sharing genomic DNA samples from such large collections with the scientific community.

Production of Quality Microarrays

The successful production of quality microarrays is as much an art as it is a science. Although laboratories exist that own and operate their own printing equipment, most labs will have to resort to either partly or completely outsourcing the production process to dedicated production facilities that have the necessary liquid handling and printing robotics.

In that context, the following sections on microarray construction will focus on the early stages of microarray design and production, prior to the actual microarray printing. These methods will focus primarily on spotted microarrays, as they are the most widely used platform for MCGH.

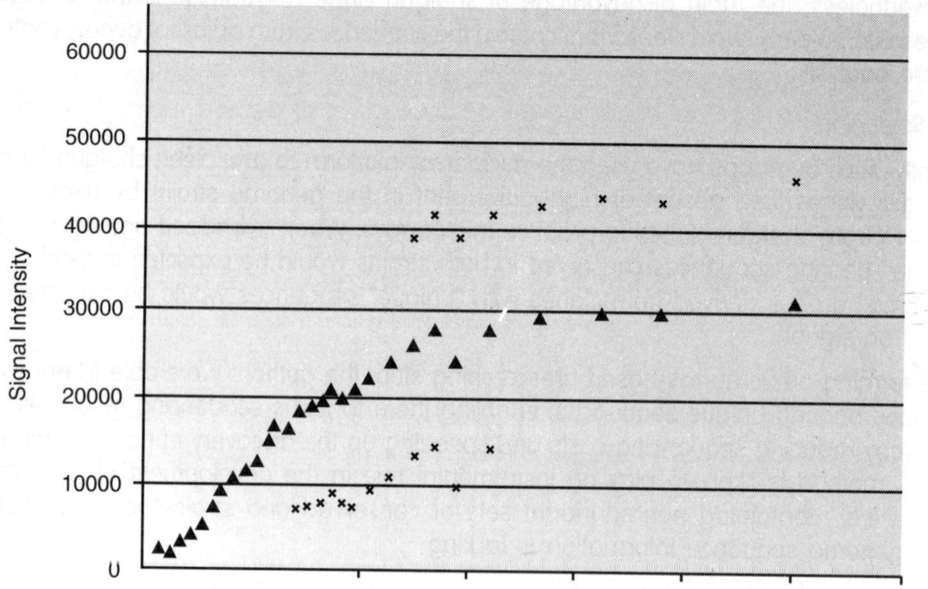

Figure 14.2: Relationship between amplicon length and microarray signal. The average signal intensity obtained from the control channel using a Campylobacter jejuni *amplicon array was tracked more than 150 hybridizations. Despite normalization of amplicon concentrations prior to arraying, decreasing signal yield is observed with decreasing amplicon length, affecting spot dynamic range and yielding erroneous log-ratios. Signal saturation for 16-bit scanning systems occurs at 65,535 units (A* = median; x = +1 standard deviation).*

Amplicon or Oligo Arrays

Whether amplicon- or oligo-arrays are used, gene-specific primer-pairs or long-oligos will have to be selected for every gene to be represented in the array. Selection of oligonuclotides or primer pairs on a genome scale can be automated and a number of free and commercial programs are available that select oligos based on complete genome sequences and optimize them for a range of criteria (i.e., similar T_m, lack of secondary structure, and so on).

A very important consideration is avoidance of potential cross-hybridization between microarray probes. Some programs incorporate steps that reject potentially cross-hybridizing probes. One hitherto largely ignored parameter in amplicon-array design is that of probe size. The typical approach of many groups has been to maximize amplicon length given the ORF size.

The resulting array is generally a collection of amplicon probes of vastly different lengths, which is exacerbated by the large number of small ORFs in bacterial genomes. This can cause significant problems upon subsequent acquisition and analysis of the data because of the direct relationship between amplicon length and average signal yield.

Laser scanning parameters that are optimal for short probes may be inadequate for longer probes owing to signal saturation, which may compromise the dynamic range of the array. With the benefit of hindsight, we currently advocate standardizing amplicon length within a relatively narrow range to minimize the considerable length effects on signal intensity.

This results in more homogeneous signal yields and dynamic range across the array. Although, ideally, the average amplicon length would be selected so that it is similar to that of the shortest ORFs of interest, a complicating factor is the abundance of very short ORFs (i.e., 200 by or less) in bacterial genomes.

We have observed that amplicons from short ORFs can generate anomalous hybridization data and, thus, a potential compromise could be to design amplicons in the 300 to 400 by range for longer genes and as close to full-length as possible for shorter ORFs. Homogeneous amplicon lengths also simplify amplicon concentration normalization prior to spotting.

Typical Bacterial Genome

Given current densities achieved with spotted microarrays and the average gene content of a "typical" bacterial genome (2000-4000 genes), several key features are highly advisable from an array-design standpoint:

1. A large number of positive and negative controls: these should be dispersed across the entire array to detect any potential spatial biases in foreground and background signal.
2. Multiple spot replicates: spot replicates not only increase the robustness of subsequent statistical analyses but are invaluable in case of individual spot failure. Many arrays are designed with two or three spot replicates that are printed side-by-side. If space allows, we advocate complete replicate arrays to be printed per slide so that some spot replicates are in a different physical location.
3. The inclusion of a "universal reference sample:" sheared whole-genomic DNA from the source strain used to make the array can act as an excellent positive control, and can be used to balance the signal from both strains being compared in the CGH experiment. This can greatly simplify image acquisition and analysis.

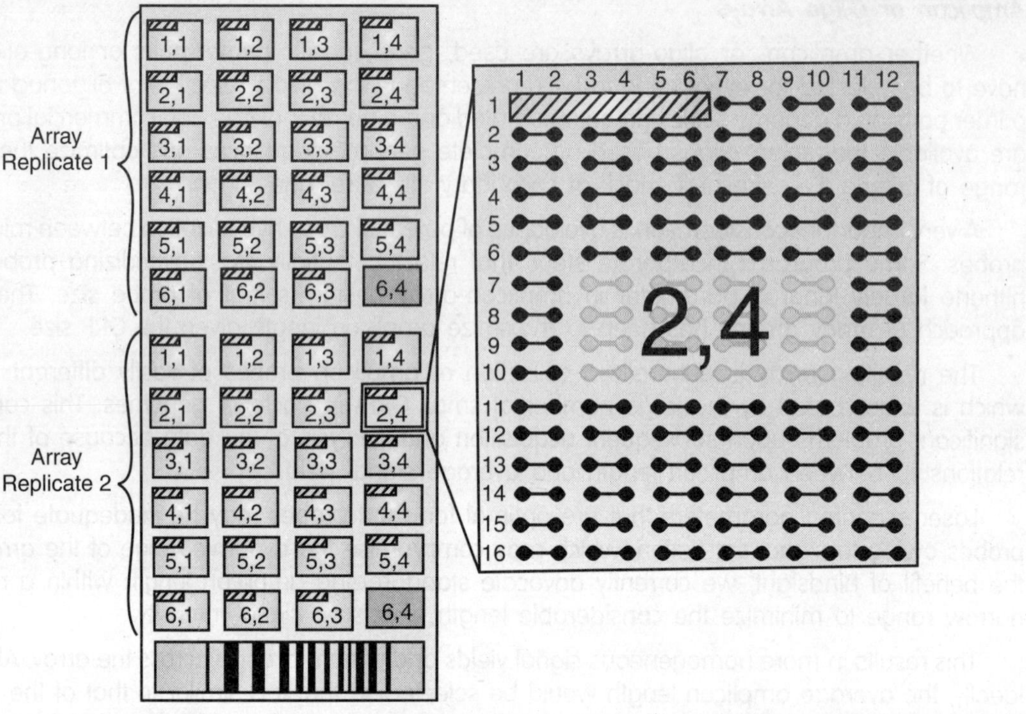

Figure 14.3: An example of key features in a successful microarray design. These include: control spots of serially diluted control DNA on each subarray (cross-hatched boxes); two full array replicates; adjacent replicate spots (black circles). A complete subarray on each array replicate (dark gray boxes) is dedicated to control spots that can be used to quickly establish balanced scanning conditions without requiring scanning of the complete array surface.

4. Normalized concentrations of amplicons for all the probes on the array.

REQUIREMENTS

Production of Microarray

Amplification

1. Forward gene-specific primers.
2. Reverse gene-specific primers.
3. PCR master mix (per 96-well PCR plate).
   a. 6.9 mL MIlli-Q water.
   b. 1.0 mL $MgCl_2$ (25 mM).
   c. 1.0 ml, dNTP mixture (2 mM per dNTP).
   d. 1.0 mL 10X PCR buffer.
   e. 40μL Taq polymerase (5 U/μL).
   f. 20μL Template genomic DNA (100 ng/μL).

Robotic Arraying

1. 96-Well Multiscreen PCR filter plate (Millipore, cat. no. MANU3050).
2. Millipore Filtration Manifold (Millipore, cat. no. MAVM0960R).
3. Milli-Q water.

Microarray CGH

Random Priming

1. Genomic DNA.
2. BioPrime Plus Array CGH Indirect Genomic Labeling System (Invitrogen, cat. no. 18096-011).

Direct Chemical Coupling

1. Genomic DNA (fragmented to average 1 K).
2. Mirus Label IT labeling kit (Mirus Bio Corp., Madison, WI; Cy-3, cat. no. MIR3600 and Cy-5, cat. no. MIR3700).
3. SigmaSpin Post-Reaction Clean-Up Columns (Sigma, cat. no. S-0185).
4. Qiaquick PCR Purification Kit (Qiagen, cat. no. 28106).

Hybridization

1. Microarray (stored at room temperature in a dessicator until use).
2. DIG Easy hybridization solution (Roche, cat. no. 1-796-895).
3. Yeast tRNA (10 mg/mL).
4. Salmon sperm DNA (10mg/mL).
5. High-humidity chamber.
6. Heat block (65 °C); hybridization oven (37 °C).

Washing

1. 20X SSC, 0.2-μm filtered (Sigma, cat. no. S-6639).
2 Washing buffers (prewarmed to 50'Q:
   a. Buffer 1: 1.0X SSC, 0.1% SDS.
   b. Buffer 2: 0.5X SSC.
   c. Buffer 3: 0.1 X SSC.
3. Hybridization oven (50ºC).

METHODS

Production of Microarrays

Amplification

1. Although the yield of each individual primer will vary, resuspend primers thoroughly in a

common volume of Milli-Q water to an average concentration of 10μM.

2. Gene-specific amplicons are generated in 100-μL reactions in 96-well reaction plates: dispense 96 μL of master mix, 2 μL of forward primer, and 2 PL of reverse primer.
3. Perform reactions in a thermal cycler. Cycling conditions will vary but the cycling should include a "touchdown component" to compensate for differences in the melting temperature among the primers in the reaction plate.
4. Visualize amplicons by gel electrophoresis and assess for the amplification of expected product sizes. Reactions yielding wrong sized or multiple amplicons may require redesign and synthesis of new primers. If low failure rates are achieved (<5 per 96-well plate), flag the failed amplicons and carry the plate through purification and arraying steps and simply ignore spots from failed probes during subsequent analysis.

   Generate additional amplicons resulting from resynthesized primers in additional plates (these can be purified subsequently). In case of higher failure rates, consolidate correct amplicons into fresh plates prior to probe purification to avoid large numbers of "failed spots" in the final microarray.

Preparation for Robotic Arraying

Filtration using 96-well Multiscreen filter plates (Millipore) is an efficient method to remove unincorporated nucleotides and primers.

1. Wash off surfactants in the filter membranes by drawing 200 μL of Milli-Q water through the filter plates on the vacuum manifold filtration apparatus at the pressure recommended by the manufacturer.
2. Load reaction products onto the filter plates and filter through until the membranes are dry.
3. Add 50μL of Milli-Q water to each well and filter through until the membranes are dry.
4. Repeat step 3.
5. Remove the plate from the apparatus and add 100 μL of Milli-Q water to each well. Seal the plate (precut sealing film or foil is available from many vendors), then place on a shaker and shake vigorously for 5-10 min to complete resuspension of the DNA.
6. Carefully remove the purified products using a multichannel pipet and transfer to new 96-well plates.
7. If access to a liquid handling robot and 96-well spectrophotometer is available, measure amplicon concentrations and transfer a common amount of each amplicon to fresh plates. Dry-down amplicons.
8. Prior to printing, resuspend amplicons in the appropriate printing buffer at a concentration of approx 0.2 μg/μL (concentration may vary depending on slide surface chemistry).

Microarray CGH

Production of Labeled Target DNA

1. Use 2μg of genomic DNA and label target DNA using the BioPrime labeling kit according to the manufacturer's instructions.

Labeled Target DNA by Direct Chemical Coupling

1. Bring up 5 μg of sheared genomic DNA to 40 μL with Milli-Q water, then add 5 μL of Buffer A, and 5 μL of dye. Incubate in a light-tight container at 37 °C for 4-16 h.
2. To perform buffer exchange and remove the bulk of unincorporated dye, pass each labeling reaction through a SigmaSpin column according to the manufacturer's instructions.
3. Purify the labeled DNA from residual dye using Qiaquick column according to the manufacturer's instructions. (This additional purification step can reduce background signal. A second elution step can increase yields by an additional 10%.)

Hybridization

1. Calculate dye incorporation efficiency of probes by calculating the A_{260}, A_{550} (for Cy-3), and A_{650} (for Cy-5). Tester and control probes should be matched so that samples with equivalent yields and incorporation efficiencies are cohybridized.
2. Pool matched tester/control targets and ethanol precipitate or dry-down in a vacuum centrifuge.
3. Prepare hybridization solution (per 100 μL : 90 μL of DIG Easy solution, 5 μL of yeast tRNA, and 5 μL of salmon sperm DNA). Warm the solution to 65 °C prior to its use to resuspend components that have precipitated out of solution.
4. Prepare high humidity chambers by quickly heating in a microwave to generate a small amount of water vapor. The humidity generated prevents the microarray from drying out. Place chambers at 37 °C until ready to use.
5. Carefully resuspend pooled labeled targets in appropriate volume of hybridization solution. If bubbles are introduced during mixing, remove them by quick centrifugation. Denature targets at 65′C for a minimum of 5 min until ready to use.
6. To set up hybridization: place a clean cover slip onto a dust-free surface; pipet-labeled targets onto the cover slip without introducing bubbles; touch the microarray array-side down onto the liquid whereas allowing it to spread by capillary action; flip the microarray and place on a humidity chamber.

   After closing the lid to the chamber, proceed with the preparation of the next hybridization, working quickly to maintain humidity and temperature in the chamber. Carefully place chamber back into hybridization oven and hybridize the arrays for 16-24 h.

Washing

1. Remove microarray from the humidity chamber (if using the chamber for more than one array, place it back into hybridization oven) and place it into a small container with enough prewarmed Buffer 1 to cover the entire array. Use gentle agitation to displace the cover slip.
2. Empty buffer and replace with fresh prewarmed Buffer 1. Incubate for 5 min at 50°C with gentle agitation. Repeat.
3. Empty buffer and replace with prewarmed Buffer 2. Incubate for 5 min at 50°C with gentle agitation. Repeat.

4. Empty buffer and replace with prewarmed Buffer 3. Incubate for 5 min at 50 °C with agitation. Allow the microarray to come to room temperature in Buffer 3. Arrays can be left in Buffer 3 until other microarrays have been processed.
5. Using forceps, carefully remove the microarray from the buffer. Shake excess buffer onto a lint-free paper towel until the array is nearly dry. It is important to avoid leaving large drops of buffer on the array surface to prevent the buffer from crystallizing onto the array surface.
6. Store arrays in a light tight slide holder until laser scanning. Repeat for additional arrays.

Microarray based Methods

As with all other microarray-based methods, the MCGH detection and analysis process involves:

1. the acquisition of images for both control and tester channels using (typically) a confocal laser-based scanning system;
2. analysis of the images to extract signal; and
3. analysis and interpretation of the signal data.

Microarray Puzzle

Image acquisition is an extremely important, and often neglected, piece of the microarray puzzle and improper scanning methods can seriously compromise subsequent data analysis.

An assumption made in MCGH studies is that the amounts of DNA from tester and control samples hybridized to the array are approximately equal for the majority of spots in the array. This stems from the expectation that most genes will be highly conserved; several lines of evidence including both genome sequence data and largescale MCGH studies would indeed suggest that this assumption is solidly founded.

Thus, in a typical MCGH microarray experiment one should expect the majority of spots to generate similar signal and this prior knowledge provides an invaluable guiding principle during image acquisition. More specifically, laser-scanning parameters (laser power, photo-multiplier tube-PMT, detector gain) can be chosen which generate similar average signal levels in both control and tester channels (i.e., "balanced scans") for the highly conserved genes, which represent roughly 80% of the spots on the array.

This minimizes any potential bias in the signal intensity data prior to subsequent analysis. Although data normalization techniques can remove some of this bias, Cui et al. have shown that whereas data from unbalanced scans is very sensitive to the normalization method used (i.e., different normalization methods can yield very different results), data from balanced scans is quite robust when different normalization techniques are applied.

Analysis of Microarray Data

Analysis of microarray data requires the use of specialized software to quantify the signal from images acquired for each channel after laser scanning. Signal extraction is generally carried out in two steps: in the first step, spots are identified, the raw foreground and background signals are calculated for every spot on the array and spots with unreliable signal (e.g., bad spot morphology, very low signal-less than three to five times background) are filtered out from further analysis; in the second step, the data from all good quality spots is "normalized" by adjusting it for biases intrinsic to microarray technology.

Although sophisticated mathematical approaches can be used to "clean-up dirty data," a far more sensible approach is to reduce any potential sources of bias when possible (i.e., through measures such as uniform probe lengths, matching of labeled targets according to concentration and dye incorporation efficiency, and balanced scanning practices) to minimize the extent of data adjustment necessary to remove any additional data bias.

Early methods of data normalization tended to be linear in nature, such that a similar adjustment was made to all spots. Newer approaches recognize that biases in the data can vary with respect to spot intensity (e.g., dye bias is intensity dependent) and with spatial position on the array (e.g., hybridization gradients, arrayer print-tip effects).

Most array analysis programs now include the "printtip loess" normalization method, one of the most popular and robust intensity- and spatially dependent normalization methods. This method will generally produce excellent results from typical microarray data. Two-channel microarray data is typically transformed by calculating the base 2 logarithm of the ratio of the channel 1 signal to channel 2 signal (i.e., $_{1092}${channel1/channel 2}), generally referred to as the "log-ratio" such that genes with differential signal have log-ratios that deviate significantly from 0.

Excellent ways to visualize microarray data are the MA-plot (sometimes also referred to as the Ratio-Intensity or RI-plot), in which the logratio is plotted against the mean signal intensity of both channels, and the histogram of logratio frequencies. Both methods enable the visualization of potential outliers in the data, which can be seen as spots that are removed from the main body of the log-ratio distribution.

Analyzing MCGH Data

A number of programs, both freely and commercially available, have been developed to analyze microarray data for an excellent freely available microarray analysis suite developed by TIGR). However, in our experience, no single analysis platform has been developed specifically to deal with the idiosyncrasies of bacterial MCGH. We will instead focus our discussion on overall approach for analyzing and visualizing data.

Microarray Based Transcript

Because of the technical variability inherent in microarray experiments and the biological variability in complex biological systems, statisticians have strongly emphasized the importance of generating sufficient replicates for microarray-based transcript profiling experiments.

To accurately measure gene expression, researchers typically perform several replicates (including technical, biological, and dye-swaps) and subject the expression ratios to rigorous statistical analysis to gain confidence based on the concordance of the results. A significant advantage of MCGH vs transcript profiling experiments is the fact that the biological sources of noise are essentially absent: genomic DNA is not subject to the short-term compositional fluctuations observed with mRNA.

This fact alone overcomes the main analytical hurdle faced in microarray analysis because (with careful experimental design and technique) biological noise is generally a far greater source

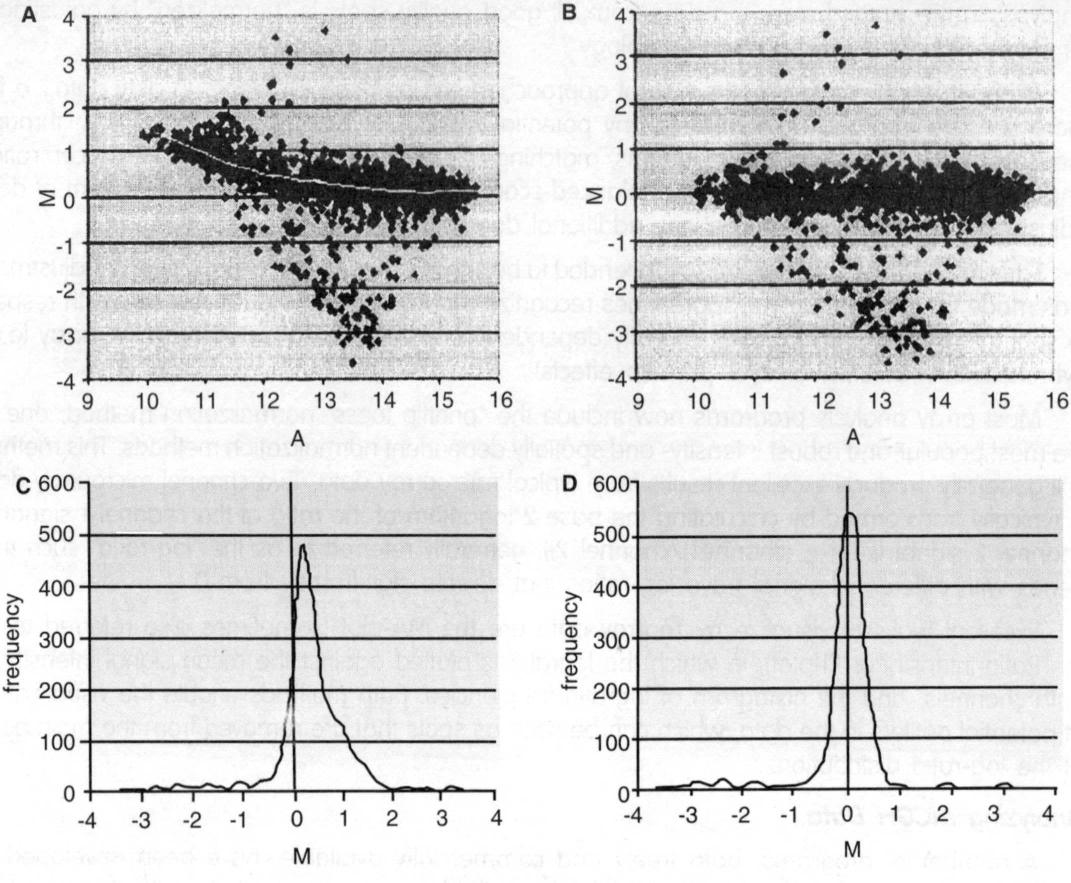

Figure 14.4: Effects of normalization on log-ratio distribution. The log-ratio data from an MCGH experiment is shown using the "MA plot" (A,B) and "log-ratio histogram" (C,D) visualizations. Raw data is shown in A and C. Data after Loess normalization is shown in B and D. Note the increasing bias in the distribution in the low intensity range and the removal of the intensity-dependent bias in the normalized data.

of variability than experimental noise. Replicates are important, and we encourage their inclusion in MCGH studies.

However, the number required is much lower than with transcript profiling. Typically, we perform each experiment in duplicate or triplicate, with occasional dye-swaps to ensure that dye bias is not corrupting the data.

MCGH Outlier Determination

Outlier determination is possibly the single most significant component of microarray data analysis. MCGH outlier determination has generally depended on using a cut-off for assigning genes into conserved or divergent categories. In practice, this cut-off could be an empirically determined value based on analysis of a strain known to be missing (some) genes; or a cut-off derived from other published cut-off values (e.g., the twofold signal difference used in much of the early microarray literature).

One problem with these approaches is that they apply the same cutoff to multiple arrays in a dataset despite the fact that technical variability and the number of divergent genes can greatly influence the shape of the logratio distribution. Although normalization discussed in elsewhere in this chapter ensures that data within an individual microarray has been adjusted for bias, multiple arrays in a dataset might retain different degrees of bias with respect to one another such that not every array will display the same dynamic range and such that the same log-ratio value may not necessarily be "equivalent" in different arrays.

To analyze multiple arrays with a fixed cut-off value, it is imperative that arrays be normalized against each other. A different approach altogether has been to develop methods that define dynamic cutoff values that are array-specific. Kim et al. have developed the program GACK, which determines thresholds specific to each array based on the deviation between the log-ratio distribution and an idealized normal distribution derived from the original log-ratio distribution, improving the sensitivity of divergent-gene detection.

It is important, however, to acknowledge the limited reliability of sequence divergence estimates from hybridization data. We have found that although a trend between log-ratio and the level of sequence conservation is observable, the log-ratio distributions of genes with different levels of sequence conservation are partially overlapping, making accurate prediction of gene divergence levels extremely difficult and, thus, advocate great caution with the interpretation of gene divergence from hybridization data.

Overall, we believe it is important to analyze data at two levels: a "high-sensitivity mode" where all potentially divergent genes are detected and a "high-accuracy mode" where those genes with the greatest likelihood of being highly divergent or absent are identified.

At present, it is very important for groups to verify and carry out an empirical assessment of MCGH results (for example, by determining false-positive and false-negative rates based on MCGH characterization of other sequenced strains if available, or by PCR-based screening to confirm absent genes) to determine suitable parameters for analysis. The future is likely to bring advances in analytical algorithms and improvements in the accuracy of MCGH data analysis should be forthcoming.

Gene Conversation Profile

From the point of view of CG, the main goals of visualization and high-order analysis are to compare and contrast the gene conservation profile of the strains being analyzed, and to reveal local and global trends in gene conservation. Although MCGH cannot directly address gene order, or synteny, direct comparison of genome sequence data has shown that many closely related strains do indeed preserve synteny to a large degree.

Thus, arranging MCGH data using the gene order from the reference strain used to build the array provides a useful starting framework for displaying gene conservation profiles. To compare the gene conservation profiles of two strains, simple ideogram plots can be used to display genomic regions with potential copy number differences, as seen in Figure elsewhere in this chapter. To compare the gene conservation profiles of multiple strains, a number of programs that implement different data clustering algorithms have been developed for the analysis of gene expression data and work extremely well with MCGH data.

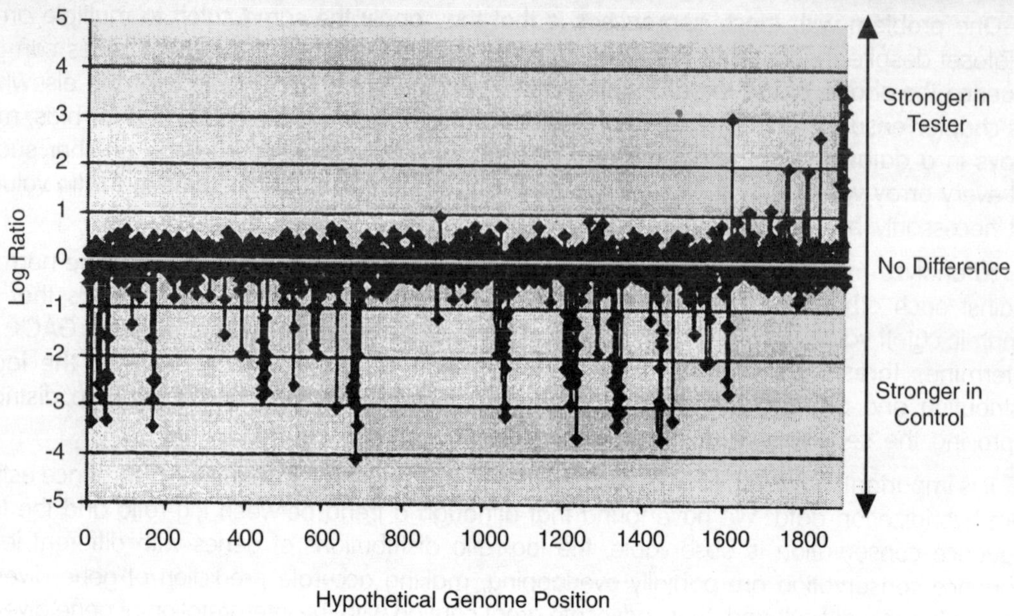

Figure 14.5: Ideogram plot depicting the MCGH log-ratio data comparing a "Tester" Campylobacter jejuni *strain against the "Control" strain. The data for each gene has been plotted using a hypothetical location based on the original genome location in the reference strain.*
Negative log-ratios, indicating stronger hybridization signal in the Control strain are observed throughout and represent possible divergent/absent genes. Positive log-ratios can be observed beyond genome position 1634, indicating stronger hybridization signal in the Tester strain. These genes are known to be absent in the Control strain but may be present in the Tester strain.

Despite the increasing sophistication of the various clustering methods, even simple techniques such as hierarchical clustering have been extremely effective at uncovering strain relationships based on whole-genome gene content, enabling genome-based typing or genomotyping. In addition, because genome content is a valid form of phylogenetic signal, MCGH data can be used in for genome-based phylogenetic analysis (or phylogenomics).

To visualize global gene conservation trends for the species, the percentage of strains divergent at each gene can be plotted. Using this approach, numerous MCGH studies have shown the recurring theme of plasticity or hypervariable regions, spanning several adjacent genes that appear to be absent in many strains. Interestingly, many of these regions contain functionally related genes and tend to be over-represented with strain- or species-specific genes of unknown function.

Final Goal

Despite some shortcomings, which are technical and analytical in nature, MCGH has proven a very reliable technique in the study of bacterial CG and it currently remains the most accessible means of performing large-scale CG.

A major limitation is the limited "genome coverage" achieved by microarrays because only those genes included in the array can be tested for. Typically, microarrays have been based on the genome of a single strain, which undoubtedly places limitations on the resulting data because

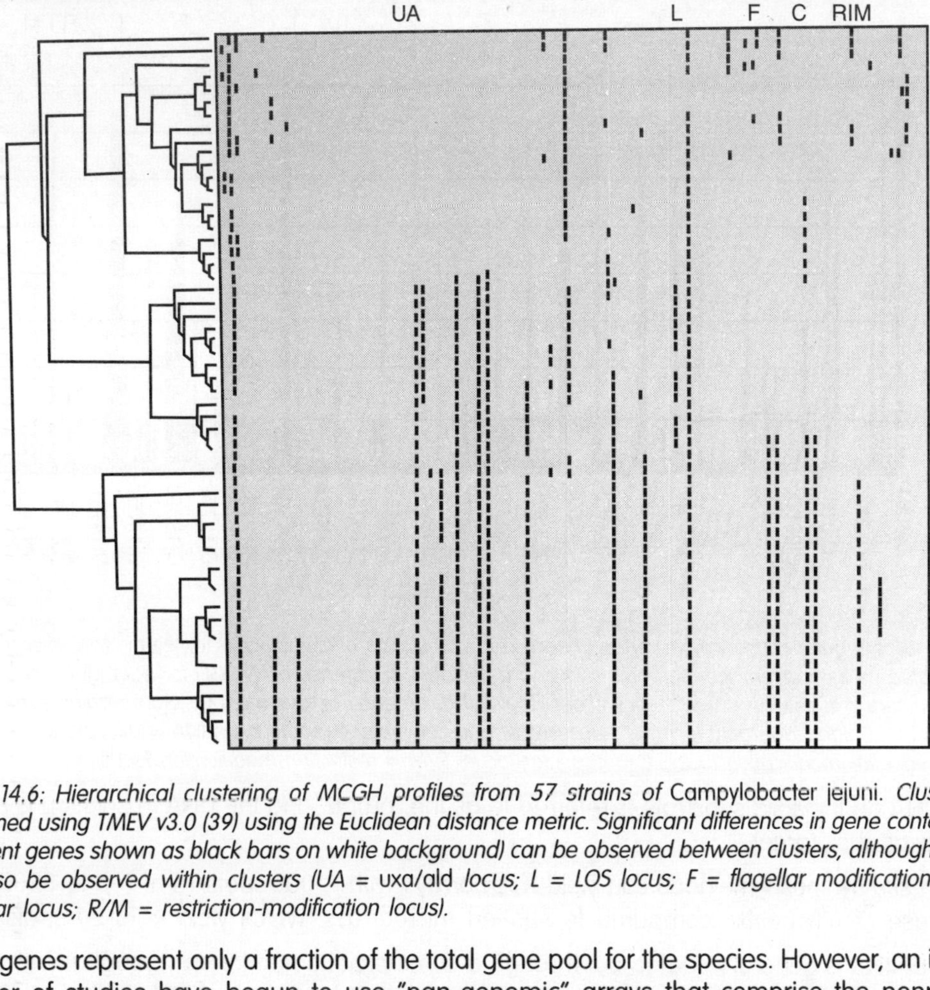

Figure 14.6: Hierarchical clustering of MCGH profiles from 57 strains of Campylobacter jejuni. *Clustering was performed using TMEV v3.0 (39) using the Euclidean distance metric. Significant differences in gene content (absent/ divergent genes shown as black bars on white background) can be observed between clusters, although differences can also be observed within clusters (UA = uxa/ald locus; L = LOS locus; F = flagellar modification locus; C = capsular locus; R/M = restriction-modification locus).*

those genes represent only a fraction of the total gene pool for the species. However, an increasing number of studies have begun to use "pan-genomic" arrays that comprise the nonredundant gene complement of two or more strains, though again the gene coverage achieved is limited by the availability of additional genome sequences.

It is interesting to note that among the most promising avenues for increasing the effective genome coverage of pangenomic microarrays is the discovery of novel genes by MCGH-based screening of shotgun-arrays. Similarly, some of the most promising approaches for rapid whole-genome sequencing are to be found with ultra-high resolution MCGH using oligo-arrays.

In the short term, much can be learned about the CG of a bacterial species using single genome-based microarrays. When coupled with novel MCGH-based methods for gene discovery and resequencing, MCGH can provide an integrated approach to bacterial CG.

POINTS TO REMEMBER

1. In this chapter, we will conform to the nomenclature proposed by Duggan, in which the

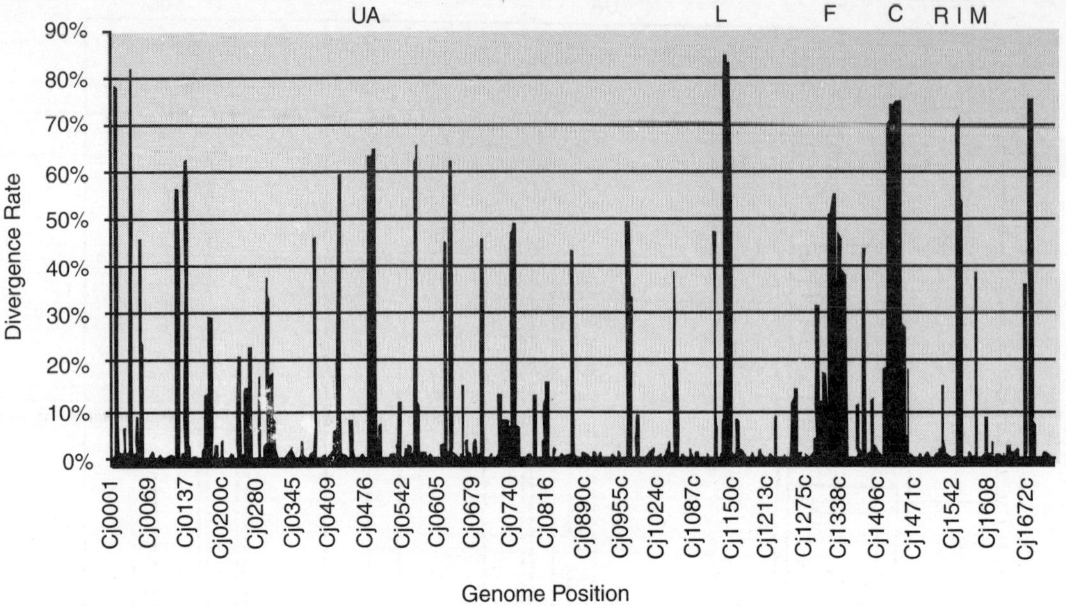

Figure 14.7: Global gene conservation trends observed in 150 strains of Campylobacter jejuni. *The divergence rate was calculated as the percentage of strains showing detectable divergence on MCGH for each gene on the array. Note the presence of several loci with significant levels of divergence across most* C. jejuni *strains surveyed. Among the most highly divergent loci are those involved in cell-surface modification* (UA = uxa/ald locus; L = LOS locus; F = *flagellar modification locus; C = capsular locus; R/M = restriction-modification locus).*

DNA on the solid substrate is referred to as the "probe" and the DNA in solution is referred to as the "target."

2. *In situ* synthesis methods can yield microarrays with ~1000 spots/mm^2 (up to 8200, in the case of Affymetrix), compared to spotted microarrays, which yield < 10-20 spots/mm^2.
3. Spotting is generally carried out through a commercial operation or through the increasing number of microarray production core facilities that have been established at a number of academic or governmental research institutions.
4. The specifics of the platform used, such as the slide surface chemistry, may require modifications to buffers and others. Some slide surface chemistries, for example Corning Ultra GAPS, work equally well with amplicons and longoligos and require virtually no changes in hybridization or washing conditions. Protocols that have been optimized for your platform of choice are generally available from the production facility or slide vendor.
5. The software program "primer3" is a freely available and well-respected primer design tool with the ability to select primers using a number of different parameters and with extensive reporting capabilities about the primer sets picked. Its interface, however, is designed for selecting one primer pair at time.

   To use it to generate the thousands of primer sets required for whole genome amplicon -based microarrays, we have written a Perl program (make-primers) as a controlling shell to primer3. Make primers will submit large sets of genes to primer3 using a number of userdefined parameters and produces a report of the primer pairs selected and expected amplicons generated for each gene in the submitted dataset.

Make-primers is controlled by a graphical interface, and is available free from our group (http://informatics/bio.nrc.ca/software/make_primers). It will run on all platforms for which Per] and primer3 are available (typically Linux/Unix, PC, and Mac).

6. Dynamic range: the difference between the maximal possible signal intensity (i.e., signal saturation) and nonspecific background signal.

7. Experimental design for most MCGH applications is such that two samples are compared by cohybridization to the microarray. Because data analysis is generally carried out by calculating the signal ratio of the two samples, "normalization" of amplicon concentrations yields prior to spotting is seldom considered because it has long been assumed that any concentration effects would be negated by the ratio calculation.

   Nevertheless, signal intensity does affect the dynamic range of a spot, affecting signal ratio values. Reamplification tags can be incorporated into genespecific primers and used to perform a second round of PCR to reamplify the gene specific amplicons obtained in the first round of PCR. Unlike the first round of amplification, which can generate very different yields for each amplicon, the reamplification reaction generally results in similar amplicon yields for all genes because common primers are used.

8. For high-throughput synthesis, oligos are synthesized in 96-well format with Forward and Reverse primers in matching plates. They are typically shipped in lyophilized form and should be reconstituted in Milli-Q water to the working concentration.

9. Most laboratories use random primer-based labeling to generate labeled targets. Although we have also used it in the past, because of its simplicity and robustness, the chemical coupling protocol described in elsewhere in this chapter is currently our method of choice. For random priming, we prefer to use the BioPrime indirect labeling system (Invitrogen). The indirect method is preferable over direct methods that directly incorporate dye-labeled nucleotides during synthesis to avoid biases as a result of the differential incorporation rate of Cy-3 and Cy-5 nucleotides by DNA polymerase.

10. Fragmentation can be achieved by digestion with restriction enzymes or sonication but we have found that the nebulization method of Bodenteich et al. using a modified asthma nebulizer, and which is routinely used in shotgun library construction, works efficiently and yields an excellent distribution of fragment sizes.

    Briefly, the nebulizer is modified by discarding the mouthpiece and sealing the large opening on the top cover using the cap from a 13-mL culture tube and sealing with Parafilm. Twenty-five micrograms of genomic DNA are suspended in 2 mL of 35% glycerol: 10 mM Tris: 1 mM EDTA (pH 8.0) and loaded on the nebulizer chamber. The nebulizer is connected to a pressurized gas line at 15 PSI (i.e., compressed air, $C0_2$) using the tubing provided in the nebulizer kit.

    While on ice, the DNA mixture is nebulized for 1 min. The nebulizer is placed in a cushioned rotor bucket (use paper towel, Styrofoam) and the DNA mixture is collected at the bottom of the nebulizer by low-speed centrifugation (-1000g for 5 min). The DNA is recovered by ethanol precipitation.

11. Although many labs eventually opt for automated microarray hybridization stations to standardize hybridization conditions, the entrylevel setup described can produce satisfactory results.

    A high-humidity hybridization chamber can be made using light-tight plastic microscope slide boxes containing a small amount of water and into which clean plain microscope slides have been placed to form a raised platform for the arrays. Each chamber can hold two or three arrays. Special microarray hybridization chambers are also available from several vendors including Corning and Bio-Rad.

12. A typical touchdown PCR profile will have 10 cycles over which the annealing temperature is decreased by 0.5'C in each cycle, starting at the maximum T_m of any primer in the plate, followed by 20-25 cycles at an annealing temperature 3°C below that of the lowest Tof any primer in the plate.

13. Low yield reactions may require pooling of multiple reactions. In our hands, with automated primer design, relatively few reactions failures are owing to poorly designed primers. The majority of these failures are because of primer synthesis failure or errors during the preparation of the PCR reactions. In the vast majority of cases, repeating the PCR reaction of a failed sample, or resynthesizing the primers has resulted in the expected amplicon being produced.

14. Dyes are reconstituted in resuspension buffer provided in the kit. The dye should be aliquoted immediately into single use amounts into O-ring cryotubes. Dyes should be stored at -20°C in a light-tight container. It is preferable to use freshly resuspended dyes whenever possible as coupling efficiency decreases with time.

15. To calculate dye incorporation efficiency, calculate the nucleotide to dye ratio:

    a. pmol nucleotides = [A260* volume (µL)* 50 ng/µL* 1000 pg/ng] /324.5 pg/pmol.

    b. pmol Cy-3 = A_{550} * volume (µL)/0.15.

    c. pmol Cy-5 = A_{650} * volume (µL)/0.25.

    d. nucleotides/dye ratio = pmol nucleotides/pmol Cy-dye.

16. It is of crucial importance that the microarray be prevented from drying during either hybridization or washing steps to avoid unspecific background from probe precipitation or wash buffer crystallization.

17. The volume required depends on the size of the array, and the cover slip type and size and should be accurately determined by setting up mock hybridizations on standard microscope slides; both insufficient and excess volume can adversely affect the hybridization.

18. The stringency of washing conditions has to be empirically determined based on the length of probes, the GC content of the organism, and so on.

19. To eliminate any residual buffer on the array surfaces, microarrays can be spun dry. Dedicated bench-top mini-centrifuges for microarrays are available from several vendors.

20. Despite the fact that Cy-3 and Cy-5 differ in their chemical characteristics (i.e., absorption spectra, extinction coefficient, quantum yield, and so on) influencing signal yields, equivalent signal yields from both channels can be expected for the majority of spots if control and tester probes have been matched according to concentration and dye-incorporation efficiency.
21. Although more sophisticated methods exist and continue to be developed, in our view, if they become absolutely necessary it may be more judicious to simply repeat the experiment because, unlike some gene expression experiments on materials that may be in extremely limited quantity, genomic DNA for additional MCGH experiments should be easily obtainable.

Chapter 15

Cloning in Trichomonas

Expressed sequence tags (ESTs), genome sequence surveys (GSSs) and many gene cloning *strategies* frequently result in partial sequence data. *Fortunately*, numerous *polvinerase* chain reaction (PCR)-based approaches have been developed to clone gull-length open reading frames (ORFs) based on the information from *partial* sequences.

These include methods such as rapid amplification of cDNA ends (RACE), rapid *amplification* of genomic ends (RAGE), and inverse PCR Such approaches can be successful using small quantities of *complementary* DNA (cDNA) or genomic DNA as starting material and avoid the need to go through the complex and tedious process of constructing and screening gene libraries.

A major advantage of these methods is that they use less starting material than the more demanding traditional cloning strategies, and this is particularly relevant for work with unicellular organisms in which RNA and DNA may be in short supply. We have used both the RACE and the RAGE approaches to clone partial and full-length ORFs from *amitochondriate* parasitic microbial eukaryotes.

A RAGE cloning approach was used to complement a RACE cloning strategy when this did not allow us to obtain a complete ORF with flanking *regions* and, hence, infer the likely starting codon of a gene of interest. The *amitochondriate* microbial eukaryotes we are investigating are of interest from both *evolutionary* and *biomedical* perspectives.

We describe below gene cloning results from work with two human parasites that were formally members of the *Archezoa* kingdom. The kingdom *Archezoa* was created to encompass a heterogeneous grouping of *amitochondriate* microbial eukaryotes, which were *hypothesized* to represent early *branching* and primitively mitochondriate eukaryotes.

They would have diverged from the main eukaryotic trunk prior to the acquisition of the bacterial endosymbiont that gave rise to mitochondria. Data from several *laboratories* have shown clearly that none of the known eukaryotes are Archezoa as defined originally.

All extant investigated eukaryotes possess genes that suggest the past presence of mitochondria, and it is possible that all possess some form of mitochondria that may currently lie cryptic

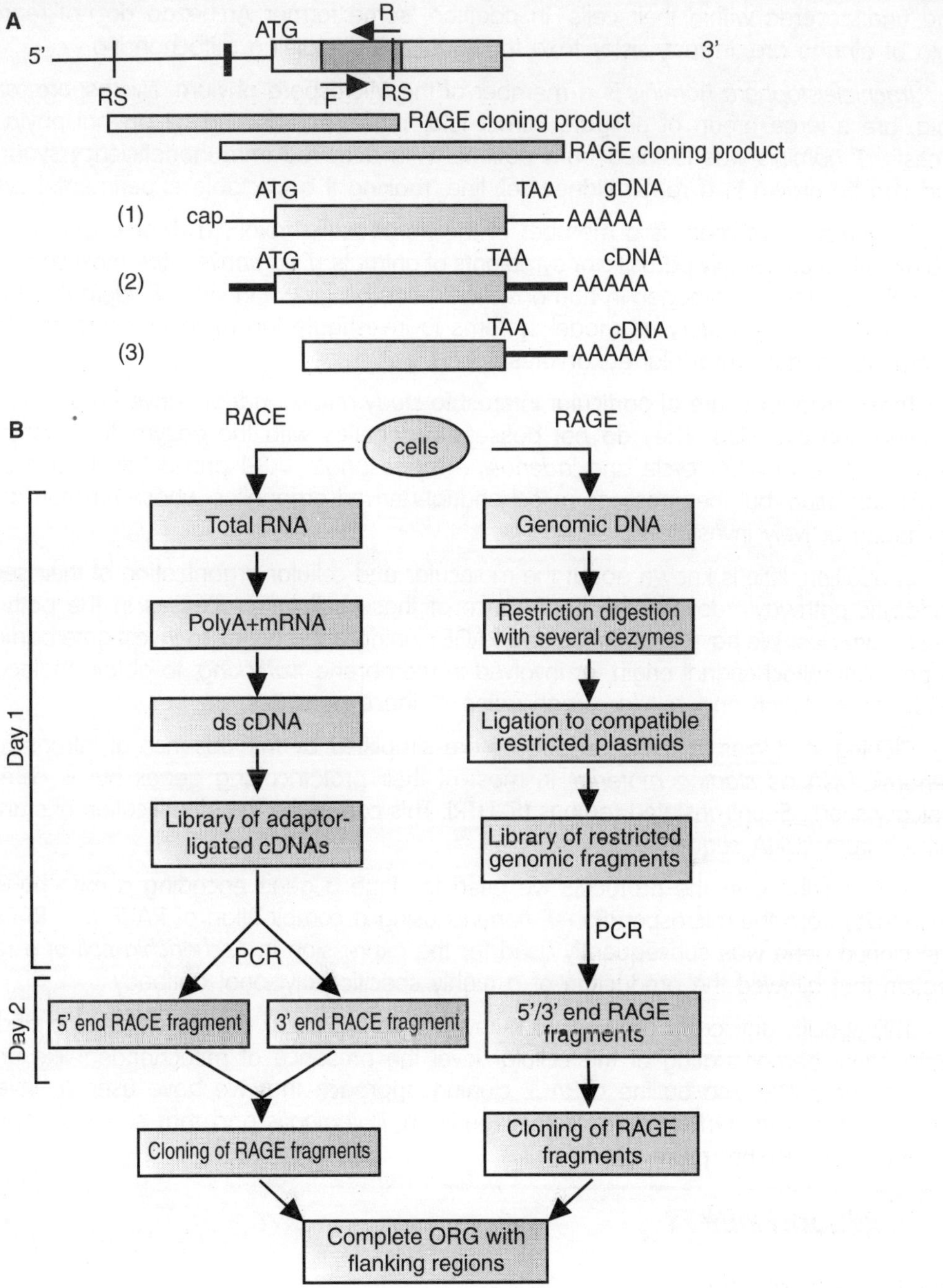

Figure 15.1: Overview of RACE and RAGE cloning strategies. (A) Schematic organization of a protein-coding gene (genomic DNA: gDNA) with the ORF (boxed), the start codon (ATG), and the stop codon (TAA). (B) The major steps are indicated for the RACE and RAGE approaches, from nucleic acid extraction to cloning. Both cloning approaches can focus on either the 5'- or the 3'-end of an ORF.

and undiscovered within their cells. In addition, some former Archezoa do not represent early taxa at all and are, in fact, sister taxa to organisms containing mitochondria.

Trachipleistophora hominis is a member of the Microspora phylum. Microspora, or microsporidia, are a large group of obligate intracellular parasites infecting all animal phyla and some protists. *T. hominis* was isolated from patients with acquired immunodeficiency syndrome (AIDS) and can be grown in a rabbit kidney cell line, making it a tractable experimental system.

Trichomonas vaginalis is a member of the Parabasalia taxon, a diverse group of flagellated protozoa that are mainly parasitic or symbionts of animals. *T. vaginalis* is the most common nonviral sexually transmitted pathogen in humans, and it can be grown in vitro. *T. vaginalis* and *T. hominis* are used in our laboratory as model systems to investigate the molecular and cellular evolution of amitochondriate microbial eukaryotes.

These organisms are of particular interest to study mitochondrial-derived organelles and gene function and evolution. They do not possess organelles with the enzymatic machinery for the *tricarboxylic* acid (TCA) cycle and *adenosine triphosphate* (ATP) production based on *oxidative phosphorylation*, but they possess mitochondrial-derived organelles whose function and structure are being actively investigated.

In addition, little is known about the molecular and cellular organization of their secretory and endocytic pathways, despite the importance of these cellular processes in the pathogenicity of these parasites. We have used RACE and RAGE cloning approaches to investigate particular genes of potential mitochondrial origin, or involved in membrane *trafficking*, to obtain molecular cues to study the evolution and cellular organization of these parasites.

Cloning strategies in these organisms are simplified by the absence of introns (when using genomic DNA as *starting material*) in most of their proteincoding genes but is complicated by typically short _5'-untranslated regions (5'-UTR). This can make the identification of starting codons difficult from cDNA sequences.

We describe later the protocols we used to clone a gene encoding a mitochondrial Hsp70 (mitHsp70) from the microsporidian *T. hominis* using a combination of RACE and RAGE methods. The cloned gene was subsequently used for the expression in *Escherichia coli of* a recombinant protein that allowed the production of a highly specific polyclonal antibody.

This specific antiserum allowed us to investigate mitHsp70 protein expression and its cellular localization, demonstrating at the cellular level the presence of mitochondrial-like organelles in microsporidia. We also outline a RACE cloning approach that we have used to investigate the diversity of a small GTP-binding protein family in *T. vaginalis* and that could be applied to the cloning of most other protein families.

REQUIREMENTS

Isolating Nucleic Acid

DNA from T. hominis

1. Saturated phenol solution, pH 7.6 (BDH Laboratory Supplies, cat. no. 4367546).
2. Chloroform (BDH Laboratory Supplies, cat. no. 10077W).

3. Ribonuclease A (RNAase A) from bovine pancreas (Sigma, cat. no. 83831).
4. Microcon 100 filtration columns (Millipore, cat. no. UFC7PCR50).
5. Tris-EDTA (TE) (10:1), pH 7.5. 10 mMTris-HCl, 1 mMEDTA, pH 8.0, pH adjusted to 7.5 with 1 M HCl.
6. 10% Sodium dodecyl sulfate (SDS) solution (Sigma, cat. no. 05030).
7. 0.10- to 0.11-mm diameter glass beads (Braun Biotech International, cat. no. 854140/0).
8. Mikrodismembrator U. B. Bead beater (Braun Biotech International, type 853172/2).

Total RNA from T. hominis

1. Lysis buffer: 62.5 g guanidine thiocyanate, 4.4 mL of 0.75 M sodium citrate. 30% (w/v), 2.2 mL Sarkosyl, 77.5 mL DEPC-dH_2O.
2. Saturated phenol solution, pH 6.6 (BDH Laboratory Supplies, cat. no. 4367546).
3. Chloroform (BDH Laboratory Supplies, cat. no. 10077WF).
4. Agarose gel electrophoresis equipment and 10-mg/mL ethidium bromide stock.
5. DEPC-dH_2O.
6. Phosphate-buffered saline (PBS), 10X concentrate (Sigma, cat. no. P7059).
7. Agarose (Sigma, cat. no. A4804).
8. 1 X TBE: 10.8 g Trizma base, 5 g boric acid, 4 mL of 0.5 M EDTA, distilled H_2O to 1 liter.

mRNA from T. hominis

1. Dynabeads mRNA purification kit (Dynal, cat. no. 610.06).

Hsp70 Gene of T. hominis

1. Marathon cDNA amplification kit (Clontech, cat. no. K 1802-1).
2. PLATINUM *Taq* DNA polymerase (Invitrogen, cat. no. 10966).
3. Gene-specific oligonucleotide primers.
4. Microcon 100 filtration columns (Millipore, cat. no. UFC7PCR50).

mitHsp70 Gene of T. hominis

Extraction of Vector

1. Hind III restriction endonuclease and appropriate lox buffer (New England Biolabs, cat. no. R0104T).
2. pPCR-Script Amp cloning kit (Stratagene, cat. no. 211188).
3. Shrimp alkaline phosphatase and appropriate 10X buffer (Promega, cat. no. 9PIM 820).
4. Microcon 100 filtration columns (Millipore, cat. no. UFC7PCR50).
5. T4 DNA ligase (Stratagene, cat. no. 600011).

DNA Amplificatior Product of T. hominis

1. Hind III restriction endonuclease and appropriate 10X buffer (New England Biz-labs, cat.

no. R0104T).

2. *Taq* pzlymerase, 5 U/μL (Bizline, cat. no. M95801B), supplied with 10X NH_4 reaction buffer and 50 *mM* $MgCl_2$.
3. Gene- and vector-specific primers.
4. Microcon 100 filtration columns (Millipore, cat. no. UFC7PCR50).

Genetic Engineering

1. *Taq* polymerase, 5 U/μL (Bioline, cat. no. M95801B), supplied with 10X NH_4 reaction buffer and 50 *mM* $MgCl_2$.
2. BSA, 10 mg/mL (Promega, cat. no. R3961).
3. dNTP 400 mM, (dATP, dGTP, dCTP, and dTTP) (Invitrogen, cat. no. 10297-018).
4. Centriflex gel filtration cartridges (Edge Biosystems, cat. no. 42453).
5. QIAquick gel extraction kit (Qiagen, cat. no. 28706).
6. pGEM-T-EASY vector system (Promega, cat. no. A1360).
7. T4 DNA ligase (Stratagene, cat. no. 600011).
8. Maximum-efficiency DH5a-competent E. *coli* cells (Life Technologies, cat. no. 18258012).
9. X-gal (5-bromo-4-chloro-3-indolyl-(3-D-galactopyranoside), 50 mg/mL (Promega, cat. no. V3941).
10. IPTG (isopropyl (3-D-thiogalactopyranoside), 100 mM (Sigma, cat. no. 15502).
11. Ampicilin sodium salt (Sigma, cat. no. A2804).
12. Material and solution to grow DH5a E. *coli.*
13. Material and solution for standard TBE agarose gel, including a gel imaging system, e.g., UVP imager (Ultra Violet Products, Cambridge, UK).
14. Hyperladder DNA ladder (Bioline, cat. no. HYPL1200).
15. BigDye Terminator v 3.0 sequencing standard (ABI PRISM, cat. no. 4390303).

PROCEDURES

Here, we describe how to perform extraction of genomic DNA and RNA from the eukaryotic organisms used in our research. We then give detailed *protocols* for the generation of cDNA from mRNA and the *amplification* of the ends of selected cDNAs. This approach is complemented by the description of a method for the amplification of the ends of genomic DNA fragments, using a *T. hominis* gene as an example. Amplification products may be cloned and sequenced as appropriate.

Isolation of Nucleic Acids

DNA from T. hominis

T. hominis cells are maintained in cultured rabbit kidney cells and spores purified from these cultures, as previously described. Purified spores are used as *starting* material for both mRNA and genomic DNA purification. Because spores are *surrounded* by a thick spore wall and, therefore, are difficult to break, we use bead beating (glass beads are added to the sample, which is shaken at high speed) to disrupt the cells prior to nucleic acid *extraction.*

1. Centrifuge purified spores at 3500g for 5 min (we used between 1.8×10^8 and 2.25×10^8 spores per extraction).
2. Discard supernatant and resuspend spores in 1 mL TE 10:1, pH 7.5.
3. Transfer spore suspension to a sterile 2-mL screw-cap tube and centrifuge at 3500g for 5 min. Discard the supernatant.
4. Add 400 pL TE 10:1, 10 µL 10% SDS, 2µL of RNAase A, and 400 µL of 0.10-to 0.11-mm diameter glass beads to the tube. Bead-beat (as described above) for 45 s at 2000 shakes per minute.
5. Add 800 µL of phenol, pH 8.0, to the bead-beaten cells, mix by inversion, and centrifuge at 10,000g for 10 min.
6. Recover the supernatant and add to it 400 µL of chloroform. Centrifuge at l0,000g for 10 min.
7. Recover the supernatant and concentrate the samples using microcon 100 cartridges, according to manufacturer's instructions. Test a small aliquot for presence of DNA using an agarose gel. This material can be used for either PCR or the RAGE protocol described in elsewhere in this chapter.

Total RNA from T. hominis

To prepare cDNA we first need to purify total RNA, from which mRNA is purified. The latter *is* used for cDNA synthesis that will be the template DNA for the RACE PCR reaction.

1. Centrifuge culture medium of infected cells with abundant spores for 10 min at 200g so that spores and other cells form a pellet. Wash twice with double-distilled sterile H_2O and resuspend in PBS and use directly, or store at -80°C until RNA extraction.
2. Wash spores with RNAase-free sodium phosphate (20 mM sodium phosphate at pH 8) and then bead-beat in 400 L of lysis buffer for 1 min at 2000 shakes per minute. Centrifuge sample at 10,000g for 10 min on a benchtop centrifuge.
3. Remove supernatant and rinse beads with 300 L of RNase-free 20 mM sodium phosphate at pH 8.
4. Bead-beat spores again and centrifuge (as in step 2) and pool the supernatants.
5. Add 700 gL of phenol, pH 6.6, to the pooled supernatant and invert to mix the two solutions. Centrifuge for 10 min at 10,000g and retain the aqueous layer.
6. To this, add 700 µL of phenol-chloroform-isoatnyl alcohol (25:24:1 v/v) and centrifuge the mixture at 10,000g for 5 min.
7. Keep the aqueous layer and add 700 µL of chloroform-isoamyl alcohol (24:1 v/v), mix, and then centrifuge at 10,000g for 5 min.

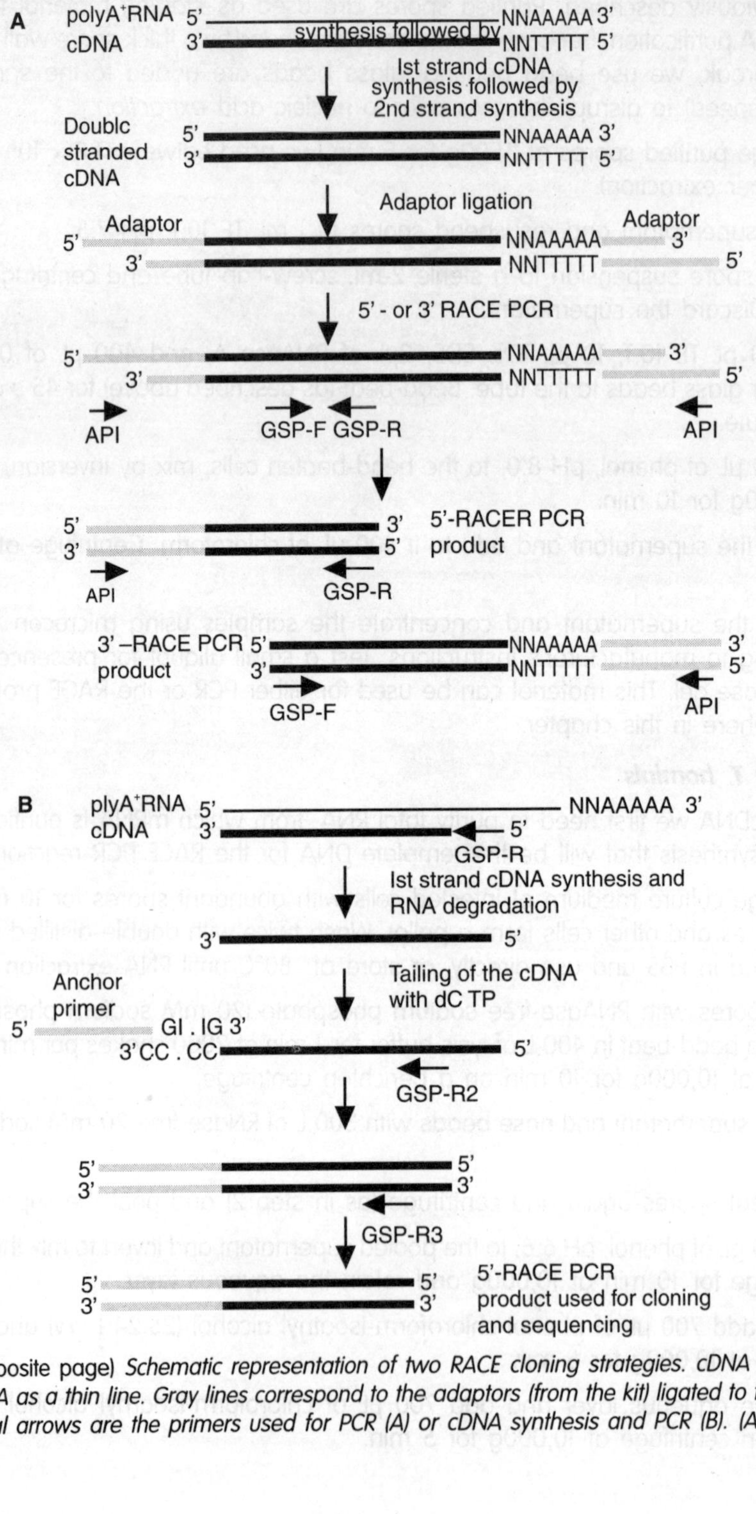

Figure 15.2: (Opposite page) *Schematic representation of two RACE cloning strategies. cDNA are shown in black thick lines, mRNA as a thin line. Gray lines correspond to the adaptors (from the kit) ligated to the double-stranded cDNA. Horizontal arrows are the primers used for PCR (A) or cDNA synthesis and PCR (B). (A) The first approach*

(Caption Contd.)

8. Precipitate the RNA by adding 2.5 vol of 100% ethanol to the extract and placing it at -20°C for 1 h. Then centrifuge this at 10,000g for 30 min at 4°C.
9. Wash the pellet with 70% (v/v) ethanol and resuspend in 50 µL TE 10:0.1 pH 7.5 (or DEPC-dH2O).
10. Run a small aliquot on a 1.2% 1X TBE-agarose gel to check for the presence of RNA.
11. Preferably, one should proceed immediately to the next step to reduce chances of RNA degradation. Alternatively, store at -80°C until use.

Extraction of mRNA

1. Extract poly-A+ mRNA from total RNA using Dynal Dynabeads° following the manufacturer's protocol.
2. Proceed directly to the cDNA synthesis step described below to avoid mRNA degradation.

Hsp70 Gene of T. hominis

To clone a mitHsp70 gene, we used the Clontech Marathon cDNA amplification kit for RACE cloning, following carefully the manufacturer's protocol, allowing both 5'- and 3'-RACE from the same cDNA preparation. In brief, this involves the following steps:

1. To generate the cDNA, reverse-transcribe the mRNA using an oligo d(T) primer (provided in the kit), which binds to the polyA tail of the 3'-end. The second strand is produced from this according to the Gubler and Hoffman method.
2. Ligate the adapters from the kit by blunt-end ligation to each end of the cDNAs using the protocol provided by the manufacturer. The library of adaptor-ligated cDNA is now ready for RACE PCRs.
3. Design gene-specific primers from previously sequenced areas of the gene that you are attempting to amplify. Design a nested primer to use in case no product is amplified using your first PCR primer. In our work, gene-specific primers were designed to a fragment of the *T. hominis* mitHsp70 gene, obtained by PCR from genomic DNA, using degenerate primers as described previously.

   The gene-specific primers (GSPs) were ThHsp70-421F (GSPF1), ThHsp70-542F (GSPF2),

---

(Caption 15.2 Contd.) *corresponds to the methodology implemented by the Marathon cDNA amplification kit. This approach has the advantage of allowing both 5'- and 3'-RACE cloning from the same cDNA preparation. It can lead to the cloning of a complete ORF if the mRNA transcripts are not degraded during their purification, and if cDNA synthesis reaches the 5'-end. If the mRNA is characterized by a long 5'-UTR, then it makes the identification of the start codon more likely.*

As T. hominis and T. vaginalis both have short 5'-UTR, and Trichomonas has high levels of nuclease activity, many RACE clones do not contain the start codon. (B) A second approach, as implemented by the Invitrogen Life Technologies 5'-RACE kit, increases the chance of cloning the 5'-end by focusing the cDNA synthesis on the particular gene of interest by using a GSP (GSP-R) to initiate the cDNA synthesis. Performing shortened cDNA synthesis also reduces the likelihood of inhibition of cDNA synthesis because of potential severe secondary structures along the mRNA.

A primary PCR is performed with an anchor primer and a GSP (GSP-R2). A secondary PCR typically is needed to obtain a PCR product (using GSP-R3). It is useful to sequence several clones to obtain as much sequence data as possible from the 5'-end. This approach was used successfully to clone several full-length ORFs in T. vaginalis.

ThHsp70-864F (GSPF3) for the 3'-RACE, and ThHsp70-652R (GSPRl) and ThHsp70-258R (GSPR2) for the 5'-RACE. Use these primers in combination with the primer APl, complementary to the adaptor, provided in the kit.

4. If no PCR product can be detected in an agarose gel from the first amplification, then it may be necessary to carry out a nested PCR, which requires the design of two additional GSPs used with the AP2 primer (provided with the kit), an "internal" (in relation to APl) adaptor-specific primer. For cloning of the *T. hominis* mitHsp70 a two-step amplification procedure was needed, because both primary 5'- and 3'-RACE PCRs (GSPF/Rs and APl) did not yield detectable PCR products. A second round of successful nested 5'- and 3'-RACE PCR was carried out using the "internal" GSPs (GSPF/R2) with the "internal" AP2 adapter primer provided by the kit.

5. Carry out PCR amplifications using Invitrogen high-fidelity Platinum *Taq* and a Perkin Elmer GeneAmp Systems 9600 thermal cycler using the cycling conditions described below. Mix all ingredients as detailed in the Marathon kit and start the cycles. The Marathon kit suggests a touchdown PCR for primers with high *Tm* (>_*70°C), to reduce back ground. This may help to increase the specificity of the PCR, but it was not needed for our work on *T. hominis* mitHsp70.

   RACE cloning PCR conditions:

Initial denaturation:	30 s at 94°C
Followed by 25-30 cycles:	5 s at 94°C
	4 min 68°C

6. Electrophorese an aliquot of the PCR products on 1% TBE agarose gel to check for the presence of PCR product. If there is a detectable band, then electrophorese the remaining material on a gel and elute the band of interest using the QIAquick gel elution kit as described in the user manual. Quantify the eluted material on a gel by running a small aliquot of the eluate in parallel to a molecular size maker with a known amount of DNA for comparison.

7. Ligate the eluted DNA band to the pGEM-T-EASY vector using the following protocol, and mix:

T4 DNA Ligase 10X buffer	1 µL
pGEM-T-EASY vector, 50 ng/µL	1µL
PCR product (28-85 ng)	x µL (molar ratio 3:1-1:1 to the vector)
T4 DNA ligase (1 Weiss unit/µL)	1µL
Analar water to	10µL

   Perform a negative control ligation without PCR product; from this, no colonies or very few-should be obtained here. As a positive control, one can use a control insert provided by the pGEM-T-EASY kit. Incubate the ligation overnight at 4°C.

8. Use 3µL of the ligation mix to transform competent *E. coli* DH5a*. The remainder of the ligation is stored at -20°C if this step has to be repeated. Incubate plates overnight at

37°C to allow E. *coli* colonies to grow. Typically, two dilutions of cells are plated 1:10 and 9:10 to provide well-spread colonies.

9. Pick up colonies for plasmid minipreps and perform restriction digestion or PCR directly from bacterial colonies to identify clones with appropriate inserts; 3-5 clones (per PCR product) inserts are typically sequenced.

The combination of the sequence data for the 5'- and 3'-RACE PCR products provide an ORF encoding a potentially complete mitHsp70 but no 5'-UTR. Sometimes it is not possible to obtain any 5'-UTR regions and hence infer the position of the potential start codon. In this case, a complementary RAGE approach may be used to provide the missing data.

5'-End mitHsp70 Gene of T. haminis

To complement the 5'-end sequence data obtained by the RACE, we have used a RAGE cloning approach. The protocol given here broadly follows that of Mizobuchi and Frohman. The result of the RAGE cloning approach is shown in Figure 15.4. A two-step amplification procedure is often needed to obtain a PCR product, and direct sequencing is required to confirm that the product contains the relevant sequence. The combined cDNA and genomic DNA sequence data allowed us to obtain the complete ORF for *T hominis* mitHsp70.

Cloning Vector

1. Select an appropriate cloning vector and digest with a restriction enzyme. This enzyme should cut your known fragment of sequence at a single site. In our experiments, Stratagene pPCR Script was selected as vector and was digested with HindIII for 80 min at 37°C.
2. Digest approx 500 μg of plasmid with 100 U of restriction enzyme, adding appropriate restriction enzyme buffer to a final concentration of 1X buffer. Add an appropriate volume of double-distilled H_2O so that the restriction enzyme is not more than 10% of the total reaction volume. Check the level of digestion on an agarose gel. A single band (linear DNA) should be obtained for a linearized plasmid.
3. Remove the restriction enzymes from the mix using a Microcon microconcentrator, according to the manufacturer's instructions.
4. To prevent self-religation of the vector ends, treat the cut vector with alkaline phosphatase. Incubate approx 1.2μg of the digested vector with 0.7 U of shrimp alkaline phosphatase (SAP) for 1h at 37°C. In our experiment, 1.2 μg of vector corresponded to a volume of 12μL, and a 20-μL reaction was set up by adding 0.7 μL of SAP, 2 μL of the provided 10X reaction buffer and 5.3 μL of H_2O.
5. Heat-inactivate the phosphatase by incubation for 30 min at 65°C and remove the restriction enzymes using a Microcon 100 filtration column.
6. Assess the success of this treatment by ligation of the vector to itself using untreated cut vector as a control. Transform the ligation products into DH5α. No colonies should be seen in the alkaline phosphatase-treated vector in the absence of insert.

DNA Amplification Product of T. hominis

Restriction enzyme sites were identified in the mitHsp70 fragment that had been previously

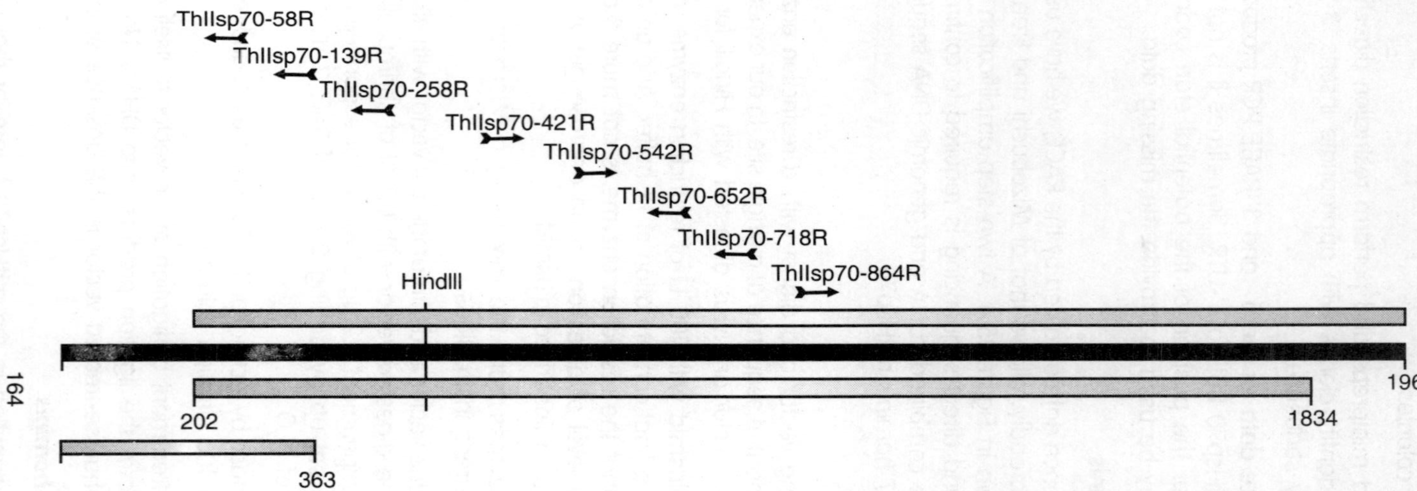

Figure 15.3: The position of primers used for PCR relative to the sequenced T hominis *genomic DNA and cDNA encoding mitHsp70. In addition to the GSPs (some are not shown here for simplicity), cloning vector-specific primers were used for double-strand sequencing of the entire ORF and flanking regions. The black box shows the total DNA fragment amplified and sequenced from genomic DNA (see GenBank entry AF492453 for full sequence including the 5'- and 3'-UTRs). The light gray box shows the complete ORF with a total of 1632 base pairs. The white box shows the DNA fragment amplified and sequenced from cDNA using both 5'- and 3'-RACE. The darker gray box shows the segment sequenced from 5'-RAGE cloning. The position of the restriction site HindIII, used to digest the genomic DNA, is also shown. For Hsp70 primer sequences, see Note elsewhere in this chpater.*

amplified and sequenced. This fragment has a HindIII restriction site 340 bps 3' of the most 5' by in the known sequence.

Because of the small amount of DNA that can be extracted from spores and of the smaller size of the typical microsporidian genome relative to that of mammals, only 0.5 μg of DNA was digested with HindIII for a RAGE cloning attempt.

1. Digest genomic DNA (0.5 g) to completion with 3.5 μL (35 U) of HindIII over-night at 37°C.
2. Remove the restriction enzyme with a Microcon 100 filtration column.
3. Ligate the genomic DNA to the restricted and phosphatase-treated vector over-night at 4°C using T4 ligase using a 1:1 ratio of insert to vector.
4. Carry out PCR amplifications on the ligation mix the following day. Two rounds of PCR may be needed. PCR conditions are given in polymerase chain reaction. For the first round use an external GSP combined with the plasmid-specific primers (T3 or T7 for pPCR Script); we used either ThHsp70-258R (GSP1) or ThHsp70-139R (GSPIb). The plasmid-specific primers sit either side of the HindIII cloning site.

   It is necessary to use primers from both sides of the cloning site, as the potential insert orientation is unknown, making up a total of four primer combinations. In our case, we had: GSP1-T7, GSP1-T3, GSPIb-T7, and GSPIb-T3.
5. If no PCR product can be detected by gel electrophoresis, purify potentially amplified material from this first round of amplification using a Microcon 100 column to remove former primers and dilute 1/100 in Analar water. For the second round of PCR, use an internal gene-specific primer and 1μL of the purified primary PCR.

   Here, we used two GSP primers, either ThHsp70-139R (GSP1b) or ThHsp70-58R (GSP2) with the plasmid-specific T3 or T7 primers and the diluted primary PCR reaction as template DNA. In our experiment, one combination of primers, including ThHsp70-139R, produced a PCR product that was shown by direct sequencing to correspond to the 5'-end of the mitHsp70 ORF.
6. Electrophorese PCR products on a 1.2% 1 X TBE agarose gel and gel-extract any band obtained by PCR using the QIAquick gel extraction kit. Sequence PCR products directly or clone using, for example, pGEM-T-EASY before sequencing.

Polymerase Chain Reaction

Here, we present the general protocol we use to perform the hot-start PCR mentioned in elsewhere in this chapter. It is assumed that the user is familiar with the danger of DNA contaminations that can lead to spurious PCR products; hence, a negative control containing no template DNA is always run in parallel with the samples.

1. Mix all components as listed below:

   5 μL *Taq* 10X reaction buffer

   4 L MgCl2 (100 mM)

   1 μL BSA (10 mg/mL)

   1 μL dNTPs (400 mM)

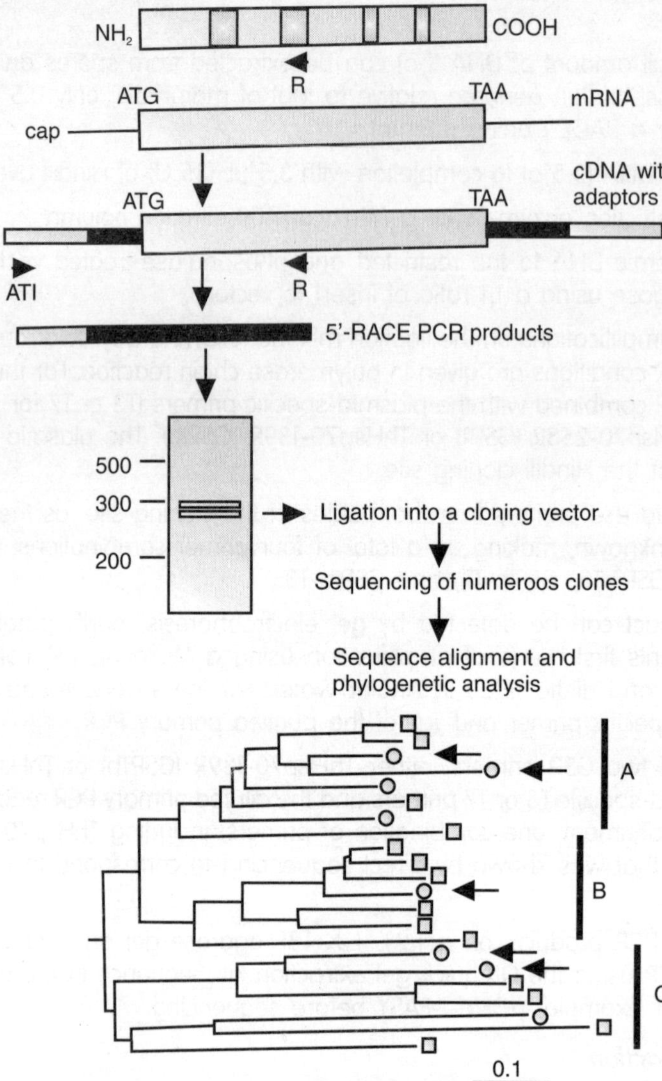

Figure 15.6: A RACE cloning strategy used to clone members of the Rab gene family in T. vaginalis. A schematic representation of a Rab GTP-binding protein is shown with conserved domains in gray boxes. Primers can be designed to hybridize to one or more conserved domains and used in a RACE cloning approach.
Shown here is the 5'-RACE cloning approach as used for Rab proteins. Total mRNAs were purified and used for cDNA synthesis using the Marathon cDNA amplification kit. The primer API, specific for the ligated adaptor. was used with a degenerate reverse primer complementary to the conserved domain of the G-2 domain of Rab proteins. The resulting RACE PCR amplicons of various sizes (ca. 250-300 bp), reflecting a combination of heterogeneity in paralog length and variable success in cDNA synthesis, were ligated into the pGEM-T-EASY plasmid.
Sequencing of the clone fragments allows identification of Rab-encoding cDNAs by comparison to known sequences. A phylogenetic tree is used to differentiate between paralogs. The tree shown here suggests three major paralogs, A, B, and C, with the new sequences highlighted with arrows.

6 μL primers (3.2 pmol each)

6 μL template DNA

0.5 μL *Taq* (5 U/μL)

H_2O to 50μL

2. Heat mix 5 min 96°C (hence hot start)
3. Add 0.5 μL *Taq* (5 U/μL) and start cycles:

   1 min at 94°C

   1 min at 56.5°C 35 cycles

   1.5 min 72°C

   30 min at 72°C 1 ×
4. Run a small aliquot, typically 10% of the PCR reaction volume, on an agarose gel to check for the presence of a PCR product.

Cloning of Gene Family

Numerous genes form complex families with many paralogs. Standard PCR can be used to investigate their diversity, in which case two conserved domains allowing the design of degenerate primers typically are needed. We have used a RACE cloning approach, as used for mammals, to investigate *in Trichomonas* the members of the Rab small GTP-binding protein family to complement an ongoing *Trichomonas EST* project.

These two approaches have identified with more than 30 distinct cDNAs so far, an unexpectedly large Rab gene family for a singlecelled organism (Hirt et al., unpublished data). Figure 15.6 glves an outline of the approach. We used the strategy described by Chavrier et al. in combination with the Marathon cloning kit described above.

1. Choose an appropriately conserved protein region for your gene family of interest that allows the design of a degenerate primer for PCR. The primer properties might be better for either a 5'- or 3'-RACE. If two conserved regions or primer directions are promising, then it may be useful to try alternative options to increase the chance of success in surveying the gene family diversity.
2. Perform eithor 5'- or 3'-RACE PCR (5'-RACE was performed for the Rab genes) and clone products into a plasmid vector as described in previous subheadings. (PCR products of expected size range were obtained for the chosen Rab-specific primer and amplicons were gel-purified and cloned into pGEM-T-EASY as described earlier.)
3. Prepare plasmid minipreps to analyze random clones by restriction digestion.
4. Sequence several clones that show distinct restriction patterns.
5. Compare the sequence by aligning them and perform phylogenetic analysis to differentiate them.

We could obtain, without difficulty, numerous pGEMT-RACE clones with distinct restriction fragments and identified among these by sequencing 10 distinct (partial sequence) Rab paralogs.

The partial sequence data can then be used to clone full-length cDNA using, for example, a combination of 5'- and 3'-RACE cloning approach as presented earlier. Of these 10 5'-RACE cDNA clones, 8 have been confirmed by further cDNA cloning and sequencing, validating the approach.

POINTS TO REMEMBER

1. For RNA purification, one can use a standard phenol extraction protocol. However, in some organisms, numerous nucleases are released, and these must be inhibited in some way. When working with the parabaslid *T. vaginalis,* one needs to take extra care to inhibit the numerous nucleases released during cell lysis and, hence, use methods that will efficiently destroy their activity.

   We have used either a guanidium thiocyanate-based method or an RNA extraction kit. For the guanidium thiocyanate-based method, it is crucial to use fresh solutions as described. Excellent results were also obtained for *Trichomonas* using the SV total RNA isolation system from Promega, cat. no. *TM048.*

2. Primers are key *for any* PCR-based cloning approach, and we recommend that researchers new to the technique carefully read more details in the literature. We have used OLIGO v. 5.0 for Macintosh (National Biosciences) and Primer3 on the Web http://www-genome.wi.mit.edu/cgi-bin/primer/primer3_www.cgi to design primers that are typically 20-30 nucleotides in length. The great majority of primers designed using these methods resulted in success.

   Specific information for primer design for RACE PCR is also given in the Marathon cDNA amplification *kit.* For degenerate primer design, the chance of success may be increased if you are able to take into account any available information about the codon usage of your organism. This may allow *you to* reduce primer degeneracy. For additional information on degenerate primer design see, for example, ref. elsewhere in this chapter.

3. Because of the small amount of available material for *T. hominis* preparations, we did not attempt to quantify either the mRNA or the cDNA. The extracted mRNA was processed directly for RACE PCR. In the case of nonlimiting starting material, one should use the optimal amount of polyA+ mRNA (1 μg) for the Marathon kit. This will provide an optimal amount of cDNA, and the more template DNA that can be used in the RACE PCR reactions the better. This might be a critical parameter for rare mRNAs. This was not a problem for *T. hominis* mitHsp70.

4. In the case of the *T. hominis* Hsp70 gene, this is probably to the result of short 5'-UTRs in microsporidia, which can have short intragenic sequences. Also, it may have been degraded during the purification, removing the most 5'-end of the mRNA including the start codon.

5. Mizobuchi and Frohman suggest the use of at least three digestion enzymes and 5 μg of DNA for each digestion as detailed here. However, because of the small amount of DNA purified from spores, we made an attempt with a reduced amount of DNA. This experiment was successful using a single digestion enzyme and only 0.5 μg of genomic DNA for the digestion.

This success may be because of the fact that microsporidia have relatively small genomes sizes *(18,26)* and, therefore, a DNA extraction contains more copies of genome per microgram of DNA, compared to a DNA extraction from mammalian cells.

6. The sequences of the Hsp70 oligonucleotide primers (22-30-mers) used in PCR amplification were as follows:

ThHsp70-58R	5'-GGCCACGTTGTCCTGCATTATAGAAAT-3'
ThHsp70-139R	5'-CGGTTTGCCAACTATCACGTTTTCACCCGA-3'
ThHsp70-258R	5'-AGTTCTCCGTTGTCGTCTATAA-3'
ThHsp70-421F	5'-TACACAGCGGGAGGAGACTAAAA-3'
ThHsp70-542F	5'-ACTTAGGCGGTGGTACGTTTGATA-3'
ThHsp70-652R	5'-GCCTAATTTTCGCCAGATCAATGTCAC-3'
ThHsp70-718R	5'-CAATCGTCACGGTTTCCTGTGTTGATAG-3'
ThHsp70-864F	5'-TTGCTCCACTGATAAAAAGAAC-3'

Chapter 16

Phylogeny of Bacteria

The discovery of new genera and species is at the core of understanding biological diversity within the framework of a stable classification system. The natural environment, particularly the marine biosphere, provides a rich source of many new microbial species. Sampling of such environments suggests that the diversity of prokaryotes is probably 100 times higher than previously expected.

By contrast, the discovery of completely novel bacterial species of human origin has become an increasingly difficult and *demanding* task, and the emergence of new species of proven clinical significance is now a comparatively rare event. In recognition of these difficulties, microbiologists are encouraged to proceed in taxonomic studies by use of the "*Candidatus*" concept, which entails publication of descriptions of novel but as yet *uncultured* organisms, provided they have been well characterized by *molecular* techniques and by electron *microscopy*.

Some recent examples using this approach include *Helicobacter suis, Helicobacter bovis,* and *Cam-pylobacter hominis.* Interestingly, a culture method was subsequently developed for the latter species. For the *nonspecialist, taxonomy* can sometimes prove a confusing subject, because basic terms may not necessarily have been agreed upon, while the interpretation of some initially *unambiguous* and precise definitions may have changed with time, as a result of new developments and the availability of new *information.*

To assist the reader to overcome these potential difficulties, key terms that are generally consistent with the sense and views expressed by Cowan will be defined at the outset. *Systematics* is the scientific discipline that embraces the many approaches involved in the characterization of the diversity of organisms and their arrangement in an orderly scheme of relationships. Central to systematics is *classification,* the process by which organisms are ordered on the basis of their *relationships* into groups (taxa) within a hierarchical scheme, from the basic unit, the species, through the different ranks of *genus, family, order* and *class.*

Phylum and kingdom, and likewise subspecies and variety, have no formal standing in bacterial *nomenclature,* so are not recognized by the International Code of Nomenclature of Bacteria, which provides a formal set of Rules and Statutes, agreed upon by the International Committee of

Systematic *Bacteriology*, to standardize the naming of bacteria. The Code is used in conjunction with the Approved Lists of Bacterial Names, published in 1980, and subsequent notifications of new approved and validated names are published on a regular basis in *the International Journal of Systematic and Evolutionary Microbiology*.

The key requirements of any classification are that it should be stable, reproducible, scientifically based, and have high information content. However, the purpose of classification varies among different groups of microbiologists. For example, in practical medical bacteriology, the main objective is to facilitate a better understanding of relationships at the species and genus level, whereas relationships at higher levels remain very much the domain of more *academic/theoretical taxonomists*.

Apparently "natural" classifications may well be an ideal to aim for, rather than the artificial classifications developed for special purposes. However, the distinction between these extremes may not necessarily be clearcut, and may lead to different *interpretations* of taxonomic analyses, and the development of possibly *incompatible* nomenclature.

Taxonomy is the scientific theory of classification and, in practice, is applied in conjunction with nomenclature (the process of allocating names), and with identification (the means by which unknown microorganisms are allocated to previously described taxa). Taxonomy has an important function in clinical *bacteriology* in providing a stable framework for the description and identification of bacterial species that cause disease, although it is recognized that names must change occasionally as new and improved information is gathered, despite the short-term *confusion* it may cause.

It is also recognized that the ecological role of bacteria can, in certain cases, take priority in deciding species status. Consequently some medical organisms associated with defined clinical symptoms may continue to bear names that are not in accord with their genomic relatedness. Despite the numerous detailed systematic investigations undertaken since the publication of the Bacteriological Code, there are no formal procedures for the classification of bacteria.

The process of classification, although dynamic, continues to be subjective and dependent on the particular type of organism under study, as well as the current state of the art of relevant methods for strain characterization. *Recommendations* have been made from time to time to *reconcile* and *reevaluate* approaches to bacterial systematics.

Attempts have also been made at standardization by the publication of minimal standards for particular genera, but these are by no means comprehensive, and only provide guidelines and are not *mandatory* for the description of *novel taxa*. The aim in this chapter is to highlight possible paths that provide strategies when considering the description and validation of a new bacterial species *originating* from human clinical specimens.

Taxonomy, like many aspects of clinical microbiology, has moved into the postgenomic era, so it is essential to encompass data on individual genes as well as on complete genome organization, and to assess how such *developments* might impact on the more traditional concepts of bacterial speciation. These aspects will be considered here by reference to the *Campylobacteriaceae* and related organisms that include *helicobacters*, as they provide a good model group to illustrate the application of different *taxonomic* approaches.

Another new area that needs to be factored into contemporary taxonomy is the growing information on the population structure of individual species, with its importance in relation to the species definition. Production of trees to illustrate *relationships* or similarities is one of the basic strategies involved in defining new species.

As investigators use ever enlarging datasets, and a growing number of analytical tools, taxonomists have a key role to play in interpreting such analyses for the benefit of the scientific community. The purpose here is to consider possible paths through the veritable forest of potential microbial trees, and to identify those that might bear fruit.

PHENETIC CLASSIFICATION

A classification derived from detailed study of the current observable or expressed properties of the organism (the phenotype) is described as *a phenetic* classification. It is important to recognize that the term *phenetic* does not have any evolutionary implications except in the sense of showing the present-day product of evolution.

The use of phenotypic traits, such as the biochemical tests employed for the characterization of aerobic metabolically active bacteria, such as *Salmonella, Staphylococcus,* or *Pseudomonas,* has been a traditional approach, but it has a number of inherent limitations, both conceptual and practical.

These are particularly evident when applied to the characterization of more fastidious and intrinsically metabolically inactive microorganisms, such as slowgrowing bacteria (e.g., mycobacteria), microaerobes (e.g., *Campylobacter* spp.), or strict anaerobes, such as *Clostridium* spp.

The taxonomic validity of species and genera defined on traditional characters might be open to question, as such tests sample only a minute fraction of the properties encoded by the genomes of such *microorganisms.*

Description of Species

Until the mid-1960s, classification of bacteria largely relied on the description of species based on readily observable phenotypes, represented by morphological (microscopic, cellular, and cultural), physiological, and *biochemical* characteristics. This tradition followed in the footsteps of the work of the *nineteenth-century* founders of medical *microbiology,* such as Robert Koch.

The application of this systematic observational approach, familiar to generations of microbiologists, is illustrated extensively in classic bacteriology texts, such as Bergey's *Manual of Determinative Bacteriology.* Some strain features, such as antibiotic susceptibility and host pathogenicity, are generally not regarded as valid taxonomic features at the *species level,* although until recently, animal tests were used for determination of strain *pathogenicity* for some species (e.g., *toxigenic* variants of *Corvnehacterium diphtheria* and *Escherichia coli).*

Similar approaches were adopted in other fields, such as insect pathology and plant pathology, where particular variants (*pathovars*) were typically described on the basis of the disease caused by infection of a *susceptible* host.

Despite the limitations of traditional test characteristics, they nevertheless continue to provide the starting point for the formal description of new bacterial genera or species, which should include as many basic phenotypic properties as possible. Typical test categories include culture medium requirements, gaseous requirements, growth temperature range, Gram stain, cell shape and size, cell motility, and enzymic indicators of biochemical activity, such as *cytochrome oxidase, catalase, nitrate reduction,* and ability to ferment different carbohydrates (*sugars*).

Such tests form a key part of the formal genus and species descriptions, and are used in the development of *dichotomous* keys and tables for practical identification purposes in medical microbiology. Although it is difficult to apply standard *criteria* to all bacteria, there are now *minimal standards,* drawn up by relevant experts, recommended for defining new species for a growing number of bacterial *genera, including,* for example, *Campylobacter* and *Helicobacter.* Such standards emphasize the importance of basing descriptions on a number of individual isolates that represent species diversity in terms of ecological, host, and geographic origins.

Numerical Taxonomy

The rapid developments in computing technology in the 1960s facilitated the foundation of numerical taxonomy, which now provides a rigorous *mathematical* basis for the analysis of biological data. The basic stages in performing numerical *phenetic* taxonomy for bacteria comprise:

- Strain selection
- Test selection
- Recording of results for qualitative and quantitative tests
- Data coding
- Numerical analysis to produce a similarity matrix and hierarchical clustering
- Interpretation-notably definition of clusters

The principal outcome from any such numerical analysis is one or more treelike structures (*dendrograms/phenograms*), which provide a two-dimensional *representation* based on degrees of similarity or distances between different strains, defined as *operational* taxonomic units (OTUs). Trees form a part of mathematics called graph theory, and in simple terms are composed of lines (*branches*), that intersect at internal nodes and *terminate* at external nodes at the tips of branches that represent the OTUs.

The tree topology is determined by the order in which the different branches bifurcate. A tree is described as rooted if there is a particular node (the root) from which a unique directional path leads to each OTU. By making *comparisons* of large numbers of characters (ideally in excess of 50), it is possible to define groups (*phena*) with the greatest overall similarity of characters.

The advantage of the numerical taxonomic approach is that it should avoid subjectivity if every character is given equal weight. The introduction of numerical (*statistical*) methods based on overall similarity or distance marked an important departure from the traditional approaches to taxonomy, which were more subjective and essentially concerned with the definition of monothetic taxa, in which certain characters were heavily weighted.

Initially, the data used for such numerical analyses relied mainly on traditional bi- or *multistate phenotypic* features derived from basic attributes of the organism such as *morphology,* and *physiological* and *biochemical* reactions. However. a problem of performing numerical analysis based only on conventional tests, is to have an adequate number of discriminatory tests.

This problem is exemplified by metabolically unreactive organisms such as *Campylobacter* spp., which typically give mostly negative results in metabolic tests. An attempt was made to circumvent this practical difficulty by using a range of alternative tests, such as determinations of carbon-source utilization, which in the case of *Campylobacter* have proved a specialized but nevertheless useful basis for numerical phenetic analysis.

Molecular Taxonomic Parametus

Because a phenetic classification is based on current observable characters, it can make use of macromolecular data, including genetic information, wherever available. From the mid-1960s, there were rapid developments in analytical chemical *techniques* and *instrumentation,* so that many new *methodologies* became available for the detailed chemical *analysis* of *microorganisms.*

The application in systematics of these new types of data is broadly referred to as *chemotaxonomy,* and the topic has been reviewed and documented extensively. The main categories of strain characteristics, particularly those based on the comparison of the structural properties of component *macromolecules* of cells are listed in Table 16.1.

Study of the chemical structure of individual cellular components has provided many valuable *chemotaxonomic* markers, at both the genus and species level. Such markers have included components of the cell wall (e.g., *pepti-doglycan, polysaccharides*) and cell membranes (e.g., *polar lipids, isoprenoid quinones*), and *metabolic* end-products such as cellular fatty acids.

Of these chemotaxonomic approaches, data derived from cellular fatty acid profiles have proved generally useful and applicable across many different *bacterial genera,* and *databases* have been developed that allow for rapid species identification.

However, in most instances, such data, while important in *descriptive* studies of taxa, do not

Table 16.1: Main Phenetic Characters Used in Bacterial Systematics and Their Applicability for Use in Numerical Analysis.

Conventional microbiological tests
One-dimensional (1D) electrophoretic protein profiling
Whole-cell fingerprinting (PyMS and/or GLC)
Fingerprinting of cell surface components (MALDI)v
Cell wall analysis
Cell membrane analysis
End products of metabolism (cellular fatty acid analysis)
DNA % G+C content and genome size
DNA-DNA hybridizationc
RAPD/ribotyping/PCR-RFLP fingerprints

lend themselves readily for use in numerical taxonomy and the generation of taxonomic trees. Nevertheless, certain types of chemical data have been applied in *numerical* taxonomy with the successful *generation* of phenetic trees as indicated in Table's 16.1 and 16.2.

Electrophoretic Analysis

One-dimensional SDS-polyacrylamide gel electrophoresis (PAGE) of whole-cell proteins has provided a widely used approach for *phenetic* analysis, particularly at the level of bacterial species. The technique, if performed under carefully *standardized* conditions, provides excellent data for *numerical* analysis.

Protein profiles typically comprise 30 to 40 discrete bands of varying intensity, which, after *staining* and drying of gels, can be scanned with a densitometer. The *absorbancies* are digitized, then stored electronically, and numerical analysis is performed with similarities between profiles being calculated using a correlation coefficient (Pearson *product-moment*).

Trees are constructed by a distance method, such as the *unweighted* pair group method with arithmetic mean (UPGMA). An example to illustrate the *application* of this approach used in the description of *H. pullorum,* a new species from human feces, is shown in the form of a rooted dendrogram in Figure elswhere in this chapter.

Dedicated analysis software packages, such as GelCompar (Applied Maths, Korttrijk, Belgium), are now available, that allow corrections for background and for *distortions* in the electrophoretic gel profiles.

Fingerprinting

The direct analysis of components of whole bacterial cells is another approach that can be used to generate an organism fingerprint. Favoured analytic methods for this purpose have included pyrolysis mass spectrometry (PyMS) and, more recently, Fourier-transform infrared (FT-IR) spectroscopy and *matrix*-assisted laser desorption time of flight (MALDI TOF) *spectrometry*.

Basically, the techniques involve whole-cell pyrolysis to generate volatile low molecular weight products, or ionization of cell-surface components, after the organisms are grown under carefully controlled conditions. Each method provides a complex fingerprint for each strain. The data are then subjected to numerical analysis, usually by ordination methods because of their complexity, to determine the key *discriminatory* features of OTUs that can be used in identification of taxa at the species and *subspecies* levels.

Ordination methods produce *multidimensional* diagrams and are non-hierarchic; they use multivariate analyses such as principle component analysis, and *canonical variate analysis*. While these approaches may have some merit as rapid means of strain identification, particularly in situations where high throughput of samples is required (e.g., in the analysis of food contaminants), they have not been developed and established as primary methods for taxonomic research or in clinical *microbiology*.

Genomic DNA Analysis

The second source of information, and the most fundamental for use in systematics, is provided by the investigation of cellular DNA and RNA-the socalled *genomic* approach. Development of this approach was greatly facilitated by the availability of rapid techniques for the isolation of bacterial

Table 16.2: Numerical Taxonomic Analyses Commonly Used to Infer Taxonomic Structure of Bacteria.

Phenetic analyses	Comment
• Dissimilarity/similarity matrix determined (Simple matching or Dice coefficients) + hierarchical cluster analysis by UJPGMA + representation (dendrogram)	Distance method applicable to conventional phenotypic test data
• Dissimilarity/similarity matrix determined (Pearson Product Moment Correlation Coefficient) + hierarchical cluster analysis by UPGMA + representation (dendrogram)	Applied to electrophoretic band profiles, e.g., protein or AFLP patterns
• Dissimilarity/similarity matrix determined (either as above) + multivariate analysis (principle component analysis) represen- (ordination diagram)	Usually applied to complex data sets, e.g., PyMS but also applicable to further detailed analysis of taxotation nomic structure evident from dendrograms to investigate relationships between individual OTUs
Phylogenetic inference analyses	
• Pairwise distance methods from multiple sequence alignments such as neighbor joining (NJ) + representation (rooted or unrooted tree)	Most widely used approach in bacteriology. Trees are constructed from a series of matrices and distances are directly calculated to internal nodes.
• As above, but with clustering by UPGMA equally distant from the root	Rarely used as the clustering assumes all taxa are
• Maximum parsimony methods	Rarely used in bacterial phylogenetics
• Maximum likelihood	

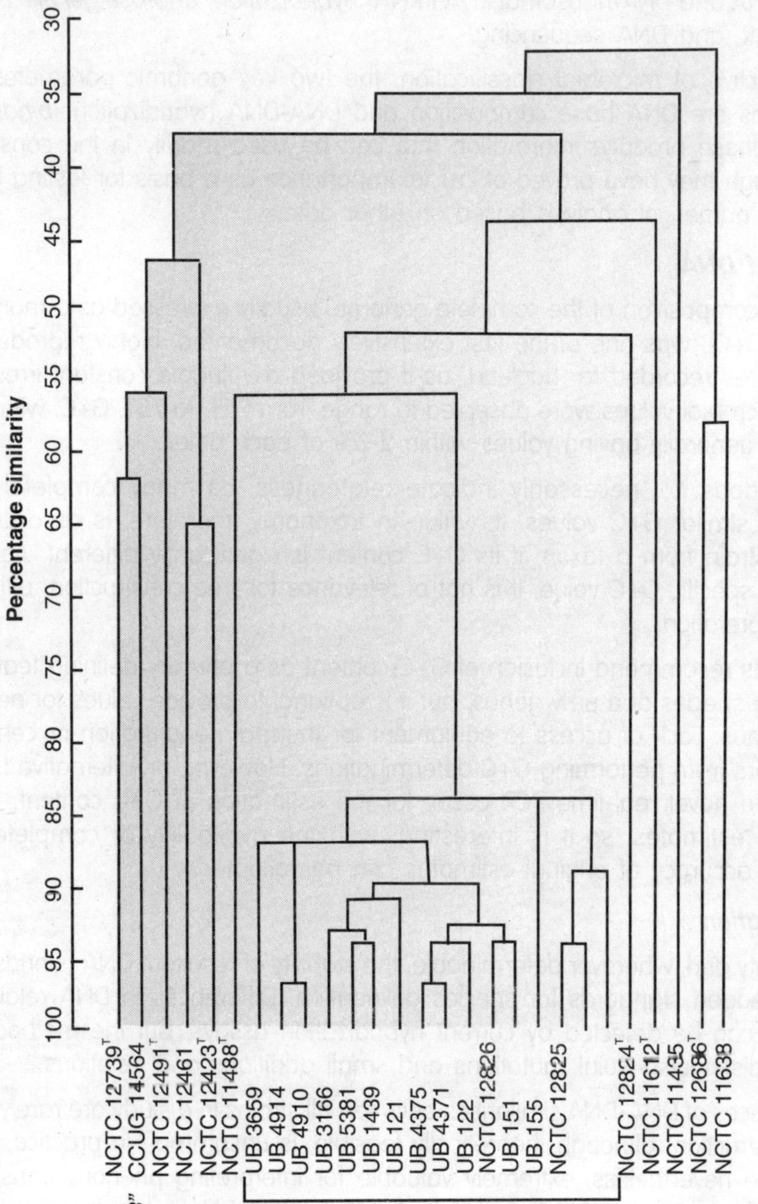

Figure 16.1: Dendrogram of numerical analysis of the total protein pattern of H. pullorum isolates and reference strains of other helicobacters and closely related organisms.

DNA and its analysis by a variety of methods, including buoyant density *centrifugation*, thermal denaturation, DNA-DNA and DNA-ribosomal RNA (rRNA) *hybridization, agarose gel electrophoresis, blot hybridization*, PCR, and DNA sequencing.

From the perspective of microbial classification, the two key genomic parameters used in phenetic classifications are DNA base composition and DNA-DNA hybridization (*binding*) levels. However, neither of these produce information that can be used readily in the construction of phenetic trees, although they have proved of crucial importance as a basis for testing the validity of taxa defined from numerical analysis based on other data.

Base Composition of DNA

The overall base composition of the complete genome, usually expressed as a mol% guanine plus cytosine (mol% G+C), was one of the first extensively documented, highly reproducible, and stable genomic features recorded for bacteria, as it provided a valuable constant irrespective of growth conditions. Microbial values were observed to range from 25% to 75% G+C, with members of the same species generally having values within 2-3% of each other.

The same G+C does not necessarily indicate relatedness, as many completely different organisms can have similar G+C values. Its value in taxonomy, therefore, is as a criterion on which to *exclude* a strain from a taxon, if its G+C content is significantly different. As individual strains have a single specific G+C value, it is not of relevance for tree construction, although it is of value in tree interpretation.

Minimal standards recommend inclusion of G+C content as a primary defining feature of the type strain of the type species of a new genus, but it is optional to provide values for new species of an established genus. Lack of access to equipment for thermal denaturation or centrifugation may deter investigators from performing G+C determinations. However, an alternative is provided by the availability of a novel real-time PCR assay for the estimation of G+C content. Most G+C values are empirical estimates, so it is interesting, with the availability of complete genome sequencing, that the accuracy of original estimates can be validated.

DNA-DNA Reassociation

DNA-DNA similarity and, wherever determinable, the stability of resultant DNA hybrids, currently provide the acknowledged standards for species delineation. Differences in DNA relatedness of between 5 and 10% can be detected by current hybridization assays, but the methods are not sensitive enough to distinguish point mutations and small additions and deletions.

As complete matrices of DNA-DNA similarities between all strains in a study are rarely available, their use in tree construction, although theoretically feasible, is uncommon in practice. DNA-DNA hybridization data are nevertheless, extremely valuable for interpreting phenons constructed by numerical methods. There are no definite rules to correlate DNA-DNA relatedness and levels in the taxonomic hierarchy, but the guidelines indicated below are now generally accepted:

- Hybridization levels of 70% or more, and low levels of mismatching (high thermal stability) indicate almost identical sequences and that strains should be classified in the same species.
- Hybridization levels of less than 20% indicate strains should belong in different species,

or even in different genera.

- DNA-DNA hybridization data between 20% and 70% are more difficult to interpret, and may depend on the method used and the particular group of *organisms.*

DNA-rRNA Hybridization

Saturation hybridization between rRNA and rRNA cistrons, which form a minor part of the total DNA (generally up to 0.4%), was an important tool for the elucidation of bacterial relationships at the intergeneric and *suprageneric* levels, prior to the widespread availability of DNA sequencing. In this technique, the thermal stabilities of the DNA-rRNA hybrids formed on membrane filters were used to establish *relationships* based on the parameter $T_{m(e)}$, the thermal melting point of the elution profile of the hybrids.

The values were used mainly for clustering different bacterial groups, with *groupings* presented typically in the form of two-dimensional similarity maps with Tm(e) *plotted* against % rRNA binding. The data were not suitable for use in numerical analyses, but were extremely valuable for *interpretation* of relationships inferred by such methods. The methods were complex and have now *largely* been replaced by direct sequencing of rRNA cistrons.

DNA Based Criteria

A wide range of DNA-based criteria have been used in identification of predefined taxa, or in the typing of strains within species, but most of these do not provide information of direct value in *systematics* and, with a few *exceptions,* are not suitable for numerical analysis. The two most widely used approaches from which *fragment* patterns (*fingerprints*) lend themselves for analysis by numerical methods are ribotyping and *amplified* fragment length polymorphism analysis (AFLP). Another fingerprinting approach that has proven taxonomic application is restriction fragment length *polymorphism* (RFLP) analysis of the 16S rRNA and 23S rRNA genes, where profiles may provide criteria for species-specific *identification,* although they have limited application in *numerical* analysis. The strength of RFLP-based approaches, however, is generally in sub-specific *discrimination* (strain typing), rather than in establishing interspecific relationships. The application of numerical methods to such data is considered elsewhere.

EVOLUTIONARY PATH

Phylogenetics is the study of the ancestry of organisms over long time scales, involving many thousands or even millions of years. In 1859 in *The Origin of Species,* Darwin described a tree-like structure to illustrate diversification of species. Then, in 1866, Haeckel proposed possibly the first *phylogeny* and classification of life illustrated in the form of a tree. *Traditionally* such studies in biology have *relied* on the availability of fossil records, but in the case of *bacteria* no such evidence was available.

Nevertheless, that did not deter occasional early speculative attempts based on purely descriptive data. For instance in 1921, Orla-Jensen proposed a phylogenetic scheme based on physiology, while in 1936, Kluyver and van Neil developed a scheme based on *morphology.* However, it was not until the 1960s that a *conceptual* leap was made, when Zuckerkandl and *Pauling* introduced the idea that macromolecules such as proteins can act as documents of evolutionary history, containing a record of changes since *divergence* from a common *ancestor.*

This concept marked the birth of molecular systematics in biology, which may now be defined as the study of phylogenies and processes of evolution by the analysis of DNA or *amino-acid* sequence data. This original idea was taken forward in 1980 by Kimura, who introduced the concept of the evolutionary clock, and it was some 7 yr later that Woese proposed the idea of an independent evolutionary clock *embedded* in the genotype for *prokaryotic* organisms, for the first time giving *microbiologists* the capacity to infer evolutionary histories and relationships.

Central to this approach was the selection of a suitable molecular *chronometer* that was not restricted to particular taxonomic levels or bacterial groups, and had not undergone *subtle* functional changes. As there is no fossil record for microbes, *phylogenetic* inferences depend almost entirely on the *indirect* evidence provided by the analysis of sequence data that are *considered* to express rates of genetic change with time.

It was proposed by Woese that the most useful molecular chronometers for microbial *taxonomy* were rRNAs, as they were universal, constant, often present in *multiple copies,* and had highly constrained functions that were not changed by the organism's environment. Furthermore, there were three types of rRNA present in all bacteria, namely 16S rRNA, comprising approx 1500 nt and found in the small (30S) *ribosomal subunit,* and 5S and 23S rRNA, comprising about 115 and 2900 nt, respectively, and both found in the large (50S) *ribosomal* subunit.

These macromolecules were easy to isolate, could be sequenced directly, and in the case of 16S and 23S rRNA, contained considerable taxonomic information that was typically present in multiple copies in the genome. As each *nucleotide* in the RNA represents a multiple-character test with four alternative states (namely the bases, A, G, C, and U), the sequences could be subjected to *rigorous* mathematical analysis according to the methods developed for *numerical* taxonomy.

Another advantage of rRNA is that, because of functional constraints, it is a nonlinear chronometer. The primary structures are alternating sequences, invariant, more or less conserved to highly variable regions. In addition, rRNAs are capable of forming higher-order structures (*helices*) by short- and longdistance *intramolecular* interactions involving inverse complementary sequence stretches.

The frequencies of compositional changes at different positions in the molecules vary greatly, so it is *feasible* to investigate phylogenetic relationships from the domain to the species level. Slowly evolving positions provide information on the deepest groups, whereas positions of higher *variability* may indicate more recent branchings.

Gathering and Sequencing Data

Since the 1980s, phylogenetic analysis based on oligonucleotide catalogs, then on sequences of 16S rRNA, and, more recently, on high-quality sequencing of 16S rRNA genes (rDNA) has become widely accepted as the reference method for defining phylogenetic lineages. In 1987, Woese presented the first *comprehensive* 16S rRNA phlyogenetic tree, and provided evidence for division of living organisms into three *primary domainsthe Eucarya,* the *Bacteria,* and the *Archaea.*

The phylogenetic trees based on 16S rRNA have led to the division of *Bacteria* into 18 major lines of descent (traditionally called phyla, but more recently, divisions), which fall into two primary and structurally distinct groups-the Gram-positives, which appear to be ancestral and structurally allied to the *Archaea,* and the Gram-negatives.

With the availability of PCR-based amplification/sequencing methods, the number of sequences has increased dramatically, with publicly accessible databases containing over 16,000 sequences of small-subunit sequences (http:/ /rrndb.cme.msu.edu/rrndb/ . Furthermore, species-specific probes and PCR assays targeting 16S rRNA genes are also widely used in detecting and identifying species in clinical samples.

Another major use of 16 rRNA gene sequences has been in exploring and surveying the diversity of environmental sources, by isolation of target sequences from bulk samples of environmental DNA. The importance of such analyses is that they provide the only means of determining the *phylogenetic* position of nonculturable organisms.

Architecture of 16S rRNA Tree

Why construct trees? The simple answer is that a tree is the most convenient way currently available of *representing* cluster analysis-derived relationships in two-dimensional space, particularly for the analysis of large amounts of complex data. No single tree is ideal, so it is important to appreciate, therefore, that trees are drawn to illustrate and to improve understanding of the biological relationships between the OTUs, although some information is lost in the *reductive* process.

This applies equally to both phenetic and phylogenetic trees. Trees are *essentially* attempts at the reconstruction of phylogenies, and should be viewed only as models of possible events in evolutionary history, as they are based on indirect *evidence* from sequence data with a number of assumptions. As there are numerous strategies for constructing *phylogenies,* many different trees may be obtained, depending on the *algorithm* used and the sequences included.

Furthermore, the number of possible trees increases dramatically as the number of taxa increases; it can be calculated that while four taxa could give three unrooted trees, fifty taxa could give 3×10^{74} possible trees.

The first step in extracting phylogenetic information from 16S rRNA genes is to make multiple alignments, *preferably* of high-quality nucleotide sequences. These may be new sequences as well as sequences downloaded from a public database. An ideal alignment procedure would ensure that homologous positions were placed in correspondence, in order to *maximize* their overall similarity. *Homology* is assumed when two sequences are similar due to common ancestry, but it should not be overlooked that two sequences could also be similar because they have evolved *independently* by chance (*homoplasy*).

In general, two main types of analysis are used to infer phylogenetic trees from sequence alignments, namely evolutionary distance methods such as UPGMA and Neighbour Joining, which are algorithmic in that they use a specific series of calculations to estimate a tree, and character-based (tree searching) methods, that include maximum likelihood, maximum parsimony, and Bayesian methods.

More detailed overviews of the methods and programs for analyzing DNA sequence data can be obtained elsewhere. In the case of 16S rDNA phylogenetic analysis, the most commonly adopted approach is to construct a distance matrix from all aligned sequences, where distances are expressed as the fraction of sites that differ in the alignment.

Corrections for multiple base changes are applied, as the number of observed differences between DNA sequences may be underestimated, due to a number of superimposed substitutions that may have occurred during the evolution of the lineages from a common ancestor. In a simple model with only one parameter, such as that of Jukes and Cantor, it is assumed that all four bases occur at equal frequencies, and that all types of substitutions are equally likely.

The phylogenetic trees are then constructed from the distance matrix; in the case of 16S rRNA gene sequences, the technique usually applied is neighbour-joining, in which the tree is reconstructed from a series of matrices. An example of this approach for type species of the genus *Helicobacter is* shown in Figure elsewher in this chapter.

Reliability of Gene Trees

The reconstruction of evolutionary trees using sequence divergence from a common ancestor is based on a number of assumptions about substitution rates, irrespective of the analytical methods. A key question for any particular method and dataset is how reliable a representation is the tree in terms of its topology? It is important that the tree should allow us to see the order in which existing taxa diverged from a hypothetical common ancestor.

Branch lengths are then proportional to evolutionary distance, and distances are typically expressed as the number or fractions of substitutions per site. Tree reliability is usually assessed by means of nonparametric bootstrapping, in which data are resampled to give a series of pseudoalignments.

Typically 100 to 1000 bootstrap repetitions are required. The value at each node then indicates the number of times the Glade occurred in the bootstrap replicates, and provides a measure of confidence in the Glade.

Interpretation and Validation

The interpretation and validation of 16S rRNA phylogenies in terms of their biological relevance can be facilitated by direct comparison with other categories of data, such as those derived from phenetic analysis that could include conventional bacteriological tests, protein profiles, or DNA-DNA hybridizations. Application of a comparative approach is feasible, however, only where relevant information is available, which is usually at species and genus levels.

Comparative studies using alternative taxonomic criteria may yield results that are ambiguous or that apparently contradict the 16S rRNA phylogenies and cast doubt on the reliability of that approach in certain applications. A major limitation of 16S rRNA phylogeny is the lack of discrimination between closely related species within genera.

This problem has been noted from studies of a number of genera, and is illustrated by the analysis of the various species classified in the genus *Helicobacter,* within which there is a continuum without any evidence of a distinct phylogenetic substructure, despite clear distinctions in phenotypic characteristics and host specificities.

Another limitation in the application of 16S rRNA phylogeny is that sequences are generally available only for single isolates of each species, often the type strain of the species, which may not actually be representative of that species.

For instance, 16S rRNA phylogenetic analysis of multiple strains delineated *Campylobacter hyointestinalis* into two principal groups corresponding to subspecies *lawsonii* and *hyointestinalis,*

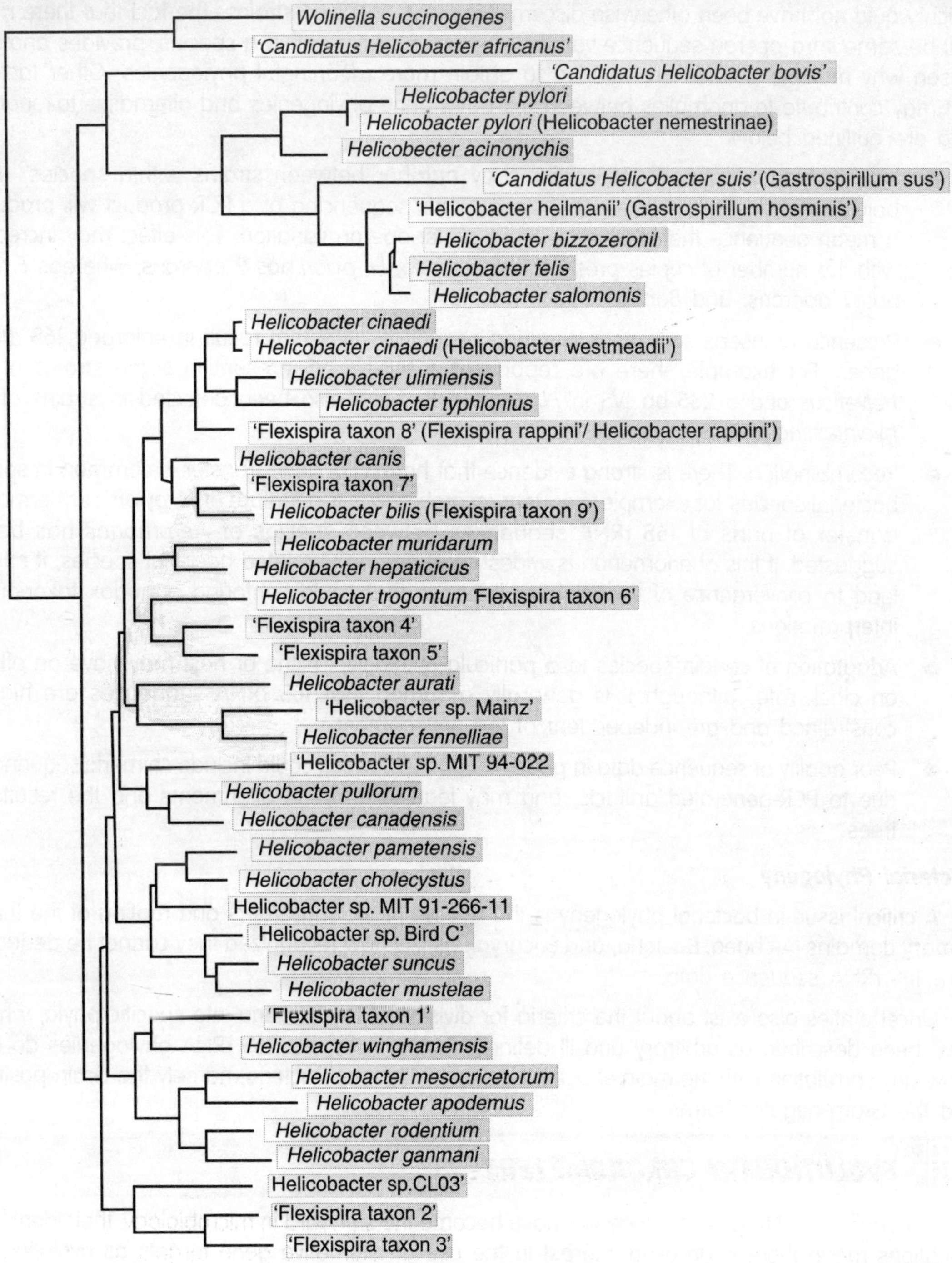

Figure 16.2: Phylogenetic tree of members of the genus Helicobacter based on 16S rDNA sequences.

which would not have been otherwise discernable using only type strains. The fact that there may well be some intra-operon sequence variation between strains within a species, provides another reason why multiple isolates are needed to obtain more meaningful phylogenies. Other factors that may contribute to anomalies between 16S rRNA gene phylogenies and alternative taxonomic data are outlined below:

- Inter-operon variation in 16S rRNA copy number between strains within species, and between strains within different species. Direct sequencing of a PCR product will produce a mean sequence that may mask subtle inter-operon variation. This effect may increase with the number of copies present. For instance, *H. pylori* has 2 operons, whereas *E. coli* has 7 operons, and *Bacillus cereus* has 12.
- Presence of inserts such as intervening sequences (IVS) that result in enlarged 16S rRNA genes. For example, there are reports of a 148-bp IVS present in some strains of *C. helveticus* and a 235-bp IVS in *H. canis.* IVSs have also been detected in strains of *C. hyointestinalis, C. sputorum* and *C. rectus.*
- Recombination. There is strong evidence that horizontal gene transfer is common in some bacterial species-for example, frequent recombination is a feature of *H. pylori.* Furthermore, transfer of parts of 16S rRNA sequences between species of *Aeromonas* has been suggested. If this phenomenon is widespread among unrelated bacterial species, it might lead to convergence of evolutionary lineages that could confound orthodox taxonomic interpretations.
- Adaptation of certain species to a particular ecological niche or host may have an effect on clock rate, although it is generally assumed that 16S rRNA sequences are highly constrained and are independent of the environment.
- Poor quality of sequence data in public databases, which could include chimeric sequences due to PCR-generated artifacts, and may lead to errors in alignments and the resultant trees.

Bacterial Phylogeny

A critical issue in bacterial phylogeny is the relative branching orders and rooting of the three primary domains (Archaea, Bacteria, and Eucarya), as it is now recognized they cannot be deduced from 16S rRNA sequence data.

Uncertainties also exist about the criteria for division of the Bacteria into specific phyla, which have been described as arbitrary and ill defined. Furthermore, the 16S rRNA phylogenies do not show any correlation with the main structural groups within the Bacteria, namely the Grain-positive and the Gram-negative forms.

EVOLUTIONARY CHRONOMETERS

Although 16S rRNA gene phylogenies have become the standard in microbiology, their identified limitations mean there is growing interest in the use of alternative gene targets as evolutionary chronometers. The expansion in datasets from the growing number of genomic sequencing projects has raised the possibility of identifying alternative markers for inferring phylogenetic relationships.

New markers might avoid the dependence on 16S rRNA and the assumption that its content reflects the evolutionary history of an organism. For example, protein-encoding genes may evolve faster than rDNA, and phylogenies based on such genes may have a higher resolution than those based only on 16S rRNA sequences. Such analyses can then be used to test whether the gene tree corresponds to the species tree.

In an attempt to overcome some of the limitations of 16S rRNA phylogeny, Gupta proposed a novel approach that involved identification of conserved inserts and deletions (indel signature sequences) in genes encoding various proteins. Results from analysis of some 60 completed bacterial genomes provide strong evidence that the phylogenetic placements and branching orders of taxa deduced by indel analysis are highly reliable and internally consistent.

The indels are in genes that encode key proteins involved in highly conserved housekeeping functions, such as glutamate semialdehyde aminomutase, the Hsp 60 and Hsp 70 heat shock proteins, 5-aminoimidazole-4-carboxamide formyltransferase, and phosphoribosyl pyrophosphate synthetase.

The rapidly expanding datasets of multilocus sequence types will provide an important source for identifying polymorphisms in housekeeping gene sequences, which may be linked to nonsynonymous mutations that might provide valuable phylogenetic as well as typing markers (http://www.mlst.net/new).

Sequencing of 23S rRNA

In theory, the 23S rRNA, with a size of about 3 kb, should be the most informative marker, as it is estimated that under optimal circumstances, it should provide a resolution of one evolutionary event per 0.2-0.3 million years. However, there are only a limited number of sequences of this gene available for analysis, compared with the many now available for the smaller 16S rRNA gene.

Nevertheless, despite the differences in the degree of information content, detailed and comprehensive phylogenies inferred from equivalent datasets of both subunit sequences are in good agreement for the three primary domains, and for phyla within the domain Bacteria. Interestingly, the branches within the 23S rRNA tree were generally longer than those of the 16S rRNA tree, indicating a higher degree of variability in the large subunit.

The taxonomic information in the large subunit sequences is demonstrated by its use in a rapid identification scheme for *Helicobacter* species based on RFLP analysis. However, 23S rRNA genes of several bacterial genera, including *Bacteroides, Salmonella, Leptospira, Yersinia, Coxiella, Helicobacter,* and *Campylobacter,* have been shown to contain IVSs, and the effect of these enlarged genes on phylogenic analysis has not yet been established.

DNA Gyrase B Subunit Gene

PCR amplification and sequencing of the DNA gyrase B subunit gene *(gyrB)* of *Acinetobacter* strains provided a greater degree of resolution than that based on 16S rDNA. Phylogenetic trees showing the relationships of various genera and species in the Enterobacteriaceae were also constructed from analysis of *gyrB,* and compared with 16S rRNA genes.

Differences between the gene trees were noted in areas of low resolution or low bootstrap

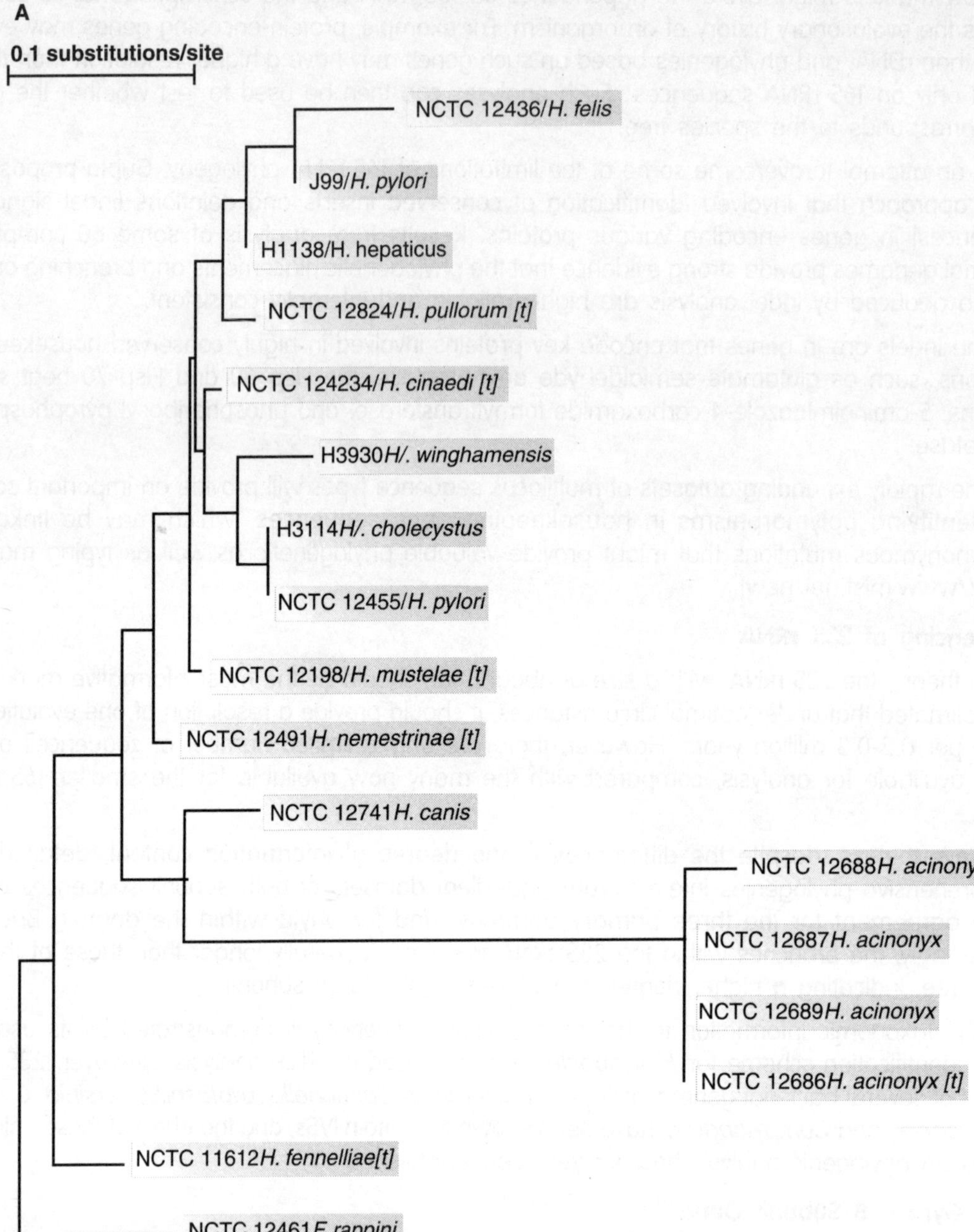

Figure 16.3: Phylogenetic trees based on ureI DNA sequence (A) and amino acid sequence alignments (B, next page) of species within Helicobacter.

B

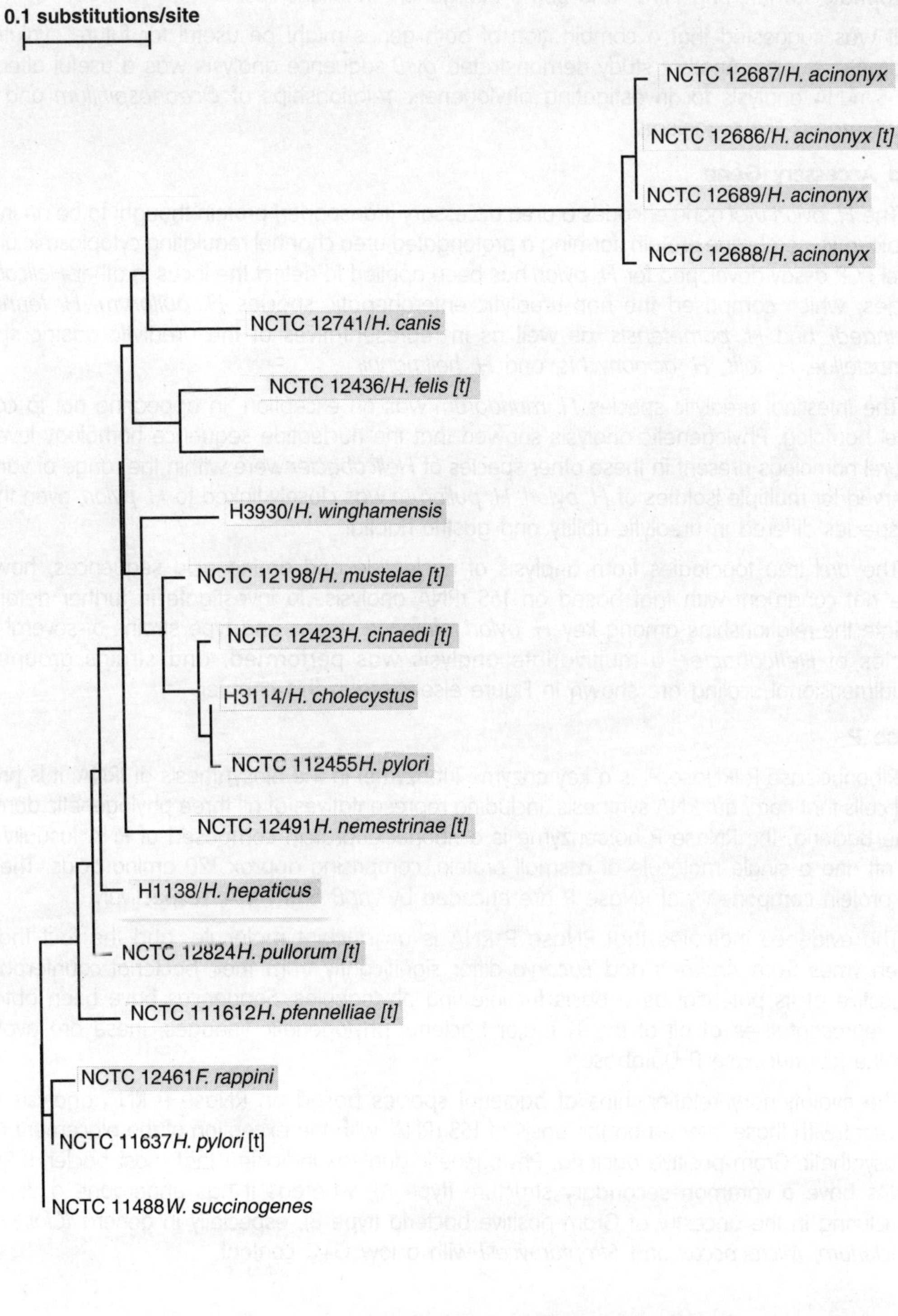
0.1 substitutions/site
NCTC 12687/H. acinonyx
NCTC 12686/H. acinonyx [t]
NCTC 12689/H. acinonyx
NCTC 12688/H. acinonyx
NCTC 12741/H. canis
NCTC 12436/H. felis [t]
H3930/H. winghamensis
NCTC 12198/H. mustelae [t]
NCTC 12423H. cinaedi [t]
H3114/H. cholecystus
NCTC 112455H. pylori
NCTC 12491H. nemestrinae [t]
H1138/H. hepaticus
NCTC 12824H. pullorum [t]
NCTC 111612H. pfennelliae [t]
NCTC 12461F. rappini
NCTC 11637H. pylori [t]
NCTC 11488W. succinogenes

values for one or both of the molecules. Overall, the 16S rRNA sequences were useful for describing relationships between distantly related taxa, whereas *gyrB* sequence comparisons were more appropriate for inferring intra- and some intergeneric relationships.

It was suggested that a combination of both genes might be useful for future evolutionary systematic studies. Another study demonstrated *gyrB* sequence analysis was a useful alternative to 16S rRNA analysis for investigating phylogenetic relationships of *Oceanospirillum* and other *Pseudomonas-like* organisms.

Urea Accessery Gene

The *H. pylori ureI* gene encodes a urea accessory (transporter) protein thought to be an integral cytoplasmic membrane protein, forming a protongated urea channel regulating cytoplasmic urease. *A ure*I PCR assay developed for *H. pylori* has been applied to detect the locus in other *Helicobacter* species, which comprised the non-urealytic enterohepatic species *H. pullorum, H. fennelliae, H. cinaedi,* and *H. pametensis,* as well as in representatives of the urealytic gastric species *H. mustellae, H. felis, H. acinonychis,* and *H. heilmannii.*

The intestinal urealytic species *H. muridarum* was an exception, in appearing not to contain *a ure*I homolog. Phylogenetic analysis showed that the nucleotide sequence homology levels for the *ure*I homologs present in these other species of *Helicobacter* were within the range of variation observed for multiple isolates of *H. pylori. H. pullorum* was closely linked to *H. pylori,* even though the species differed in urealytic ability and gastric habitat.

The *ure*I tree topologies from analysis of nucleotide and amino acid sequences, however, were not congruent with that based on 16S rRNA analysis. To investigate in further detail and validate the relationships among key *H. pylori* reference strains and type strains of several other species of *Helicobacter,* a multivariate analysis was performed, and strains grouped by multidimensional scaling are shown in Figure elsewhere in this chapter.

RNase P

Ribonuclease P (RNase P) is a key enzyme (ribozyme) in the biosynthesis of tRNA; it is present in all cells that carry out RNA synthesis, including representatives of all three phylogenetic domains. In the Bacteria, the RNase P holoenzyme is a ribonucleoprotein composed of RNA (usually 350-400 nt) and a single molecule of a small protein, comprising approx 120 amino acids. The RNA and protein components of RNase P are encoded by *rnpB* and *rnpA,* respectively.

The evidence indicates that RNase P RNA is an ancient molecule, and the fact that the holoenzymes from *Archaea* and *Eucarya* differ significantly from their bacterial counterparts is suggestive of its potential as a basis for inferring phylogenies. Sequences have been obtained from representatives of all of the 11 major bacterial phylogenetic lineages; these are available from the Ribonuclease P Database

The evolutionary relationships of bacterial species based on RNase P RNA analysis were consistent with those inferred on the basis of 16S rRNA, with the exception of the placement of the photosynthetic Gram-positive bacteria. Phylogenetic analysis indicated that most bacterial RNase P RNAs have a common secondary structure (type A), whereas it has undergone a dramatic restructuring in the ancestry of Gram-positive bacteria (type B), especially in genera *(Clostridium, Eubacterium, Micrococcus* and *Streptomyces)* with a low G+C content.

Interestingly, each genus was shown to have an RNase P RNA with a distinctive secondary structure. Conserved and variable domains were detected within the *rnpB* sequences of members of the *Prochlorococcus* group. The RNase P RNA and 16S rRNA phylogenetic trees were not fully congruent, but the application of *rnpB* sequences allowed a better resolution between clades of very closely related genotypes.

Other Protein Encoding Genes

Other protein-encoding genes have been investigated with regard to their potential for construction of bacterial phylogenetic trees; they include the genes encoding *recA* elongation factors, ATPase subunits, RNA polymerases, translation initiation factor *(infB),* and *groEL* and other heat shock proteins. As an example, *groEL* has been used in the classification of *Bartonella spp.*

THE POLYPHASIC PATH

The term *polyphasic* was introduced by Colwell to refer to classifications based on a consensus of all available data (both phenotypic and genomic), and would now be considered analogous to phenetic classification.

With the growth of molecular phylogenetics and the recognition of its key role in classification, it was subsequently suggested by Vandamme and colleagues that polyphasic classification should be reinterpreted to include phylogenetic data, although it was left open how it should be best applied to derive relationships at the different taxonomic levels.

The polyphasic and pragmatic species definition was recently reevaluated, in view of many recent innovations, and it was concluded that the present system was generally sound and that the community was served well by the bacterial species defined by Rossello-Mora and Amann as "a category that circumscribes a (preferably) genomically coherent group of individual isolates/ strains sharing a high degree of similarity in (many) independent features comparatively tested under highly standardized conditions."

CONCLUDING REMARKS

Over the past 20 yr, application of numerical taxonomy to many different categories of bacterial strain data has resulted in the generation of a multiplicity of taxonomic trees, with topologies that vary depending on the computational methods used and the input data. Significantly, 16S rRNA gene trees have provided the mainstay of efforts in reconstructing bacterial phylogeny and, as a result, have provided many novel insights and significant advances in our understanding of the classification of bacteria that were not apparent by the application of phenetics alone.

Nevertheless, there is evidence that 16S rRNA gene phylogenies have limitations, which indicate that future strategies should not be reliant on equating singlegene evolution with species evolution. A single gene may be misleading if constraints or specific selections differ between organisms. A more rigorous test should be for congruence between trees constructed with data for several genes encoding different cellular functions, which should be combined to build a "species phylogenetic tree."

This would be consistent with a polyphasic taxonomic approach. The selection of alternative

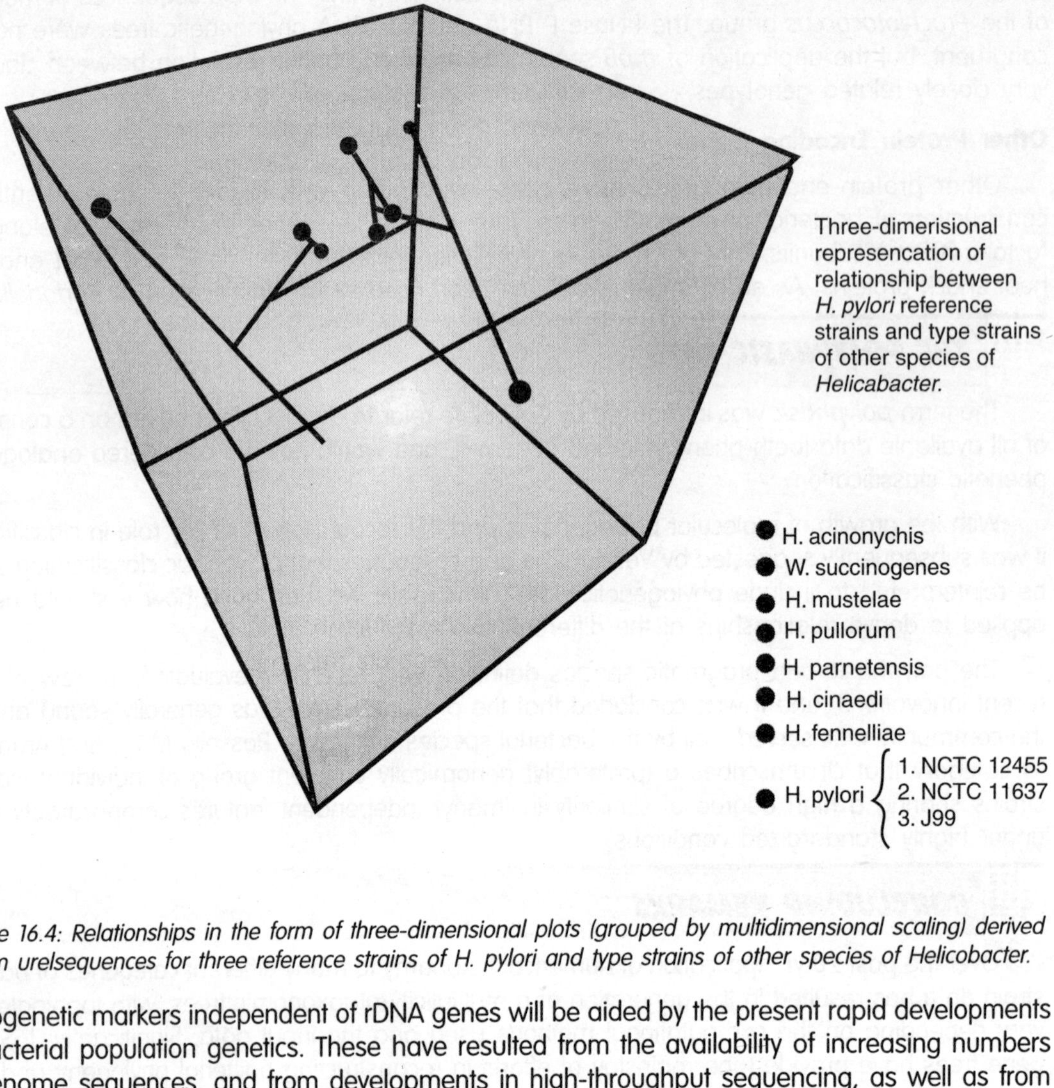

Figure 16.4: Relationships in the form of three-dimensional plots (grouped by multidimensional scaling) derived from urelsequences for three reference strains of H. pylori and type strains of other species of Helicobacter.

phylogenetic markers independent of rDNA genes will be aided by the present rapid developments in bacterial population genetics. These have resulted from the availability of increasing numbers of genome sequences, and from developments in high-throughput sequencing, as well as from new algorithms for analysis of population structure from multiple-sequence alignments, such as Markov chain, Monte Carlo, and Bayesian inference.

The availability of sequences on more strains and longer sequences, particularly of genes that have optimal rates of evolution for making inferences, will increase the robustness of phylogenetic estimates. Bacterial taxa at all levels appear to be currently defined largely by activities dependent on the presence or absence of genes relative to other taxa.

The availability of increased numbers of complete genomes, and of tools for comparative genome analysis, such as DNA microarrays, should facilitate the development of the species genome concept based on the presence of a core set of known genes. These could be used in combination with measures of the extent of polymorphism within genes and also within amino acid sequences, so that rates of synonymous and nonsynonymous substitutions are taken into account.

This approach could provide a new way of defining bacterial taxa that would avoid reliance on a single 16S rDNA sequence, usually from one strain. Robust methods of statistical analysis are an essential requirement for obtaining the best phylogenetic estimates, and although there are many software packages available, there is a need for continued development and application of improved phylogenetic methodologies.

Chapter 17

Molecular Typing

The availability of the whole genome of number of bacteria has opened the way for the evaluation of repetitive DNA motifs and their application to epidemiological investigation. Large proportion of genomic DNA consist of repeats, where DNA motifs exist in multiple copies. The repeats may vary in size, location, complexity, repeat mode, and are designated as belonging to different repeat classes such as direct repeats, dyad repeats, and inverted repeats.

Short DNA tandem repeats have been recognized as an important mechanism for controlling the expression of some bacteria outer membrane proteins. Furthermore, short DNA sequence repeats may be involved in coding or in the promoter area of genes associated in the biosynthesis of surface antigens.

Epidemiological analysis of infections, caused by pathogenic bacteria, is dependent on the precise identification at the strain level. During the last 10 yr, hypervariable minisattelite regions have been utilized to detect DNA fingerprints in the human gnome and other eukaryotes like *Candida albicans, Candida crusei, Leishmania infantrum, Sphaeropsis sapinea.*

During the last 6 yr VNTR has been used for molecular typing of several bacterial species including *Bacillus anthracis, Yersinia pestis, Mycobactrium tuberculosis, Haemophilus influenza, Francisella tuleransis, Xyllela fastidosa, Legionella pneumophilia, Staphylococcus aureus, Salmonella enterica* subsp. *enterica* serovars *typhimurium, Escherichia coli 0157,* and *Neisseria meningitidis.*

The genome of a particular bacterium can be screened for repetitive DNA with the Tandem Repeats Finder, which is available for free at www.tandem.bu.edu/tools.html. The program generates an output file giving the repeat location, the repeat segment size, the nucleotide composition, and the copy number of the array. PCR primers can be designed from the sequences flanking the repeats of the loci.

Primers may be designed by free software available at www.biotools.umassmed.edu/biopass/primer3 www.cgi. In a VNTR typing assay, the length of PCR products is measured. From this length, the number of units are deduced for each locus, and this list of values for a given strain is the genotype. It should be emphasized that some loci may have low/no polymor-phism in a genetically diverse set of isolates. Therefore, each locus may be evaluated for polymorphism before it can be used in genotyping. PCR may be run either singly, using one primer set or multiplex using more than one primer sets. The PCR-products may be pooled together into a single tube for each sample and thoroughly mixed. The PCR-mix may be run either on an agarose gel or by capillary electrophoresis.

An Internet-based resource (www.minisattelites.u-psud.fr) has recently been developed, which takes advantage of available genome sequences to help develop and run tandem-repeat based strain identification. This database allows finding of VNTR regions in sequenced organisms, comparing the VNTR loci between isolates or species and identification of similar and dissimilar VNTR regions.

The website makes possible searches for primer matches in whole genome sequences or restricted sequences associated with minisatellites. This website, as an online resource, greatly facilitates the work for development of VNTR-based typing systems. We present, in this chapter, the gel-based VNTR typing system for *Neisseria meningitides* and multiple-locus variant analysis of *Salmonella typhimurium* using capillary electrophoresis.

REQUIREMENTS

VNTR Analysis

1. Chocolate agar plate for growth of the bacteria (Oxoid), store at 4°C.
2. TE-buffer: 1 M Tris-EDTA, pH 8.0, store at 4°C.
3. dH_2O: store at 4°C.
4. PCR-primers: the oligionucleotide primers for the PCR reactions can be ordered from any commercial primer supplier, store at -20°C.
5. PCR kit (Applied Biosystems): the kit contains premixed dNTPs, DNA polymerase, PCR-buffer, ddH_2O, store at -20°C.
6. Gene Amp 9700 PCR system (Applied Biosystems).
7. 5X TBE buffer: TBE powder (AppliChem) is dissolved in dH_2O, autoclaved at 121°C for 15

Table 17.1: PCR Primers for VNTR Analysis of *Neisseria meningitidis*.

Primer	Primer sequence (5'-3')	Annealing temperature
VNTR01F	GACGGGTCAAAAGACGGAAG	59°C
VNTR01R	GGCATAATCCTTTCAAAACTTCG	
VNTR02F	CTCCCCGATAGGCCCCGAAATACC	57°C
VNTR02R	AAAGCGGCGGGAATGACGAAGAGT	
VNTR08F	GACCCTGACCGTCGGAAACC	57°C
VNTR08R	ATACCGCCCTGCTGTTGTGC	

Table 17.2: PCR Primers for VNTR Analysis of *Salmonella typhimurium*.

Primer name	(Dye name) Sequence (5'-3')	Annealing temperature
STTR3-	(HEX)-CCCCCTAAGCCCGATAATGG	61°C
STTR3-R F	TGACGCCGTTGCTGAAGGTAATAA	
STTR5-F	(HEX)-ATGGCGAGGCGAGCAGCAGT	58°C
STTR5-R	GGTCAGGCCGAATAGCAGGAT	
STTR6-F	(6FAM)-TCGGGCATGCGTTGAAA	54°C
STTR6-R	CTGGTGGGGAGAATGACTGG	
STTR9-F	(6FAM)-AGAGGCGCTGCGATTGACGATA	54°C
STTR9-R	CATTTTCCACAGCGGCAGTTTTTC	
STTR10-F	(TET)-CGGGCGCGGCTGGAGTATTTG	62°C
STTR 10-R	GAAGGGGCCGGGCAGAGACAGC	

min, store at 4°C.

8. Loading dye: dissolve 100 mg Evans blue (Sigma) in 10 mL dH_2O and 75 mL glycerol (Sigma). Store at 4°C.
9. Agarose: SeaKem ME agarose (Cambrex Bio Science), store at room temperature.
10. Gel Electrophoresis apparatus and Power Supplier Pack 3000 (Bio-Rad) and gel tray.
11. DNA molecular weight marker: 4μL 1 kb DNA-ladder (InVitrogen), 20μL loading dye and 76 μL dH_2O, store at 4°C.
12. Ethidium-bromide (Sigma): prepare 10 mg/mL by mixing undiluted ethidium-bromide in dH_2O, store at room temperature.
13. Photography apparatus: Polaroid MP4+(Polaroid). Polaroid Film should be stored at 4°C.
14. Ultraviolet (UV)-illumination apparatus: TM36 Model (KEBO-LAB, Norway)
15. Software (Applied Maths BVBA, Sint-Martens-Latem, Belgium) for data analysis and generation of dendrogram.

Capillary Electrophoresis-Based VNTR

1. Luria-Bertani (LB) agar plate (Difco).
2. ddH_2O.
3. Oligonucleotide primers: the oligionucleotide primers for the PCR reactions can be ordered from any commercial primer supplier. The forward (F) primer is 5'-labeled with a fluorescent dye (HEX, 6FAM, or TET).
4. Multiplex PCR-kit (Qiagen) the kit contains pre-mixed dNTPs, and DNA polymerase, PCR

amplification buffer, and ddH_2O.

5. ABI-310 Genetic Analyzer, POP4-polymer for 310 genetic analyzer, 10X Genetic Analyzer buffer with EDTA, 310 capillary 47 cm (all Applied- Biosystems), Geneflo-625-TAMRA internal size marker (CHIMERx), Formamide (Applied-Biosystems).

PROCEDURES

Gel-Based VNTR

Isolation of Bacterial DNA

1. Plate meningococcal isolates on chocolate agar and incubate overnight at 37 °C in a 5% CO_2 atmosphere.
2. Remove a loopful (10μ) of bacterial growth from the plate and suspend in 100 μL of 1 M TE buffer, pH 8.0, in an Eppendorf tube.
3. Boil the bacterial suspension for 10 min.
4. Centrifuge the boiled suspension at 6000 rpm for 10 min.
5. Transfer the supernatant to a new tube and use for further analyses.
6. The supernatant containing DNA may be stored at -70 °C.

Gel Electrophoresis

1. Make a list of samples, including negative and positive controls.
2. Prepare a reaction mixture of a volume 49 μL. The reaction mixture contains: lox PCR buffer 5 μL, deoxyribonucleoside triphosphate (dATP, dTTP, dCTP, and dGTP), 200 mM: 4 μL, Primer forward 100 pM: 1 μL, Primer reveres 100 pM: 1 μL, AmpliTaq polymerase 2.5 U/μL: 0.5 μL, dH_2O: 37.5 μL. Depending on the number of samples including negative and positive samples, prepare a master-mixture and dispense 49 μL to each sterile microfuge tube, which is appropriate to your thermocycler. Label the microfuge tubes.
3. Add 1 μL of supernatant containing DNA.
4. Turn the thermal cycler, place the microfuges and run: initial denaturing at 94 °C for 5 min, and 32 cycles of: denaturing: 94°C for 1 min, annealing: °C depending on primer set for 1 min, extension: 72°C for 1 min 30 s, and finally an extension step of 72°C for 5 min.
5. Mix 5 μL of each of the PCR products, resulting from the amplification of the three VNTR for each isolate, together into a single tube. Using the micropipet with a clean tip and add 5 μL of loading dye to the each PCR-product pool.
6. Seal both ends of the gel tray with tape or stoppers.
7. Make sure that one comb is in place at the negative electrode (black end of the gel).
8. Mix 4 g SeaKem ME agarose in 200 mL 1X TBE buffer and boil in a microwave until all agarose is melted. Leave at room temperature for 20-30 min.
9. Pour the melted agarose into the gel space. Let the agarose harden, which should takes 1 h at room temperature.

10. When the gel has hardened, remove the tape or stoppers.
11. Load both the pooled PCR-reactions and standard DNA markers samples into the gel.
12. Pour 1X TBE buffer carefully so it fills the electrophoresis apparatus and just cover the gel.
13. Run the gel at 40 V for 18 h (overnight), at room temperature.
14. Turn off the power supply, disconnect the electrodes, and remove the top of the electrophoresis apparatus.
15. Place gel in staining tray and stain the gel in ethidium bromide for 10 min, destain in water for 20 min. Both steps occur at room temperature.
16. Take a picture of the gel under UV illumination.

Data Analysis

The gel photograph can be analyzed using the BioNumerics Software (Applied Maths BVBA, Sint-Martens-Latem, Belgium) and a dendrogram can be constructed using the Dice coefficient of similarity and cluster analysis with the unweighted pair-group method with arithmetic averages. Both the position tolerance and the optimization can be set at 1%, as recommended in the BioNumerics Software manual.

Growth and Lysis in Bacteria

1. Streak bacteria on LB agar plates and grow at 37°C overnight in an incubator.
2. Transfer fresh *S. typhimurium* cultures, grown overnight on a LB plate, with an inoculation loop, to a microcentrifuge tube containing 300 μL of ddH_2O.
3. Vortex the cells briefly until they are evenly suspended.
4. Boil the bacterial suspension for 15 min and then centrifuge at 960g for 3 min.
5. Transfer the supernatant to a new tube for further analysis.
6. The supernatant containing DNA may be stored at -70°C.

Multiplexed PCR Reactions

1. Multiplex the primers in two solutions using the Qiagen multiplex-PCR kit in a total of 50μL as described by manufacturer:

    a. Multiplex 1 (M 1): 10 pmol of the STTR3 and STTR6 primer pairs.

    b. Multiplex 2 (M2): 10 pmol of the STTR5, STTR9 and STTR10 primer pairs.

2. Amplify both multiplex solutions with 2 μL of the supernatant containing DNA and under the following cycling conditions on a GeneAmp 9700 thermal cycler:

    1. Initial denaturing at 95°C for 15 min, and then 25 cycles of:
    2. Denaturing: 94°C for 30s.
    3. Annealing: 63°C for 90 s.
    4. Extension: 72°C for 90 s, and ending an extension step of 72°C for 10 min.

Sampling

1. Before capillary electrophoresis pool the two multiplexed PCR reactions as follows: mix 10 µL of M1 + 2.5 µL of M2, and then add 87.5 µL of ddH_2O to a total of 100 µL. This constitutes the run-mix (RM).

Capillary Electrophoresis

1. Prepare the samples as follows before capillary electrophoresis: 1 µL of the RM is mixed with 1 µL of the internal size-standard, Geneflo-625-TAMRA (CHIMERx), and 12 µL of formamide.
2. Denature the samples at 94°C for 2 min and then cool to room temperature.
3. For capillary electrophoresis the ABI-310 Genetic Analyzer is used. The electrophoresis is run at 60°C for 35 min using POP4-polymer with an injection voltage of 15 kV for 5 s, and a running voltage of 15 kV.

Data Analysis

The resulting electropherogram will typically look like Figure elsewhere in this chapter when all loci are present. Each peak is a PCR product and is equivalent to a "band" on an agarose gel. Loci STTR6 and STTR9 labeled with the dye 6FAM are depicted in blue, loci STTR3 and STTR5 labeled with the dye HEX are depicted in black, and locus STTR10 labeled with the dye TET is depicted in green.

Sizes in basepairs are automatically recorded by the ABI-310 instrument using GeneScan (Applied-Biosystems) software. The internal size-standard, labeled with the dye TAMRA, is depicted in red. This pattern will vary dependent on the number of tandem repeats at each VNTR locus in the strains under study, and give each isolate its own distinct DNA-fingerprint. Variation in size according to multiples of the repeat is interpreted as separate alleles.

BioNumerics v3.5 (Applied-Maths) can be used to enter PCR amplicon sizes, or sizes converted to allele numbers, recorded at each locus, and to draw dendrograms in which the pattern similarities are visualized, and matching profiles can rapidly be identified. The MLVA profiles can rapidly be shared among laboratories using email.

POINTS TO REMEMBER

Gel-Based VNTR

1. Use the micropipet and sterile tips, sterile microfuges.
2. The 10X PCR buffer, dNTP and primers are held at room temperature for 30 min and then placed on ice.
3. Keep the enzyme on the ice.
4. Keep the water at room temperature.
5. Keep the master-mix on ice.
6. Use separate room for adding DNA. Thaw the supernatant containing DNA at room temperature prior adding to reaction mixture, keep the samples on ice prior running PCR.

High concentrations of DNA may result in no PCR-products. Dilute DNA in dH_2O

7. Because loading dye contain Evans blue (toxic and may cause cancer) and glycerol (harmful), it is important to use gloves. PCR product may be reduced to 2.5 µL if the PCR products have a high intensity.
8. Use a comb allowing application of at least 25 µL volume.
9. Do not leave the melted agarose at room temperature longer than 30 min and be sure the agarose has not begun to harden.
10. Do not touch or move the gel until it is hardened.
11. Load the first and the last well with standard DNA marker and be sure you keep track of which samples are loaded into which wells.
12. Let the gel run until the dye migrates approximately to the other end of the gel.
13. Because ethidium bromide is mutagenic, it is important to use gloves.
14. Use safety glasses to protect your eyes from UV light.

Capillary Electrophoresis-Based VNTR

15. Use micropipets, sterile filter tips and sterile microfuge tubes.
16. Thaw the PCR-multiplex kit and primer solutions on ice.
17. Keep the bacterial lysate on ice.
18. Use a separate room for adding the bacterial lysate containing the DNA.
19. Dispense the formamide into microfuge tubes and store at -200C until needed, discard any unused formamide.
20. Use chemically resistant disposable gloves, lab coat, and safety glasses when handling the POP4 polymer.

Other Points to Remember

21. Extra bands. Some VNTR profiles can be difficult to read because of extra bands/peaks occurring in the gel/electropherogram. These bands can have several causes. They may be real bands indicating a duplicated segment of DNA, or in the case of locus STTR10 in *S. thyphimurium* the presence of more than one *pSLT* plasmid. These are however, rare problems. The extra bands may also be artefactual bands, which may result from:
    a. The differential migration of the two DNA strands in denaturing acrylamide gels.
    b. The "+A" activity of the *Taq* polymerase generating x + 1 bands.
    c. Slippage of the *Taq* polymerase during polymerization generating "stutter" or "shadow" bands.
    d. Nondenatured secondary structures adopted by the PCR products.

    Using capillary electrophoresis, problem A is not seen as only one strand is labeled with fluorescence and can thus be visualized. In the case of *N. menin-gitides* and the *S. typhimurium* VNTR method, using the previously mentioned protocol, it has also been shown that problems B and D do not result in misinterpretation.

Problem C does however occur in some isolates, and will manifest on the gel/electropherogram as either a row of minor intensity peaks each repre-senting a repeat unit of the VNTR locus forming a "saw" pattern before the correct band/peak or a larger band/peak that appears some VNTR repeat units short of the correct band/peak.

In the case of capillary electrophoresis, it is relatively easy to find the correct peak in the "saw" pattern of artefactual bands, as the correct band will be the last and most intense (i.e., highest peak) in a row of small peaks leading up to it. In the case of a larger more predominant band, the correct band can be found by running a dilution of the PCR products, as the incorrect band will in this assay be greatly reduced in intensity or even disappear.

22. Loss of some peaks. Some isolates will be negative for some of the VNTR loci used in the *S. typhimurium* method. This is normal, and especially locus STTR 10 and locus STTR6 are prone to show no amplification in a subset of isolates. This is still valuable information and is regarded as a "0" allele in comparisons. Loss of bands is a rare problem in *N. meningitides.*

23. Loss of all or almost all peaks. If all bands (gel-based method)/peaks (including the red peaks from the internal size-marker) (capillary electrophoresis-based method) are absent, this points to a problem with the electrophoresis instrument. In the case gel-based method the thermocycler should also be inspected for a possible malfunction.

    In the case of capillary electrophoresis-based method, the capillary, the running buffer and POP4-gel should be replaced. If the problems persist, the readings for the laser (capillary electrophoresis-based method) and the power supply (in both methods) should be checked to ensure they are working within operational parameters.

    If all bands/peaks are absent, the cause of this problem may be faulty PCR amplification, and all conditions and mixes used for the PCR amplification should be checked, and new primer mixes should be prepared. The Thermocycler should also be inspected for a possible malfunction.

    If just one or two peaks appear, this indicates that the isolate in testing is not a *N. meningitides* or *S. typhimurium* isolate and should be inspected by other methods.

Chapter 18

Protein Expression Screen

Protein expression and purification have traditionally been time-consuming, case-specific endeavors, and are considered to be the greatest bottlenecks in most proteomics pipelines. *Escherichia coli (E. coli)* is the most convenient and cost-effective host, although *optimal* conditions for the expression of different proteins vary widely.

Proteins vary in their *structural stability, solubility*, and *toxicity* in this environment, resulting in differing rates of protein degradation, formation into insoluble inclusion bodies, and cell death, thus affecting the amount of soluble protein that can be obtained from *E. coli* grown in culture. A variety of *E. coli* strains designed to address many of these problems have been made, by allowing for disulfide bond formation in the *E. coli* cytoplasm, more efficient usage of codons that are rarely found in *E. coli* genes, and a reduction of specific *E. coli* proteases (Novagen, Stratagene).

The T7 expression system has also helped to alleviate many problems associated with recombinant expression by allowing for tightly regulated induction of expression. However, variability in the recombinant expression of proteins still exists. Induction strength and time, and growth temperature are just a few parameters that can have drastic effects on the amount of soluble protein that can be obtained from the *E. coli* host.

Adding further complexity to this problem, affinity tags that are used as a means to standardize protein purification have varying effects on protein stability and solubility. The hexahistidine tag, which allows for protein to be adsorbed to Nil'-charged resin, is often used because it is small and often does not need to be removed by protease digestion prior to downstream experimentation.

Another useful tag is the maltose binding protein (MBP) tag, which allows for the protein to be captured with amylose resin, and can stabilize and solubilize the fusion partner. A variety of affinity tags, in addition to their placement at the N-or C-terminal end of the coding sequence, must be tested for optimal protein expression.

To take full advantage of the variety of strategies developed to improve the expression of soluble protein in *E. coli,* an easy and rapid means to test many growth parameters is necessary. This chapter describes a dot-blot expression screen to test the effects of growth and induction parameters on the yield of soluble protein. The expression screen is used to detect hexahistidine-tagged proteins expressed in *E. coli,* but it is adaptable for the detection of other affinity tags or fusion partners that have suitable antibodies available.

In this example, induction time and temperature are tested; however, it can be used to test additional parameters, such as affinity tag type and placement, E. *coli* host type, and growth medium formulations.

Results of the screen may be used to segregate samples into groups for further parallel processing in an efficient highthroughput protein production pipeline.

REQUIREMENTS

1. Rosetta (DE3) (Novagen, Madison, WI) and BL21 Gold (DE3) (Stratagene, La Jolla, CA) E. *coli* competent cells or other expression host strain.
2. cDNA clones in T7 expression vector.
3. Luria-Bertani (LB) media and agar plates.
4. Chloramphenicol.
5. Kanamycin.
6. 24-well (7 mL) round bottom blocks.
7. Isopropyl-(3-v-thiogalactopyranoside (IPTG).
8. Lysis buffer: 50 *null* sodium phosphate buffer (pH 8.0), 300 mM NaCl, 10 mM imidazole. Adjust pH to 8.0 and store at 4°C.
9. Lysozyme.
10. Benzonse (Novagen, Madison, WI) or DNAse.
11. Protease inhibitor cocktail (for histidine-tagged proteins) (Sigma-Aldrich, St. Louis, MO).
12. Phenyl methyl sulfonylfluoride (PMSF).
13. Protran nitrocellulose membrane (0.2-μ pore size) (Schleicher and Schuell, Keene, NH).
14. Tris-buffered saline (TBS): 6 mMTris, 150 rnMNaCl (pH 7.5).
15. TBS T/T: 20 mMTris, 500 mMNaCl, 0.05% Tween-20, 0.2% Triton X 100 (pH 7.5).
16. Blocking buffer: TBS with 3% bovine serum albumin (BSA).
17. Penta His horseradish peroxidase (HRP) conjugate kit (Qiagen, Valencia, CA). 18. Metal-enhanced DAB substrate kit (Pierce, Rockford, IL).

PROCEDURES

Expression of Proteins

The following protocol describes the expression of proteins using the pNHis vector in the *E. coli* strains Rosetta (DE3) pLysS and BL21 Gold (DE3), based on previous results (not shown). The strain Rosetta (D.E3), used for the proteins of *Ciona intestinalis,* is optimized for the expression of

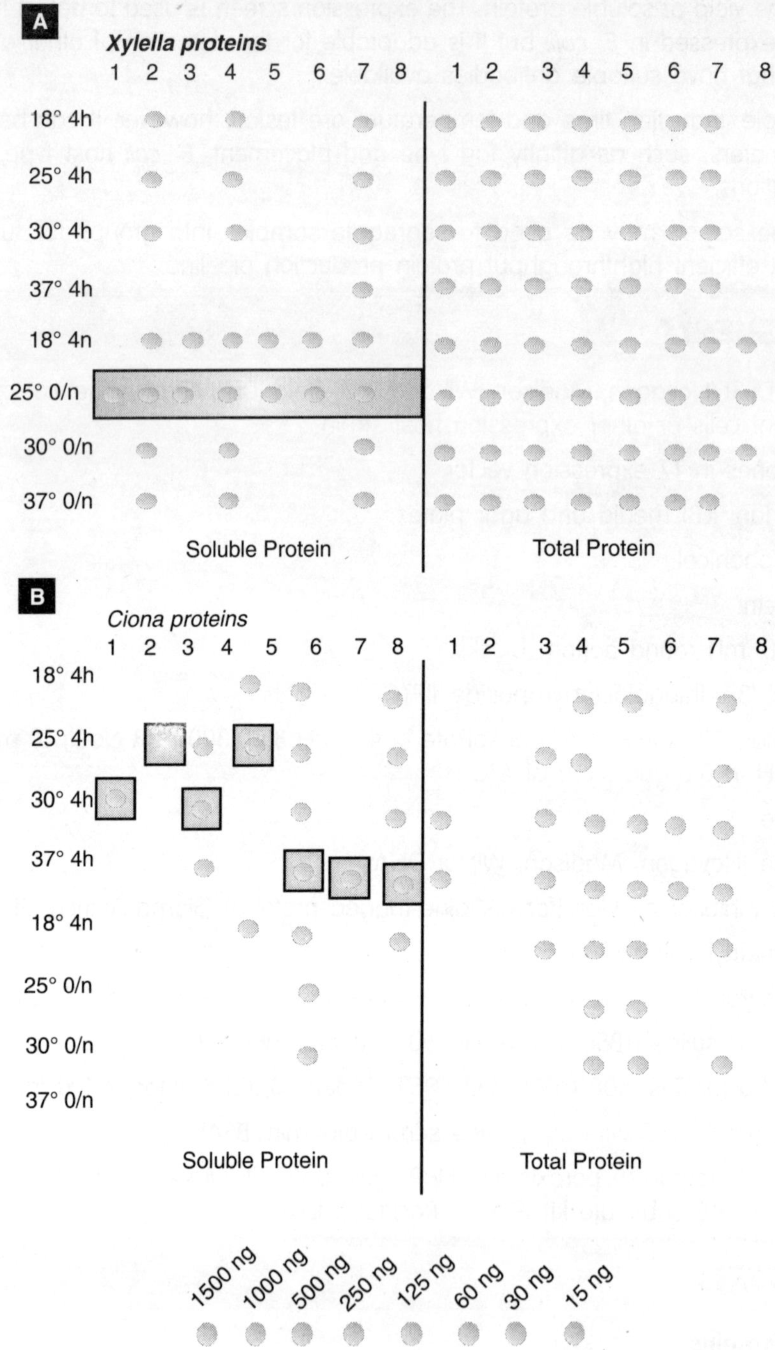

Figure 18.1: Expression screen dot blots of (A) Xylella *samples and* (B) Ciona samples. *grown for 4 h or overnight (o/n) at 18, 25, 30, and 37°C. The standard curve is shown below. The spots outlined with squares indicate samples that would be chosen for protein purification.*

proteins containing rare codons often found in genomes enriched with an abundance of either GC or AT base pairs.

The strain BL21 Gold (DE3) is used for the expression of proteins from *Xylella fastidiosa.* Note that alternative expression plasmids and *E. coli* strains may be used, requiring adjustments to the general protocol such as the use of antibiotics.

1. Transform or electroporate Rosetta (DE3) pLysS or BL21 Gold (DE3) cells with pNHis expression plasmids following the manufacturer's protocol. Plate onto LB agar plates containing 50 µg/mL kanamycin and 50 µg/mL chloramphenicol (Rosetta cells only), and grow overnight at 37°C.
2. Pick a single colony of each sample and grow 5-mL cultures (in a 24-well block) of LB containing 50 µg/mL kanamycin and 50 µg/mL chloramphenicol (Rosetta cells only) under non-inducing conditions (no IPTG) by shaking at 180 rpm at 37°C overnight.
3. Identify the parameters that will be tested, such as inducer concentration, temperature, and time of culture growth following induction of protein expression. Keep in mind that some sets of conditions will require a separate 24-well growth block. Aliquot 5 mL of LB containing 50 µ.g/mL kanamycin and 50 µg/ mL chloramphenicol (Rosetta cells only) into each well of the blocks, and inoculate with 100 RL of the overnight starter cultures.
4. Grow cultures by shaking at 180 rpm at 37°C until an optical density (O.D.) at 600 nm wavelength of approx 0.6-0.8 is obtained. Add IPTG at the desired concentration and continue to shake at the desired temperature for the desired length of time. In this example, growth temperature (18, 25, 30, and 37°C) and time (4 h or overnight) following induction with 1 mM IPTG were tested.
5. Pellet the cells by centrifugation at 4°C and freeze at -70°C or in a dry ice/ethanol bath. Prepare fresh lysis buffer by adding 1 mg/mL lysozyme, 2.5 U/mL Benzonase r'nuclease, 2 mMMgCl_2, 2.µL/mL protease inhibitor cocktail, and 1 rAPMSF. Resuspend the frozen pellets in 500 µL fresh lysis buffer and shake on a plate shaker at 4°C for 30 min to lyse the cells.
6. Remove 10 µL of crude lysate (spotted on the membrane as total protein) and remove the insoluble material by centrifugation. The supernatant (cleared lysate) and crude lysate are then used in the dot blot procedure described below to determine the amount of soluble and total recombinant protein, respectively.

Dot Blot Techniques

1. Spot 2 µL of crude and cleared lysate onto Protran· nitrocellulose membrane using a multichannel pipettor. Also spot a set of standards. This example spotted a sample protein, isocitrate dehydrogenase (IDH), in a range of 15—1500 ng. Allow the membrane to dry completely.
2. Incubate the membrane by gentle shaking at room temperature as follows: TBS, 5 min↔3; blocking buffer, 30 min; TBS T/T 5 min ↔ 3: penta His IIRP conjugate (1:1000 in blocking buffer) 30 min; TBS T/T, 5 min ↔ 5.
3. Dilute the DAB concentrate (10X stock) with stable peroxide buffer to 1 X following the

manufacturer's protocol. Submerse the membrane in the DAB solution and develop for approx 3 min. Rinse membrane with H_2O, incubate in a dark place until it is completely dry, and scan the image using a flatbed scanner.

POINTS TO REMEMBER

1. A wide variety of *E. coli* strains designed for protein expression are commercially available, derived from the *E. coli* strain BL21, that are tailored to facilitate disulfide bond formation, fine-tune protein expression levels, enhance the expression of proteins that contain rare codons, and other specific requirements.

   These strains are available as lambda DE3 lysogens, which carry a chromosomal copy of the T7 RNA polymerase gene under control of the lac promoter (inducible by IPTG). Once induced, the T7 RNA polymerase drives expression of genes that are under control of the T7 promoter, such as those cloned into the pET family of exprcssion vectors (Novagen, Madison, WI).

   Dramatic differences in protein expression can be seen when different strains are used for the production of proteins in many cases; therefore, it is useful to test several strains when producing many proteins in a high-throughput format without knowledge of their specific characteristics.

2. Conditions most commonly tested are inducer (IPTG) concentration (0.1 mM to 1.0 mM), growth temperature after induction (18°C to 37°C), and duration of time following induction (4 h to overnight). Additional parameters to consider testing include *E. coli* strain, media formulation, and placement of affinity tag (N- or C-terminal).

Chapter 19

Protein Synthesis in *E. coli*

Large collections of purified proteins have become essential to systems-biology programs for generating protein ligands and for validating protein interactions. These large protein collections will likewise benefit structural studies, and quests for protein-based therapeutics. Whether many protein targets need to be produced, such as a microbial proteome, or a few poorly expressing protein targets need to be expressed as soluble fragments (divide and conquer approach), efficient high-throughput processing can be a bottleneck.

This chapter describes a method for efficient high-throughput purification of hexahistidine-tagged proteins that are expressed in *Escherichia coli (E. coli)* using immobilized metal affinity chromatography (IMAC) in a 96-well format. This approach is particularly suitable for proteomic applications that require modest amounts of highly purified proteins to be generated very efficiently.

This approach is also very useful for identifying protein targets that are most amenable to scaled-up production for use in structural studies. The typical yield of proteins purified using this system is 50-150 μ.g, which is generally greater than that of many in vitro expression systems, and much less costly.

The method as described has been optimized for purifying approx 150 μg of hexahistidinetagged protein, but the method is flexible, so that the amount of affinity matrix and culture volumes can be adjusted for optimal binding capacity, and consequently highest purity. Although the method detailed here uses IMAC to purify hexahistidine-tagged proteins, this basic platform can be used with many other tags and affinity resins.

REQUIREMENTS

1. 24-well (7 mL) and/or 96-well (2 mL) round bottom blocks (Qiagen, Valencia, CA).
2. Genemate 96-well filter plate with GMF-5 filter (ISC Bioexpress, Kaysville, UT).
3. AirPore Tape (Qiagen, Valencia, CA).
4. 96-well plate shaker/vortex.

5. 96-well plate/block centrifuge.
6. 96-well filter-plate vacuum manifold and regulator (Qiagen, Valencia, CA).
7. Luria-Bertani (LB) media and LB agar plates.
8. Kanamycin.
9. Isopropyl-b*-D-thiogalactopyranoside (IPTG).
10. Lysis buffer: 50 mMsodium phosphate buffer (NaH_2PO_4), pH 8.0; 300 mMNaCl; 20 mMimidazole; 0.05% polysorbate 20; 1 $mMMgCl_2$. Store at 4°C.
11. Wash buffer: 50 $mMNaH_2PO_4$, pH 8.0; 500 mMNaCl; 20 mMimidazole; 0.05% polysorbate 20. Store at 4°C.
12. Elution buffer: 50 $mMNaH_2PO4$, pH 8.0; 150 mMNaCl; 500 mMimidazole; 10% glycerol. Store at 4°C.
13. Lysozyme (Sigma, St. Louis, MO).
14. Benzonase (Novagen, Madison, WI) or DNase.
15. Protease inhibitor cocktail (for histidine-tagged proteins) (Sigma-Aldrich, St. Louis, MO).
16. Phenylmethylsulfonyltluoride (PMSF).
17. Ni-NTA Superflow resin (Qiagen, Valencia, CA) or other IMAC resin.
18. BL21 Gold (DE3) competent cells (Stratagene, La Jolla, CA).
19. 3-15% polyacrylamide gels (BioRad, Hercules, CA).
20. GelCode Blue (Pierce, Rockford, 1L).
21. Agilent 2100 Bioanalyzer and Protein 200 plus lab-chip kit (Agilent, Palo Alto, CA).

PROCEDURES

First, a general method of transformation of the expression plasmid into E. *coli* will be described, as well as appropriate conditions for growth and induction of the expression cultures. Then, the 96-well protein purification method is detailed. Last, analysis of the purified proteins is described using both lab-on-a-chip technology and traditional sodium dodecyl sulfide (SDS)-polyacrylamide gel electrophoresis (PAGE).

Expression Plasmids

Expression plasmids, in our example pNHIS, containing genes of interest fused to a hexahistidine coding tag are first purified from 5-mL cultures of a non-expression strain of *E. coli,* such as DH5a or Novablue (a K- 12 strain from Novagen), grown in LB containing kanamycin for antibiotic selection. An appropriate *E. coli* strain for protein expression is then transformed with the purified expression plasmids, as described below.

The *E. coli* cell strain that is most commonly used for protein expression driven by T7 RNA polymerase is BL21 (DE3). This strain contains a chromosomal copy of the T7 RNA polymerase gene (indicated by the DE3), the expression of which is inducible by addition of IPTG. There are many derivative and mutant *E. coli* strains available that offer strategies for producing problematic proteins.

1. Add 1 μL of purified plasmid DNA to 25 μL of BL21 (DE3) competent cells in a 1.5-mL microfuge tube, gently mix, and place on ice for 10 min.
2. Place in a 42°C water bath for 30 s, then immediately back on ice for 2 min. Add 500 μL of SOC (or LB) and grow with vigorous shaking at 37°C for 30 min.
3. Plate 50 μL onto prewarmed LB agar plates containing 50 μg/mL kanamycin, and grow overnight at 37°C.
4. Next, grow starter cultures by picking a colony from the above plates into 2 mL LB plus 50 μg/mL kanamycin in sterile 24-well blocks. Cover the blocks with airpore tape and grow overnight at 37°C, shaking at 180-200 rpm.
5. Aliquot 5 mL of LB plus 50 pg/mL μg/mL into each of the wells of new 24-well blocks. Inoculate the 5 mL of LB with 100 μL of the starter cultures. Cover the blocks with air-pore tape and grow at 37°C at 180-200 rpm for approx 3 h until the optical density (O.D.) at 600 nm wavelength is 0.6- 0.8. The OD_{600} should be checked every hour by removing 100 μL from several representative wells.
6. Add IPTG to induce protein expression, transfer the blocks to a shaker incubator that has been adjusted to the desired expression temperature, and grow for the desired amount of time. Each of these parameters should be optimized by the expression screening technique described in the previous chapter. In the example in this chapter, protein expression was induced with 1 mMIPTG and cultures were grown for 6h at 25°C.
7. Pellet the cells by centrifugation at 2500g for 10 min at 4°C and then carefully decant all media. Freeze the bacterial pellets at -80°C.

Purification of Proteins

Purification of the proteins from the cultures grown in 24-well blocks can be performed in 96-well filter plates using a vacuum manifold for column forming, washing, and elution steps, as outlined below. The hexahistidine-tagged proteins in cleared cell lysates are batch loaded onto Ni-NTA agarose in a 96-well filter plate.

Columns are formed by applying low (200 mbar) vacuum pressure. These minicolumns are extensively washed, and subsequently the proteins are eluted into a microtiter plate.

1. Thaw bacterial pellets in 24-well blocks at room temperature. While waiting, add protease inhibitors, lysozyme, and Benzonase to the chilled lysis buffer: 1 mg/mL lysozyme, 10 units Benzonase/mL, 2 μL/mL protease inhibitor cocktail (no ethylenediamene tetraacetic acid [EDTA]), and 1 mMPMSF. Keep on ice.
2. Add 0.5 mL of chilled lysis buffer plus additives to each pellet and resuspend by pipetting, using a 12-channel pipet to mix a row of 6 wells with two tips in each of the 6 wells.
3. Vortex the 24-well blocks containing the resuspended pellets at 4°C on a plate shaker for 20 min. Performing the lysis at 4°C will help prevent protein degradation.
4. Centrifuge 24-well blocks at 2500g in a plate centrifuge for 20 min at 4°C. The pellets obtained can be saved to purify insoluble protein under denaturing conditions (not discussed here).

5. Equilibrate the Ni-NTA resin in lysis buffer and resuspend to obtain a 50% % slurry. Add 50 μL of the 50% slurry of Ni-NTA resin, which is enough to bind approx 150 μg of protein, into the wells of a 96-well filter-plate. Liquid may begin to flow by gravity; in order to prevent loss of material, place the filter plate on a piece of parafilm.
6. Carefully transfer the cleared bacterial lysate supernatants from each of the 24 wells into the wells of a 96-well filter plate. Cover the filter-plate top with clear plate-seal tape, to prevent cross-contamination.
7. Mix the Ni-NTA resin with the cleared lysate in the 96-well filter plate on a piece of parafilm on a plate vortex for 30 min at 4°C.
8. Remove the 96-well filter plate from the plate vortex leaving the parfilm behind; place on the vacuum manifold (with a waste collection container below) and apply vacuum (approx 200 mbar) so that each well slowly drips over several minutes until all the lysate has filtered through, and release the vacuum after the material has flowed through.
9. Add 750 μL of wash buffer into each well and apply the vacuum just as in step elsewhere in this chapter. Repeat the wash step two more times.
10. Fit a 96-well microtiter plate (200 μL) or 2-mL collection microtubes into the vacuum manifold, carefully aligning the drop-formers to the receiver plate wells. Add 100 μL elution buffer and incubate for 2 min.

    Then apply a vacuum of 200-400 mbar and release vacuum after material has flowed through. Store the plate of eluted proteins at 4°C for short term or until evaluated, and then at -80°C for long-term storage.

Isolation of Protein

Eluted material was assessed on an Agilent 2100 Bioanalyzer using a Protein 200 Plus LabChip (Caliper) to determine protein purity, concentration, and yield. However, traditional SDS-PAGE can also be used, if necessary. The LabChip system is a fluorescence-based detection system that offers the advantages of requiring only 1-4 μL of protein samples, as compared to 10 μL required for equivalent band identification on traditional SDS-PAGE gels, and results for a set of 10 samples are obtained in 30 min.

The Biosizing software provides detailed tables of raw and analyzed data and gel-like images, and is convenient for data storage and retrieval. The Agilent 2100 Bioanalyzer system provides accurate protein concentrations, alleviating the need of running additional assays and using more of the samples.

In addition, the concentration of each individual protein is determined from the fluorescent peak area, or band, on the gel-like image; thus, accurate concentrations of a protein of interest can be easily obtained in partially purified samples.

1. For SDS-PAGE, 10 μL of each protein sample were run on 3-15% , gradient polyacrylamide gel. The geis were stained with GelCode Blue and scanned on Fluor-S MultiImager.
2. For the Bioanalyzer, the protein LabChip is prepared according to the manufacturer's instructions.

3. The samples are prepared by mixing 2 µL of eluted protein, 2 µL H_2O, and 2 µL of an SDS-based denaturing sample buffer containing (3-mercaptoethanol as well as upper and lower protein mass standards. Samples are heated to 95°C' for 2 min and spun briefly.
4. Samples and ladder are then diluted to 90 µL with H_2O, and then 6 µL of each diluted sample is loaded into a well off the LabChip according to the manufacturer's instructions. Dilution of the sample may seem counter-intuitive but is necessary to reduce the SDS concentration in the denaturation buffer.
5. The LabChip is then placed in an Agilent 2100 Bioanalyzer and the Protein 200 Plus program is run.
6. Agilent Biosizing software is used to determine the sizes of the proteins of interest by normalization against two internal standards of 6 and 210 kDa. The fluorescent peak identification settings were adjusted from the default settings to improve sensitivity-0.8 for the minimum peak height, 0.2 s for the minimum peak width, and 4 for the slope threshold.

POINTS TO REMEMBER

1. A variety of *E. coli* strains have been developed for protein expression that may improve the soluble yields of particular classes of proteins. We have found that many microbial proteins express well in the traditional BL21(DE3) strain. However, eukaryotic proteins often express better in strains that provide tRNAs for codons that *E. coli* rarely use, such as Rosetta (Novagen).

   Other examples include mutant strains that may improve the proper folding of some proteins by *promoting disulfide* bond formation, such as Origami and Origami B (Novagen), and mutant strains that facilitate IPTG concentration-dependent expression induction, such as Tuner (Novagen).

   Testing several expression strains from various vendors is often a good strategy to improve the odds of producing soluble protein, and further high-lights the utility of high-throughput expression and purification techniques.
2. A variety of uniquely formulated media are available that may improve growth. For example, richer media may be used, such as Power, Turbo, or Superior broth (Athena ES, Baltimore, MD) to increase biomass and subsequent protein yield.
3. Replicates of each of the proteins can be set up in the 24-well blocks and then combined into the same well of a 96-well block to increase the yield of low-expressing samples.
4. The chapter on expression screening describes a simple method to quickly identify the optimal condition for expression induction. The amount of IPTG (0.1 to 1 mM), optimal temperature (18°C to 37°C), and length of the induction time (4 h to overnight) should be identified to improve the likelihood of producing soluble protein and improve the yields of soluble proteins. Additionally, purification from several protein expression conditions can be tested in parallel.
5. IMAC resins are available from many sources, and the binding capacity can vary. Additionally, use of metal chelation via tetradendate adsorbant (NTA) as opposed to tridendate

(IDA), ensures that much less leaching of metal ions will occur. The purity of hexahistidine-tagged proteins is optimal from purification conditions in which the amount of the target protein is equal to or greater than the binding capacity of the column matrix.

The amount of lysate and Ni-NTA Superflow agarose that we have suggested should be a guide. The expression screening method described in the previous chapter can be used to assess and then group samples with similar expression levels and then approximate the optimal amount of Ni-NTA Superflow agarose required.

Additionally, we tested several 96-well filter plates and found that the Genemate plates have dropformers that perform better than others, with less cross-contamination and optimal recovery of eluent, and are less expensive than most other plates.

6. The compositions of the buffers can be varied. Adjusting the stringency of the wash—for example, by increasing the NaCl concentration to 500 mMor the imidazole concentration to 20 mM-can improve purity; however, lowering these concentrations may improve binding of some hexahistidine-tagged proteins. The QIAexpressionist handbook (Qiagen, Valencia, CA) provides some very useful suggestions for various buffer compositions.
7. The LabChip *protocol* recommends using 4 μL of protein sample; however, we find that using 1-2 μL of each sample plus H_2O works as well or better, and uses less material. This may be due to a dilution of components that affect the sensitivity of the assay, such as salt and imidazole. Many common reagents affect the sensitivity of the LabChip assay; therefore, care should be taken when the buffer composition is altered.

Chapter 20

Gene Expression in Baculovirus

Baculoviruses are a diverse group of viruses isolated from different insect hosts. Most of the baculoviruses isolated so far are host specific, and the majority of them are isolated from the larvae of butterflies and moths (Phylum Arthropoda, Class Insecta, Order Lepidoptera). Baculoviruses are rod shaped, 40-50 nm in diameter, and 200-400 nm in length.

The length of the capsid can extend, accommodating larger double-stranded DNA. The parallel skills developed to enable the growth and multiplication of cell lines derived from fall army worm, *Spodoptera frugiperda* (Sf9, Sf21), and cabbage looper, *Trichopulsia ni* (HighFive) made it possible to propagate baculoviruses in cell culture. This opened the door to extensive studies on the molecular biology of baculoviruses. The model virus in these studies was *Autographa californica* multiple nuclear polyhedrons virus (AcMNPV).

Most expression vectors used in the baculovirus expression vector system (BEVS) at present are based on AcMNPV infection of *S. frugiperda. Bombyx mori* nuclear polyhedrosis virus (BmNPV) has been used extensively for the production of heterologous protein in silkworms. The BEVS exploits the characteristics of the baculovirus and its mode of infection in cultured insect cells. Therefore, it is important to understand the infection cycle to optimize the level of foreign gene expression.

During the Infection cycle, the parental baculovirus in cell culture displays three distinct phases: early, late, and very late. During the early phase, viral attachment occurs followed by penetration of viral genetic material, which reprograms the host cells for virus gene expression and shuts off host gene expression. The nucleus of the infected host cell then synthesizes viral DNA (nucleocapsid) in the late phase.

The viral DNA is enveloped by budding through the plasma membrane, commonly known as the *budded virus* (BV). Logarithmic production of BV occurs between 12 to 20h post infection. A decline in host protein synthesis is observed at 18h post infection. The infective viral titer remains very high during this phase. The very late phase (24-36 h post infection) is specific to viral gene expression.

During this time, large numbers of occlusion bodies containing viruses are produced. In this phase, each virus acquires an envelope within the nucleus along with a crystalline protein matrix, a polyhedrin. The virus produced in this phase is commonly known as the *occluded virus* (OV). BV is more infectious than OV.

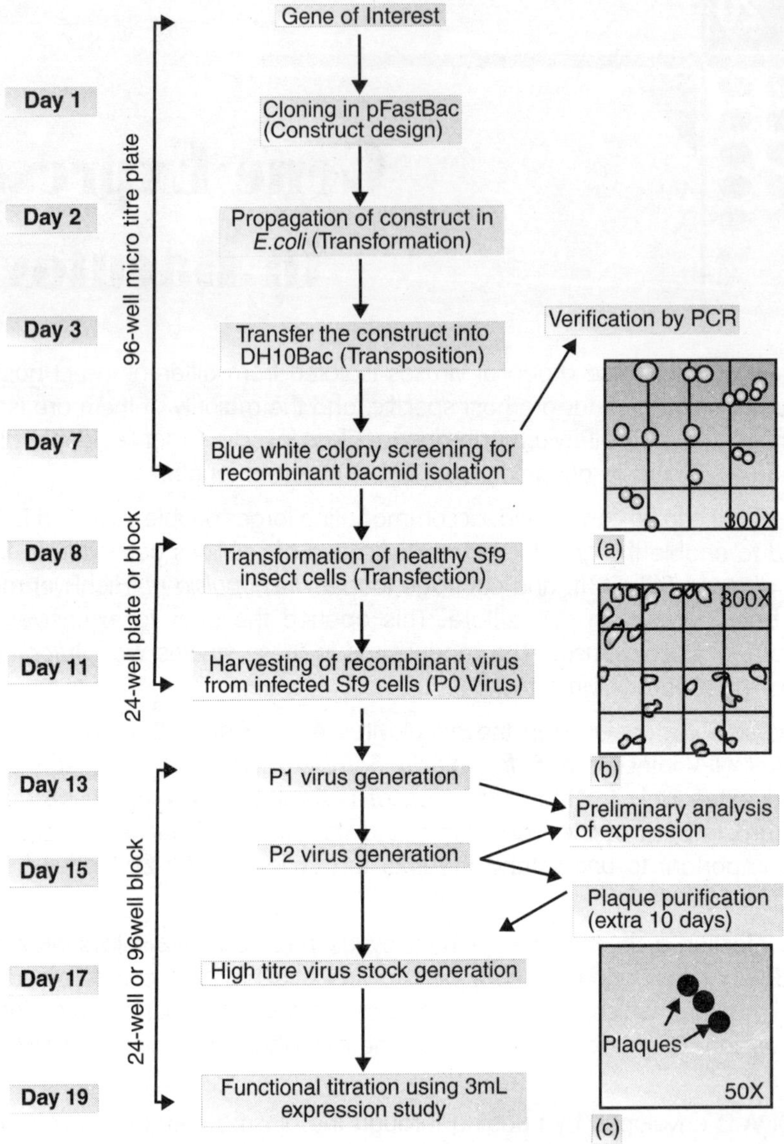

Figure 20.1: An overview of the steps involved in baculovirus-mediated cloning and expression of gene in insect cell. The images observed under the microscope help to identify (a) the healthy cells, (b) infected cells, and (c) the plaque formed by viruses.

The polyhedrin produced in the very late phase is under the control of a strong promoter called *polh,* which is responsible for the formation of OV. The *polh* promoter is responsible for very high levels of expression during the later stages of infection. The gene encoding the polyhedrin is not essential for the replication of the virus. This gene may be replaced by a foreign gene under the control of the same promoter *polh.*

Production of the recombinant protein is achieved following infection of insect cells or insect larvae with the engineered virus. Another promoter used in the BEVS is *p10,* which has been less extensively characterized but appears to be regulated in a fashion similar to the *polh* promoter. The increasing demand for BEVS over the past 20 years resulted in significant developments in using baculovirus as a prominent tool in molecular biology.

This led to the commercialization of the system, making it easier to use, especially for those with little experience in eukaryotic expression systems. The advancement of the system is reviewed by Kost et al. The wealth of information obtained from genome projects increased demand for the production and characterization of diverse groups of proteins.

The most efficient approach is to analyze several constructs and fusion tags in a variety of expression systems. Therefore, the present day challenge is to quickly screen multiple clones for desired properties, such as expression level, posttranslation modification, solubility, localization of protein, and activity. Although *Escherichia coli (E. coli)* is the first choice to achieve this using a high-throughput (HT) platform, it has limitations when expressing eukaryotic proteins.

The alternative HT system to complement the *E. coli* expression system is BEVS. One of the most commonly used systems in BEVS is the Bac-to-Bac system, which uses sitespecific transposition of an expression cassette into the baculovirus genome maintained in *E. coli.* Improvements to the throughput of recombinant eukaryotic protein in insect cells has been described by several authors.

This chapter provides a detailed methodology to achieve efficient high-throughput cloning and expression of eukaryotic proteins in insect cells. An overview of the work flow is presented in Figure elsewhere in this chapter. The rapid small-scale expression study described here yields sufficient amounts of protein to unravel the desired properties of heterologous proteins, thus addressing the challenges mentioned previously. This method is based on the Bac-to-Bac system, however, it can easily be adapted to other systems.

REQUIREMENTS

Cell Line of *E. coli*

1. *E. coli* MACH 1 (Invitrogen, C8620-03).
2. *E. coli* DH10Bac (Invitrogen, 10361-012).
3. Sf9 insect cells SFM adapted (Invitrogen, 11496-015).
4. High-Five insect cells, SFM adapted (Invitrogen, 10486-025).

Enzymes

1. Platinum Pfx DNA polymerase (Invitrogen, 11708-039).
2. Benzonase (HC, 250 units/μL) (Novagen, 71205-3).

Chemicals and Kits

1. Cellfectin reagent (Invitrogen, 10362-010)
2. EDTA-Free Protease Inhibitor Cocktail Set III (Merck Biosciences, 539134)
3. 0.4% Trypan Blue Stain (Invitrogen, 15250-061)
4. Oligonucleotides (Invitrogen, 10297-117)
   M 13 forward: 5' GTTTTCCCAGTCACGAC 3'.
   M13 reverse 5' CAGGAAACAGCTATGAC 3'
5. 96-Well DNA Purification Kits for PCR Products (Qiagen 28183 or Millipore LSKMPCR50).
6. Plasmid Miniprep Kit (Qiagen 16191 or Millipore LSKP09624).
7. Sybr-Safe DNA gel stain (Invitrogen S33102): use at 1:30,000 dilution in agarose gels.
8. Low melting-point agarose (Invitrogen, 15517-014).
9. Solutions P1 and P2 (Qiagen 19051 and 19052, respectively).
10. Isopropanol (Fisher Scientific, P/7500/17)
11. Ethanol (Fisher Scientific, E/0650DF/17)
12. Potassium acetate (FisherBiotech, BP364-500)

Plasticware

1. 96-well thin-walled PCR plates (ABgene,AB-0900).
2. Adhesive, heat-resistant film (ABgene, AB-0558).
3. 96 deep-well blocks, 2.2-mL (Thomson, 951652B).
4. 96-well filter plates (Chromacol, CLS-370011, or Thomson, 921673B).
5. Porous adhesive film for 96-well blocks (Thomson, 899410).
6. 250-, 500-, and 1,000-mL flasks with vented cap (Corning 431144, 431145, and 431147, respectively).
7. 500mL sterile centrifuge tubes (Corning, 431123).

Instruments

1. 96-well thermocycler with heated lid (Biometra).
2. Micro-express Glas-Col shaker (8 × 96-well blocks).
3. Large, chilled shaker-incubators.
4. Inverted microscope (Axiovert 25, CarlZeiss).
5. Haemocytometer, improved Neubauer (VWR International, 720-0104).

General Stock Solutions and Media

1. Penicillin/streptomycin (Invitrogen, 15140-122).
2. Kanamycin (Apollo Scientific Ltd., BIK0126).

3. Tetracycline (Fluka/BioChemika, 87128).
4. Gentamycin (Sigma-Aldrich, 46305)
5. Antibiotic stock solutions (1000X):
   a) 50 mg/mL kanamycin in water.
   b) 10mg/mL tetracycline in ethanol.
   c) 7 mg/mL gentamycin in water.
6. Bluo-gal (Invitrogen, 15519-028): 100 mg/mL in DMSO.
7. Foetal bovine serum (Invitrogen, 10082-147).
8. Heparin (Sigma-Aldrich, H3400-100).
9. L-glutamine (Invitrogen, 25030-024).
10. CHAPS (AG Scientific Inc., C-1019): 10% (w/v).
11. 5 *M* NaCl (Fluka, 71376).
12. 5 *M* Imidazole, pH 7.5 (Fluka, 56750): dissolve imidazole in 80% of the final volume, titrate with concentrated HCl to the desired pH, then adjust the volume with water, filter, and sterilize. Store at room temperature.
13. Glycerol (Fisher Scientific, G/0650/17): prepare 60% (w/v) stock, autoclave, and store at room temperature.
14. Isopropanol (Fisher Scientific, P17500/17)
15. 1 *M* HEPES (Calbiochem, 391340): dissolve HEPES in 80% of the final volume, titrate with 5 *M* NaOH to pH 7.5, adjust the volume with water, autoclave, and store at room temperature.
16. 0.5*M* TCEP (Tris[2-carboxyethyl] phosphine hydrochloride) (Fluorochem, BV 1300): dissolve the solid in water and store in 1 mL aliquots at -20°C.
17. IPTG (Apollo Scientific, BIMB 1008): dissolve the solid in water to prepare 40 mg/mL stock, filter-sterilize and store in 1 mL aliquots at -20°C).
18. TE buffer: 10 *mM* Tris-HCl, pH 7.5 and 1 *mM* EDTA.
19. Sf-900 II (1X) liquid with L-glutamine (Invitrogen, 10902-153).
20. Grace's insect medium (1X) (Invitrogen, 11595-030).
21. Express SFM (Invitrogen, 10486-025).
22. Virkon (Appleton Woods, GC383).
23. Decon90 (Decon Laboratories Ltd.)

Solutions and Media for Cloning

1. LB medium: 1% tryptone, 0.5% yeast extract, 1% NaCl.
2. SOC: 2% Bacto-tryptone, 0.5% yeast extract, 0.05% NaCl, 2.5 mM KCl, 10 mM MgCl2, 20mM glucose, pH 7.0. $MgCl_2$ and glucose added after autoclave.
3. 2X LB: LB medium prepared in 0.5 volume.

4. LB-agar plates: prepare LB medium, before autoclaving add 1% (w/v) bactoagar, and autoclave. After autoclaving, let the medium cool to 55-60'C, add antibiotics and other additives (if required), mix well, and pour into 60-mm petri dishes (approximately 10 mL/ dish).

   Prepare LB agar plates containing 50μg/mL kanamycin, 7 μg/mL gentamycin, 10μg/mL tetracycline, 100μg/ mL bluo-gal, and 40μg/mL IPTG to select for DH10Bac transformants. These plates can be stored at 4'C for up to 4 weeks, covered to prevent exposure to light.

Buffers for Protein Purification

The following solutions are made by combining general stock solutions and water; no pH adjustment is needed, but it is good practice to check and maintain the pH between 7.5 and 8.0.

1. Suspension buffer: 50 mM HEPES, pH 7.5, 0.5 M NaCl, 5% glycerol.
2. Lysis buffer: 50mM HEPES, pH 7.5, 0.5M NaCl, 5% glycerol, 1% CHAPS, 10 mM imidazole, 1 μL of antiprotease cocktail per mL buffer, 1 MM TCEP and 20U Benzonase per mL buffer, added before use.
3. Wash buffer: 50mM HEPES, pH 7.5, 0.5M NaCl, 5% glycerol, 20mM imidazole, and 1 mM TCEP.
4. Elution buffer: 50mM HEPES, pH 7.5, 0.5M NaCl, 5% glycerol, 500mM imidazole, and 1 mM TCEP.
5. Gel filtration buffer: 10mM HEPES, pH 7.5, 150mM NaCl, 5% glycerol, 0.1 mM TCEP, added before use.

PROCEDURES

Initiation of Cell Cultures

Three insect cell lines widely used in baculovirus expression systems are Sf9, Sf21, and High Five. Sf9 and Sf21 cell lines are suitable for all insect cell-based work from transfection to expression. However, we suggest starting with the Sf9 cell line. The Sf21 and High Five cell lines are suitable to optimize the expression level once the expression in Sf9 is tested. For HT work, a suspension culture is preferred over the adherent culture; therefore, the protocol is designed for cell culture in suspension.

A full description of insect cell culture methods and BEVS is described in two laboratory manuals. Basically, strict aseptic technique is required to handle and manipulate all cell lines, that is, inside a sterile laminar flow hood. Anything taken inside the hood should be surface treated with 70% ethanol to decontaminate it, unless stated otherwise.

All vessels and pipettes used with live cells or viruses must be decontaminated properly using autoclaving, Virkon (10% w/v) or diluted Decon (10% v/v). Wipe work surfaces with Virkon/bleach followed by 70% ethanol. Treat any spills immediately.

1. Before starting the manipulation, bring the Sf-900 II medium, fetal bovine serum (FBS), and antibiotic (penicillin/streptomycin) solution to 27'C. Transfer them into a laminar flow

hood. Supplement the Sf-900 II medium with a final concentration of 2% FBS (v/v), 50U of penicillin and 50μg of streptomycin per mL of medium. Mix by gently inverting the medium bottle three times. Prepare a 250-mL Erlenmeyer flask with a vented cap with 29 mL of supplemented Sf-900 II.

2. Take a cryotube containing frozen stock of Sf9 cells from a liquid nitrogen tank and thaw rapidly in a 37°C water bath. When completely thawed, rinse the outer surface of the tube in 70% ethanol, and take it into the hood.
3. Transfer the contents of the cryotube into an Eppendorf tube. Spin the tube in a tabletop centrifuge at 500*g* for 5 min. Discard the supernatant (the storage medium contains DMSO that may slow down cell growth). Resuspend the cell pellet in 1 mL of supplemented Sf-900 II medium. Use gentle pipetting in all cell handling steps.
4. Add 1 mL of resuspended cells from step 3 to the flask containing the medium from step 1. Incubate the culture at 27°C with 150rpm in a humidified shaker incubator.
5. After 24h of incubation, take out 100 μL of culture sample and place in an Eppendorf tube. Add 10μL of 0.4% Trypan blue stain, mix well and transfer a small volume into a hemocytometer. Count the cells using an inverted microscope.

   Count the number of live cells (appear white) and dead cells (appear blue). Calculate the total number of viable cells per mL of culture and, hence, the percentage viability. If the viability is less than 95%, the culture is not suitable for further work.
6. Keep a record of cell growth patterns for 24h. Once the count reaches 1.8×10^6 to 2.0×10^6 cells per mL, passage the culture with supplemented Sf 900 II medium back to 0.8×10^6 to 1.0×10^6 cells per mL, and incubate as previously.

During growth and maintenance of cultures, examine them carefully for any signs of contamination. Bacteria, yeast, and filamentous fungus often contaminate the culture. Cloudiness of the overnight culture, a pungent smell in the incubator, and swollen and disordered insect cells are the primary signs of contamination. Contamination can be verified by microscopic observation of the samples under high magnification.

However, viral contamination is hard to detect in the first instance. As the incubation continues, appearance of distorted cells and an increase in cell lysis are preliminary signs of unwanted viral contamination. All the contaminated cultures should be isolated from any others and disposed of safely.

Stock Cell Lines

Cell lines can be stored for long periods of time if grown and stored properly. We recommend backing up stocks every year in case of problems that may arise in the main stock. Preparation of ten vials of stock each with 1 mL of culture is described here. It can be expanded according to need.

1. Examine the Sf9 culture as described previously. The viability of the culture should be more than 98% and in exponential growth, that is, $\sim 1\times 10^6$ cells per mL, for the purpose of storage.
2. Prepare 10mL of freezing medium by mixing 6mL (60% v/v) Grace's insect medium, 3 mL

(30% v/v) FBS, and 1 mL (10% v/v) DMSO. Bring the medium to room temperature.

3. Transfer 100mL of cell culture (1 × 10^6 cells per mL) into two 50-mL Falcon tubes. Centrifuge at 500g for 5 min. During this time, take 10 cryovials and mark each with the name of the cell line, the date, expected cell density (i.e., 1 × 10^6 cells) and original batch number from the purchaser. Keep the vials on ice.
4. Discard the medium safely after centrifugation. Resuspend the pellets from both tubes in 10mL of freezing medium; they should give a count of ~1 × 10^6 cell per mL. Dispense 1 mL of the suspension into each cryovial.
5. Place the vials in a CryoCane and freeze gradually by placing at -20'C for 1-2 h, transfer to -80°C for 2days and finally transfer to liquid nitrogen for long-term storage. Slow freezing is a critical step for the long-term viability of insect cells.
6. After 2 weeks, thaw one vial to check viability and absence of contamination.

Cloning of Desirable Genes

The cloning of a gene into the baculovirus expression vector is a two-step process. The first step involves the cloning of the gene into a transfer vector, such as pFastBac. This contains multiple cloning sites flanked by baculovirus DNA fragments to facilitate recombination into the baculovirus genome. The second step is the transfer of the cloned fragment into a baculovirus genome.

This was done originally by cotransfection and recombination in insect cells. In the pFastBac system, the recombinant baculovirus genome is generated in bacteria using a coexpressed DNA transposition system.

When processing multiple genes in a structural genomics context, cloning based on restriction sites is problematic, since restriction sites used for cloning often are present within the gene sequence. Effective, generic alternatives include ligation-independent cloning (LIC) [25] and recombinase-based system (e.g., the Gateway system, Invitrogen).

In this chapter, use of transfer vector pFBOH-LIC-SGC is described, which is derived from the pFastBac vector by introducing LIC cloning sites. The cloning procedure used is based on method described by Stols et al. [26]. Primers for PCR include 14-base extensions that are identical to sequences at the cloning sites in the vector.

Both vector and inserts are treated with T4 DNA polymerase, exposing complementary 14-nt single-stranded tails. The vector and inserts are then annealed and introduced into bacteria. We routinely clone the same fragments in parallel into compatible bacterial expression vectors to allow comparative expression studies.

The gene fragments in the transfer vector are propagated in *a recA, endA E. coli* strain, such as MACH-1, DH5a, or DHIOB. The second cloning step is site-specific transposition. The construct cloned in pFBOH-LIC-SGC is transferred into *E. coli* DH10Bac which carries viral DNA.

The recombinant bacmid is reconstituted by recombination, and in the process, the *lacZ* cassette introduced into the system is disrupted, allowing the selection of clones by color [6]. This allows the HT generation of recombinant bacmid in *E. coli.* The procedure is described in detail:

1. Bring the DH10Bac selective medium plate to room temperature. Prepare the media plates

one day before starting the transposition experiment and allow them to dry at room temperature overnight. They can be stored for later use at 4°C protected from light for up to 4 weeks.

2. Thaw the competent DH10Bac cells on ice. Dispense 20µL into each well of a prechilled 96-well microtiter plate. Add ~0.3 µg or 2-3 µL of recombinant plasmid (construct). Mix gently using a multichannel pipette. Incubate on ice for 30 min.
3. Heat shock at 42°C for 45 s. Return to ice and chill for 2 min. Take a 96-deep-well block and dispense 900µL of prewarmed SOC medium into each well. Transfer the transformed bacterial suspension into the corresponding well of the block and cover with a porous film. Incubate at 37 °C at 700 rpm for 5 h. This incubation is essential for the transposition to occur.
4. Dilute the cultures (10µL in 90µL medium) in LB in a 96-well microtiter plate. Spread 50µL of the diluted cultures on a LB-agar plate prepared in step 1. Store the rest of the culture at 4°C.
5. Incubate plates at 37°C overnight. Check the total number of colonies the next day. If there are more than 100 or less than 10 colonies, take the rest of the culture stored at 4°C (from step 4), spread the appropriate dilution on another plate, and incubate at 37°C until the blue/white colonies can be differentiated. This usually requires 48h.
6. Streak two white colonies from each plate onto two fresh LB-agar plates containing antibiotics, bluo-gal and IPTG to verify the phenotype. Incubate for 48 h at 37°C. White colonies contain the recombinant bacmid and, therefore, are selected for the isolation of recombinant bacmid DNA.
7. Take a 96-deep-well block. Dispense 1 mL of LB containing 50µg/mL kanamycin, 7 µg/mL gentamycin, and 10µg/mL tetracycline into each well. From a single colony confirmed as having a white phenotype on plates containing Bluo-gal and IPTG, inoculate the corresponding well of the 96-deep-well block. Incubate at 37°C overnight, shaking at 700 rpm in a Glas-Col incubator. Add 330µL of 60% (w/v) sterile glycerol to each well of the 96-deep-well block. Mix well and store at -80°C.

Preparing Bacmid

1. Take a 96-well microtiter plate and dispense 50 µL of 2X LB medium into each well. The recombinant white colonies isolated after blue-white screening are used to isolate recombinant bacmid. Pick a white colony using the end of a pipette tip and suspend in the corresponding well of a 96-well microtiter plate. The suspension is used as an inoculum.
2. Take 2X 96-deep-well blocks to generate 2 mL of each culture for bacmid DNA miniprep. Dispense 1 mL 2X LB medium with 50µg/mL kanamycin, 7µg/mL gentamycin, and 10µg/mL tetracycline into each well. Inoculate each well with 20 µL of inoculum, prepared in step 1, using a multichannel pipette. Cover the blocks with porous adhesive film. Incubate at 37°C overnight at 700 rpm in a Glas-Col incubator.
3. Centrifuge the 96-deep-well blocks at 2,600g for 20min. Immediately decant the supernatant into a container for proper disposal. Invert and tap the blocks on absorbent

paper to remove residual supernatant. Freeze the pellets at -20'C for 30 min (the samples can be frozen for up to 24h prior to processing).

4. Allow the pellets to thaw at room temperature for 15 min. Vortex the blocks vigorously until the cell pellets are dispersed. Add 250µL of solution P1 (containing RNase A from Qiagen) to each well of the first block using a multichannel pipette.

   Seal the block with adhesive sealing film and mix by incubating in a Glas-Col incubator at 800rpm for about 2min. Use a multichannel pipette to resuspend the pellet, if necessary. Transfer the suspension to the corresponding wells of the second block. Seal and repeat the mixing process.

5. Add 250µL of solution P2 (Qiagen) to each well. Seal the block with adhesive sealing film. Invert gently five times and incubate at room temperature for about 10min.

6. Add 300µL of 3M potassium acetate pH 5.5, seal and mix gently, but thoroughly, by inverting five times. A thick white precipitate of protein and *E. coli* genomic DNA will form. Place the sample on ice for 20 min, followed by centrifugation at 2,600g for 30min at 4°C.

7. Transfer the clear supernatant gently into a new 96-deep-well block. Centrifuge once again at 2,600g for 30min. (Note: It is difficult to separate the clear supernatant after the first centrifugation step. Great care should be taken while transferring the supernatant. A second centrifugation step helps obtain a cleaner supernatant.)

8. Take another 96-deep-well block. Dispense 0.8 mL of isopropanol into each well. Take 0.8 mL of clear supernatant from step 7 and add it to the corresponding well. Do not try to take all the supernatant. Seal the block and invert it five times to mix and incubate on ice for 30 min. The block can be left overnight at 4°C, if necessary.

9. Centrifuge at 2,600g for 30 min. Spray the outer surface of the block with 70% ethanol and take it inside the laminar flow hood. Discard the supernatant.

10. Add 500µL of 70% (v/v) ethanol to each well. Gently tap the block to wash the pellet properly. Centrifuge at 2,600g for 30 min. Take the block back inside the flow hood. Discard the supernatant and drain the last drops of ethanol by tapping the block on layers of blotting paper or tissue paper.

11. Allow the pellet to dry inside the flow hood. Check often to avoid overdrying. It takes about 15-20 min to dry. Add 50µL of sterile TE buffer to resuspend the bacmid DNA. Separate 2 µL of each bacmid sample to measure the concentration and transfer the remaining to a microtiter plate for further use and storage.

    Make sure the absorbance ratio A260 nm/A280 nm is more than 1.7. Run a few samples on 1 % (w/v) agarose gel containing CYBR Safe DNA gel stain.

12. Use 10 µL of the bacmid DNA just prepared in each well for each Sf9 transfection experiment.

Transposition on Bacmid DNA

It has been found that colony PCR, to confirm the transposition, does not work consistently. Therefore, PCR on bacmid DNA is recommended for verifying the presence of the gene of interest.

Table 19.1: PCR master mix.

Stock solution	Concentration (100X)
10X Pfx Buffer	500 μL
10X PCR enhancer	500μL
50 mM $MgSO_4$	100 μL
25 mM dNTP mix	50L
M13-Bac forward primer (100μM)	9μL
M13-Bac forward primer (100μM)	9μL
Water	1.4 mL
Platinum Pfx polymerase	20L

Set up the PCR reaction in 96-well PCR plate(s). Prepare PCR master mix as shown in Table elsewhere in this chapter.

1. Aliquot 20μL of PCR mix into each well of a 96-well PCR plate. Take the microtiter plate containing bacmid DNA. Use a multichannel pipette to add 1 μL of Bacmid DNA template into the corresponding well of the PCR plate.
2. Perform amplification of the target genes with initial denaturation for 3 min at 94'C, followed by 30 cycles of 94°C for 45 s, 50°C for 45 s, and 72°C for 5 min. The reaction is terminated after a final extension for 7 min at 72°C.
3. Analyze 10μL of amplification product on 1% (w/v) agarose gel (containing CYBR safe DNA gel staln) by electrophoresis. Insertion of the gene into the pFastBac plasmid results in a PCR product of 2,300bp fragment of vector plus the size of the insert.

Transfection and Infection of Insect Cells

1. Grow Sf9 cells to mid log phase in Sf900-II medium supplemented with 2% (v/v) fetal bovine serum. The experimental setup is alternated between 24-deep-well blocks or 24-well TC plates and 96-deep-well blocks or 96-well microtiter plates, according to the required volumes at each step.

   To facilitate pipetting using multichannel pipettes, the convention shown in Figure elsewhere in this chapter is used in the transfer from 96-well to 24-well plates or blocks. Attach tips to every second position of a multichannel pipette for the transfer. The top four rows of the 96-well plate or block correspond to the first two in the 24-well block or plate and so forth. Alternatively, samples can be transferred one at a time.
2. Take mid-log phase Sf9 cells and dilute to 2× 10^5 cells/mL. Dispense 1 mL (2 × 10^5 cells) into each well of four 24-well TC plates. Include one for cellfectin and the other for untreated cell controls. Incubate the plate at 27°C for 1 h to allow cell attachment.

96-well	24W plate	position
A01	1	A01
A02	2	A01
A03	1	A02
A04	2	A02
A05	1	A03
A06	2	A03
A07	1	A04
A08	2	A04
A09	1	A05
A10	2	A05
A11	1	A06
A12	2	A06

96-well	24W plate	position
C01	1	C01
C02	2	C01
C03	1	C02
C04	2	C02
C05	1	C03
C06	2	C03
C07	1	C04
C08	2	C04
C09	1	C05
C10	2	C05
C11	1	C06
C12	2	C06

96-well	24W plate	position
E01	3	A01
E02	4	A01
E03	3	A02
E04	4	A02
E05	3	A03
E06	4	A03
E07	3	A04
E08	4	A04
E09	3	A05
E10	4	A05
E11	3	A06
E12	4	A06

96-well	24W plate	position
B01	1	B01
B02	2	B01
B03	1	B02
B04	2	B02
B05	1	B03
B06	2	B03
B07	1	B04
B08	2	B04
B09	1	B05
B10	2	B05
B11	1	B06
B12	2	B06

96-well	24W plate	position
D01	1	D01
D02	2	D01
D03	1	D02
D04	2	D02
D05	1	D03
D06	2	D03
D07	1	D04
D08	2	D04
D09	1	D05
D10	2	D05
D11	1	D06
D12	2	D06

96-well	24W plate	position
F01	3	B01
F02	4	B01
F03	3	B02
F04	4	B02
F05	3	B03
F06	4	B03
F07	3	B04
F08	4	B04
F09	3	B05
F10	4	B05
F11	3	B06
F12	4	B06

96-well	24W plate	position
G01	3	C01
G02	4	C01
G03	3	C02
G04	4	C02
G05	3	C03
G06	4	C03
G07	3	C04
G08	4	C04
G09	3	C05
G10	4	C05
G11	3	C06
G12	4	C06

96-well	24W plate	position
H01	3	D01
H02	4	D01
H03	3	D02
H04	4	D02
H05	3	D03
H06	4	D03
H07	3	D04
H08	4	D04
H09	3	D05
H10	4	D05
H11	3	D06
H12	4	D06

Figure 20.2: Convention for sample handling between 24- and 96-well plates and blocks

3. Bring the Grace's insect medium (serum free) and cellfectin to room temperature. Take a sterile flat-bottomed 96-well microtiter plate (that can hold -200 µL of sample). Dispense 50µL of Grace's insect medium into each well. Transfer 10 µL of recombinant bacmid DNA into each well and mix by shaking the plate gently or pipetting.
4. Mix cellfectin thoroughly by tapping the tube gently. In a 50-mL tube, mix 5 mL of Grace's insect medium and 0.3 mL of cellfectin. Mix thoroughly and add 50 µL to each well of the 96-well microtiter plate containing the recombinant bacmid DNA from step 3. Do not add cellfectin mixture to the cell control well, but add an equivalent amount of medium to it. Mix briefly by tapping the plates.

5. Cover the microtiter plate with adhesive film and incubate the mixture inside the laminar flow hood for 45 min. Dilute the solution by adding 100µL of Grace's insect medium (serum free) to each well of the 96-well microtiter plate.
6. Take out the 24-well TC plate containing cells from step 2, aspirate the medium using a multichannel pipette. Wash once with 1 mL of Grace's insect medium. Remove the medium and overlay the cells with diluted lipid-DNA complexes (0.2mL) from step 5. Add a further 0.2mL of Grace's insect medium (serum free) to each well. Incubate cells for 5 h at 27 °C.
7. Remove the transfection mixture and add 0.7 mL of SF900 II insect medium containing *2%* (v/v) FBS and antibiotics (50U penicillin and 50µg streptomycin per mL medium) to each well. Incubate cells at 27°C for 72-96h.
8. Look for signs of infection in the transfected cells 72h post transfection, by comparing with the control cells under an inverted microscope. Confluent growth of cells will be seen in control wells, whereas areas of clearing will be prominent in wells with infected cells.

   Infected cells usually are larger and deformed or elongated compared to uninfected cells. Collect viruses when the cells are well infected; this may take up to 96h or more.
9. Transfer the liquid contents from the 24-well tissue culture plate into a sterile 96-deep-well block and centrifuge at 1,500g for 20 min at room temperature. Collect the clear supernatant in another sterile 96-deep-well block. This is the PO viral stock. Store at 4°C, protected from light.
10. Further amplification of the virus can be carried out in 24-deep-well blocks in suspension culture. Dispense 3 mL of Sf9 cells (2×10^6 cells/mL) into each well of four sterile 24-deep-well blocks.
11. Infect the cells with 120µL of PO virus stocks using a multichannel pipette. Cover the block with porous seal and incubate at 27°C at 450 rpm for 48 h in a Glas-Col incubator.
12. Centrifuge the 24-well block at 1,500 g at 4°C for 20 min. Collect the supernatant and store in a 2X 96-deep-well block as P1 virus stock.

Further, viral stocks (P2 virus and P3 virus) can be amplified as necessary by repeating this procedure. The virus stocks are stored at 4°C, protected from light.

Isolation and Purification of Plaque

The first baculovirus plaque assay was described by Hink and Vail [28]. The method was modified by others in due course. However, the principle is based on the infection of cells with a very dilute virus suspension, so that each cell becomes infected by a single viral particle. The cells then are overlaid with a solid medium to limit the spread of virus particles.

After several cycles of infection, a clearing around the cell, called a *plaque,* is formed. All viruses in an isolated plaque are derived from a single viral particle and so represent a clonal population. Plaque assay is the direct method for determining the viral titer of the stock. Plaque purification allows producing a pure, clonal viral stock.

1. Take a log phase Sf9 insect cell culture and dilute to obtain a suspension of 5×10^5 cells/mL. Aliquot 2mL (1×10^6 cells) into each well of six-well tissue culture plates. Use duplicate

plates for each target. Incubate the plates at 27°C for 1 h in an incubator to allow the cells to attach.

2. During the 1-h incubation, prepare a serial dilution (10^{-1} to 10^{-8}) of the clarified baculovirus stock (either P2 virus or P3 virus stock) in Sf-900 II medium (use 0.5 mL of stock and dilute with 4.5 mL of the Sf-900 II medium). Perform the dilution in a sterile laminar flow hood.
3. After 1 h, observe the cell monolayer under the microscope. Sf9 cells should be attached and at ~50% confluency. Move the six-well tissue culture plates containing Sf9 cells to the sterile hood. Mark the wells in sequence with 10^{-4}, 10^{-5}, 10^{-6}, 10^{-7}, 10^{-8}, and blank (cells only). Remove the medium and replace each well with 1 mL of the appropriate virus dilution, using 1 mL of medium in the control well. Incubate for 1 h at room temperature to initiate the infection.
4. During the incubation period, prepare the plaque medium as follows: melt the 4% (w/v) agarose gel by heating in a microwave oven and place in a 40°C water bath until ready to use, and prewarm an empty, sterile, 100 mL container, sterile pipettes and 1.3X Sf-900 II medium to 40°C.
5. Prepare the plaquing medium. Transfer the 4% agarose, prewarmed sterile bottle and medium into a sterile laminar flow hood. Quickly combine 20 mL of 1.3X Sf-900 II medium and 7 mL of the melted 4% agarose in an empty sterile bottle. Return the bottle of plaquing medium to the 40°C water bath until ready to use.
6. Following the 1-h incubation of Sf9 cells and virus, remove the medium containing virus from the wells and replace with 2 mL of plaquing medium. Work as quickly as possible using aseptic technique. Allow the agarose to harden for 10-20min at room temperature. Incubate the cells in a 27°C humidified incubator for 7-10 days until plaques are visible and ready to count.
7. Observe the plaques under the microscope. Mark the clear plaques and use them for the amplification of recombinant virus stock. Pick each plaque using a sterile Pasteur pipette and transfer into 500 μL of Sf-900 II medium. Mix well by vortexing. Use 100μL to infect mid-log phase Sf9 cells. Follow steps 10, 11, and 12 in Sect. elsewhere in this chapter. The virus thus obtained can be used as plaque purified stock.

Recombinant Viruses

Large-scale production of virus often is necessary for the production of sufficient recombinant protein. It is recommended to keep a viral stock of 500mL for each potential protein target. Whenever possible, amplify the plaque-purified virus to keep stocks maintained. The alternative approach would be to use either P2 or P3 virus stock. The further production of virus is a two-step procedure.

1. Infect 100mL of Sf9 cells (2×10^6 cells/mL) with 200μL of plaque-purified virus or 2 mL of P3 virus. Incubate at 27°C at 150 rpm for 48 h. Centrifuge the culture at 1,500g for 25 min. Separate the clear supernatant containing the virus and store at 4°C in the dark. Analyse the pellet for possible expression.
2. Infect 500mL of Sf9 cells with 15-20mL of virus obtained as just described. Incubate and harvest the virus as specified.

Virus can be stored at 4°C for up to 6 months.in a stable condition. After 6 months, the titer may decrease. If long-term storage is necessary, aliquot the stock in smaller volumes, add 5% (v/v) FBS and store at -80°C. Frequent freezing and thawing highly reduces the viral titer. Before carrying out any expression studies using the virus stock, it is strongly recommended that a functional titration is performed.

Small-Scale Expression

This protocol has been developed using 24-deep-well blocks and is designed to rapidly analyze the expression of 96 clones.

1. Take four 24-deep-well blocks. Dispense 3 mL of mid-log phase Sf9 cells (~2×10^6 cells per mL). Infect the cells with 120 µL of virus stock to be analyzed for expression. Incubate at 27°C at 450 rpm in a Glas-Col incubator for 48 h.
2. Centrifuge the blocks at 1,500g for 25 min. Discard the supernatant safely. Use the pellet for the extraction of protein. Store at -20°C if using later.
3. Add 0.4 mL of Suspension buffer to each well of the 24-deep-well blocks containing pellets. Shake at 800 rpm in a Glas-Col shaker for 5 min to resuspend. Do not incubate longer. Transfer to ice if incubation needs to be longer. During this time, prepare lysis buffer.
4. Make sure the pellet in the 24-deep-well block is resuspended completely; if not, use a pipette to resuspend the pellet.
5. Transfer the suspension from the 24-deep-well block into a 96-deep-well block using a multichannel pipette. Add 1 mL of lysis buffer to each well. Shake the suspension at 800 rpm in a Glas-Col shaker for 30 min. Check the lysed cells under a microscope. If lysis is not complete, freeze and thaw the cells a few times.

   Take 50µL of the suspension, mix with 10µL SDS-PAGE loading dye, heat at 98°C for 10min, spin at 10,500g for 15 min in a benchtop centrifuge, and analyze the clear supernatant on SDS-PAGE (total fraction).
6. Centrifuge the block at 1,500g for 30 min at 4°C. Transfer the clear supernatant into another 96-deep-well block. Take 50µL of the suspension, mix with 10µL SDS-PAGE loading dye, heat at 98°C for 10min, spin at 10,500g for 15 min in a benchtop centrifuge, and analyze the clear supernatant on SDS-PAGE (soluble fraction).
7. Add 100µL of 50% Ni-NTA slurry preequilibrated in suspension buffer to the remaining supernatant in each well of the block. Keep on ice for about 30 min with intermittent mixing.
8. Transfer the Ni-NTA and supernatant mix into a 96-well filter block for purification. Place the block on top of a waste block (96-deep-well block) and allow the liquid to drip through the filters.
9. Discard the flow-through and wash filter plate once with 0.4mL of binding buffer and three times with 0.4 mL washing buffer. After all the liquid has dripped through the filter block, centrifuge the block at 700*g* for 1 min to remove the residual liquid.
10. Place the filter plate on top of a 96-well V-bottom microtiter plate. Add 80µL of elution

buffer to each well of the filter plate. Incubate the plate at room temperature for 5 min. Centrifuge at 700*g* for 1 min to elute the protein.

11. Transfer 16µL of the eluted fraction into corresponding positions in a fresh 96-well microtiter plate. Add 4µL SDS-PAGE loading dye. Pipette up and down to mix the sample.
12. Seal the plate and heat at 98°C for 10 min. Load 12.5 µL of sample on SDS-PAGE using a multichannel pipette. Run the gels at 150 V for 1 h 30 min. Stain the gel with biosafe Commassie blue and destain to visualize the protein bands.

This study gives a rapid analysis of recombinant baculovirus, which then can direct the expression of target proteins in insect cells and indicate which is the best to scale up.

Criteria of Expression

Expression levels of a heterologous gene may vary significantly due to different factors. The effect of several parameters can be analyzed by performing the sinall-scale expression study. This is very useful as an aid to optimize expression conditions and obtain information regarding scale-up possibilities. We recommend the following:

1. Three volumes of virus from the stock (20 µL, 40 µL, and 80 µL for 2×10^6 Sf9 cells).
2. Two concentrations of Sf9 cells (~2.0×10^6 cells per mL and ~3×10^6 cells per mL).
3. Three harvesting time points (48 h, 72 h, and 96 h).
4. Two cell lines (Sf9 and High Five).

Each experimental setup is designed in duplicate, accommodating four 24-deep-well blocks, or multiples of four, to carry out purification steps in 96-well format. The infection procedure, incubation, harvesting, affinity purification, and analysis in small-scale are as described previously.

In many cases, the purified protein obtained at this stage is enough to carry out some biochemical analysis and determine the molecular mass using mass spectrometry. With the information obtained by following this method, plaque assays may be replaced, which saves a lot of time-consuming steps. Hence, this method has been presented as a functional titration of virus.

POINTS TO REMEMBER

1. Working with insect cell lines for the first time is a challenge. Strict aseptic handling and good laboratory practice must be implemented to keep the cell lines away from contamination. It is advisable to use Pen/Strep solution, if necessary, to avoid bacterial contamination. The Sf9 cells are adapted to a serum-free medium; however, addition of 2% (v/v) FBS helps maintain the cells.

   It also avoids clumping of the cells and enhances attachment on TC plates. Such cells attach within 15 min. The routine growth and expression study can be carried out in Sf9 cells using Sf-900 II medium either supplemented with FBS or not. However, this medium is not suitable for transfection. Cells grown in Sf-900 II should be washed with Grace's insect medium prior to transfection.
2. Most insect cell culture techniques rely on adherent cultures for the initiation of growth from the stock. The idea behind this is to remove the DMSO present in the storage medium,

which hampers cell growth. We developed a different method, as described in Sect. elsewhere in this chapter.

This saves the time required for attachment and dislodging of cells in the adherent culture. It is necessary to be gentle and careful while suspending insect cells. Insect cells do not have a cell wall and hence are fragile.

3. In case a large volume of fresh cells are needed at short notice, initiate the culture using the contents of two to three stock vials in 29 mL of Sf-900 II medium and follow the routine method to grow the cells. This helps to reduce the lag period, and high cell density can be achieved more quickly.

   This also can be adapted as a routine method if sufficient stocks are available. Proper aeration is necessary and must be maintained by keeping the shaker speed as high as 180 rpm while using 250 mL Erlenmeyer flasks. The presence of 2% (v/v) FBS is vital in achieving a high-density cell culture that can be maintained at 3-4 × 10^6 cells per mL with 100% viability. Such cultures can be diluted to ~2 × 10^6 cells per mL and used directly for expression studies.

4. It is worth mentioning here that the recombinant bacmid DNA required for transfection of insect cells should be sterile and free from RNA. We found it difficult to maintain sterility with bacterial lysates purified using commercial kits. This necessitates an extra ethanol precipitation step before using the DNA for transfection.

   Furthermore, there is the possibility of cross-contamination of bacmids from one well to another. If commercial devices are preferred, make sure systems that have each well of the multi-well block constructed as an individual unit are used, that is, not sharing a large filter unit for a whole plate. We present an easy and economical method of bacmid DNA extraction suitable for transfection.

   In this simplified method, 96-deep-well blocks are used. The solutions are either ready-to-use commercial reagents or common laboratory chemicals, that is, potassium acetate and isopropanol. Cross-contamination of bacmid DNA and microbial contamination were greatly reduced employing this method.

   The yield of bacmid DNA also was higher than that obtained from the commercial kit. The user needs to be vigilant in the precise use of multichannel pipettes and tips and concentrate while performing the HT manipulation.

5. From the point of transfection, the frequent use of 24-well plates or blocks and 96-well plates or blocks may increase the risk of contamination and errors. The first problem most novice researchers encounter is the improper orientation of plates and blocks causing confusion in sample transfer from one plate or block to another. Before starting any manipulation, the plates and blocks must be labeled in proper orientation.

   Another problem arises while loading tips on multichannel pipettes. There might be an air gap, which is a cause of aerosol generation during pipetting. Aerosols can be the cause of cross-contamination within a block as well as spreading contamination across the working area and other materials.

   Ensure filter tips are used and they are fitted firmly. While dispensing liquid into deep-well blocks, touching the tips on the side wall of the well and releasing the sample gently helps avoid the formation of aerosols. The use of filtered tips is highly recommended.

6. Transfected cells should be observed daily for any signs of infection. Identification of infected cells poses another challenge for the first-time user. Cells from old culture and cells in medium with limited nutrients or insufficient aeration display morphology similar to that shown as a result of infection.

   The first sign of infection is the cessation of growth and multiplication. Therefore, each set of experiments should be accompanied by a control cell well for comparison. The success of transfection depends on the concentration and purity of the bacmid DNA used. When an experiment is set up for HT, finding out the optimum concentration for each target is not possible.

   Hence, a compromise is made to collect the virus after 72h of transfection. The amount of virus thus obtained generally is sufficient to use for follow-up amplification. If it does not work for particular targets, repeat the process for these using six-well TC plates. The DNA to cellfectin ratio may need to be optimized for particular targets to achieve successful transfection.

7. During follow-up amplification of recombinant virus, <48 h post infection is the best time to obtain an infective form, that is, a BV form. Although the gene responsible for the occlusion formation is replaced by a foreign gene, the viruses produced in the very late phase carry other cellular materials as a result of cell lysis.

   These might cause problems in subsequent infections. We observed a decrease in efficiency of infection and level of protein production when virus produced after 72 h is used.

8. If the virus stock produces slimy material, centrifuge at 17,500g, separate the clear supernatant, and titrate it for expression level. Do not filter the viral stock: Filtration results in a nearly complete loss of viral titer.

   Even with the best care, virus stocks may lose activity or get contaminated. It is best policy to store the cloned vector and bacmid DNA so that new virus can be easily generated. We recommend storing duplicate sets at 4°C and -80°C.

9. Plaque purification of virus stock is highly desirable and gives a clonal isolate. Virus stock generated from such an isolate should give uniform baculovirus clones, which avoids variation in expression products. However, the demand of HT work may lead to a compromise, so that plaque purification and functional titration or functional titration followed by plaque purification is carried out only on selected clones.

   Functional titration helps select the best clones within 48 h, while plaque purification needs at least 7 d to isolate the virus. We carry out functional titrations on a set of best possible clones directly after P2 virus generation. Any potential clone then is purified by plaque assay before generating the viral stock. This speeds up the entire process of screening by at least 1 week.

10. A small-scale study helps in identifying the best clone for large-scale production of a protein. Large-scale expression study is labor intensive, uses large amounts of consumables, and takes significant time. Moreover, the expression level may differ from

time to time and from one batch to another. To monitor such variation, we developed a procedure for rapid extraction, purification, and analysis of small-volume culture samples.

The process utilizes (1) the addition of 50μL of Insect PopCulture Reagent (Novagen 71187) in 1 mL culture samples, (2) vigorous mixing and incubation for 15 min at room temperature followed by (3) test affinity purification (e.g., Ni-NTA purification), and (4) the rapid analysis of samples performed by running them on SDS-PAGE gels for 50 min and visualizing by Coomassie staining. The overall process can be completed within 2h 30min to give an estimate about the level of expression.

Close monitoring ensures that any problems in expression of target proteins can be detected early on to avoid unnecessary downstream processing. Serum-free medium is not suitable for His-affinity purification: A component in the medium may strip-off the Ni++ from the column. If heterologous protein is expressed in an extracellular medium, buffer exchange is necessary prior to affinity purification. However, this is not a problem in small-scale expression studies.

11. Once the first expression study is carried out in Sf9 cell lines, we highly recommend analyzing the expression level in HighFive cell lines. The HighFive cell line is generally known to express a higher level of extracellular protein [30]. In our study, most intracellular proteins also were expressed in higher levels than in Sf9 cells.

    We followed the same protocol to grow and maintain HighFive cells as that used for Sf9 cells. The differences are as follows:

    (1) the speed of the shaker is maintained at 100rpm at the beginning and at a maximum of 120rpm as the cells adapt;

    (2) the medium used was Express SFM supplemented with L-glutamine, and heparin was added at 10U/mL to prevent aggregation. The addition of heparin is not necessary once the cells are adapted to suspension culture;

    (3) suspension cultures should be passaged before they reach a density of 2.0-2.5 × 10^6 cells/mL.

Chapter 21

Whole Genome

With increasing amounts of DNA sequence information available, the most direct and informative approach to studying DNA-sequence polymorphisms to assess bacterial species is sequence-based comparisons of genomes. However, it is currently not feasible to sequence whole bacterial genomes on a large scale, and hence sequencing is directed at very small regions of one or a few genes.

Therefore, whole genome analysis techniques that detect genetic polymorphisms between bacterial strains are becoming very popular with researchers. Each genotyping method attempts to identify unique characteristics of a strain which differentiates it from other strains. This could be unique DNA sequence/s, or a group of DNA fragments of unique size or sizes.

Amplified fragment length polymorphism (AFLP), first described by Vos et al. for analysis of plant genomes using radioactive labeled primers is one such technique that has been adapted successfully for DNA fingerprinting of bacterial species. Fluorescent amplified fragment length polymorphism (FAFLP) uses primers that are fluorescently labeled and simple modifications make it possible to tailor the technique for molecular epidemiological studies of bacterial genomes of any size.

The sources of AFLP polymorphisms could be point mutations within the restriction site, resulting either in loss or gain of a restriction site, or insertion or deletion between the priming sites, thus resulting in polymorphic fragments of varying sizes. FAFLP usually involves the digestion of genomic DNA with a restriction endonuclease with a six-base recognition sequence (rare-cutter) followed by digestion with an endonuclease with a four-base recognition sequence (frequent-cutter).

This is followed by ligation of the restricted fragments to double-stranded oligonucleotide adaptors, which consist of a core sequence and an enzyme-specific recognition-site sequence, to create target sites for primer annealing. A subset of these restriction fragments is amplified using primers that have sequences complementary to the ligated adaptor-restriction site. One of the primers is labeled with fluorescent dye and can be detected on an automated DNA sequencer.

FAFLP has numerous advantages over other DNA fingerprinting techniques. With or without

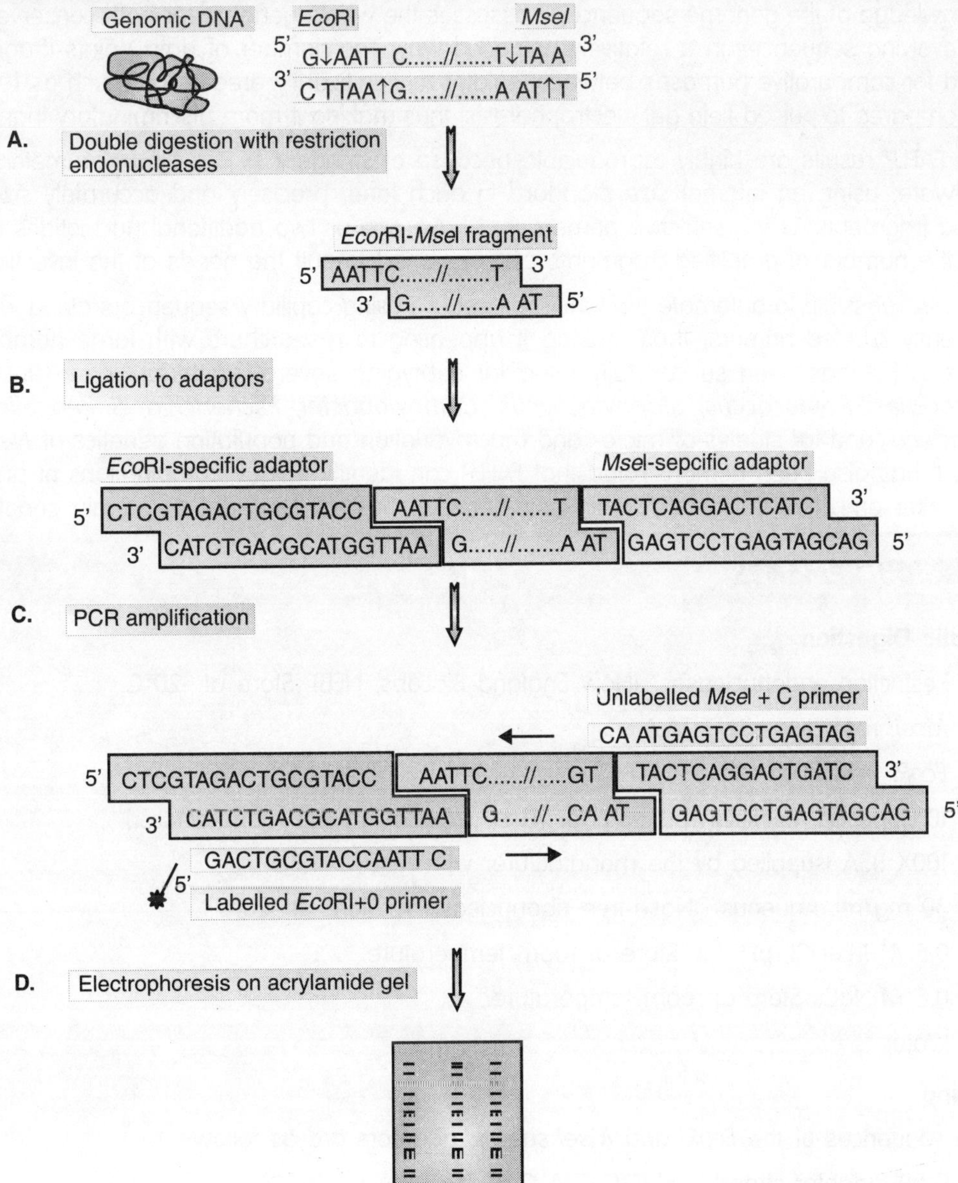

Figure 21.1: Overall scheme of FAFLP technique with EcoRI and MseI endonucleases. Total genomic DNA with EcoRI and MseI recognition sequences indicated. (A) The DNA is double-digested with EcoRI and MseI, generating EcoRI-MseI restriction fragments (as example). (B) Ligation of EcoRI-specific adaptor and MseI-specific adaptor to ends of the restriction fragments, resulting in the formation of restricted-ligated fragments which serve as template for amplification for PCR. (C) Selective amplification of these fragments with one-base selective MseI+C primer and FAM-labeled nonselective EcoRI+O primer pair (as example) by Touchdown PCR cycling conditions. (D) Resolution of amplified fragments is achieved on an automated DNA sequencer.

prior knowledge of the genome sequence, it assesses the whole genome for both conserved and rapidly evolving sequences in a relatively unbiased way. The number of data points (fragments) obtained for comparative purposes between isolates is significantly greater (as much as 10 times more, compared to pulsed-field gel electrophoresis) thus making it more discriminatory than PFGE.

The FAFLP results are highly reproducible because of stringent PCR cycling parameters, and the software, using an internal size standard in each lane, precisely and accurately sizes the amplified fragments. Using selective primers that have one or two additional nucleotides on the 3'-end, the number of amplified fragments can be varied to suit the needs of the investigation.

It is also possible to automate the whole procedure using capillary sequencers along with the fluorescently labeled primers, thus making it appealing to researchers with large numbers of samples. FAFLP has been successfully used for subtyping several bacterial genera including *Enterococcus, Streptococcus, Staphylococcus, Campylobacter, Escherichia, Salmonella,* and *Mycobacteria* and for studies of micro- and macrovariation and population genetics of *Neisseria* species. It has also been demonstrated that FAFLP can identify unique combinations of precisely sized marker amplified fragments within and between clusters, thus defining strain genotypes.

REQUIREMENTS

Enzymatic Digestion

1. Restriction endonucleases: (New England BioLabs, NEB). Store at -20°C.

   MseI: restriction-site T↓TA↑A

   EcoRI: restriction-site G↓AATT↑C
2. 10X Specific restriction buffer 2 for *MseI* (supplied by the manufacturer).
3. 100X BSA (supplied by the manufacturer with *MseI*).
4. 30 mg/mL Aqueous DNase-free ribonuclease A. Store at -20°C.
5. 0.5 *M* Tris-HCl, pH 7.2. Store at room temperature.
6. 0.5 *M* NaCl. Store at room temperature.
7. Water.

Annealing

The sequences of the *EcoRI* and *MseI-specific* adaptors are as follows:

1. EcoRI adaptor strand 1: 5'-CTC GTA GAC TGC GTA CC-3'.
2. EcoRI adaptor strand 2: 5'-AAT TGG TAC GCA GTC TAC-3'.
3. *MseI* adaptor strand 1: 5' GAC GAT GAG TCC TGA G-3'.
4. *MseI* adaptor strand 2: 5'-TAC TCA GGA CTC ATC-3'.

Ligation

1. T4 DNA ligase (NEB). Store at -20°C.
2. 10X Ligase reaction buffer (supplied by the manufacturer).

3. 2 µM EcoRI-specific adaptor.
4. 20 µM *MseI-specific* adaptor.
5. Water.

PCR Amplification

1. 1 µM Fluorescently labeled *EcoRI* primer (EcoRI+O): 5'-6FAM-GAC TGC GTA CCA ATT C-3'.
2. 5 µM Unlabeled one-base selective *MseI* primer *(MseI+C):* 5'-GAT GAG TCC TGA GTA AC-3'.
3. 10X *Taq* DNA polymerase supplied with Taq polymerase buffer and 50 mM $MgCl_2$.
4. 10 mM Each dNTPs: 10µL of 100 µM of each of the four dNTPs, and 60µL of water. Aliquots are stored at -20°C.
5. Water.

PROCEDURES

The success of the technique depends on the quality of DNA, choice of enzymes used for restriction of the genomic DNA, the sequences of the adaptors and primers, and the PCR cycling conditions used for amplification of the restricted fragments. Extraction of genomic DNA from bacterial cells should be performed using established protocols, and the DNA quantified using a spectrophotometer by standard methods or by gel comparison with known molecular weight standards.

Although it has been shown that the input template DNA concentration does not affect AFLP data, except in the case of very high or very low DNA concentrations, the quality of the DNA is very important. The choice of endonucleases and primers depends on the G+C content of the bacterial genome.

Amplification of the restriction fragments using selective nucleotides at the 3'-end ensures that the primers anneal precisely to randomly chosen sequences immediately internal to a restriction endonuclease site. This decreases the number of amplified fragments comprising an FAFLP profile thus ensuring better resolution of amplified fragments across the gel. Separation of amplified fragments can be achieved by capillary electrophoresis or on slab-based systems using the fluorescent primers.

Endonuclease and primer combinations used in FAFLP reactions for a lot of bacterial genera and species have been standardized in our laboratory. For this chapter, the most extensively used endonuclease combination, *EcoRI* and *MseI,* has been described with the following primers: a FAM-labeled nonselective *EcoRI* adaptor-specific forward primer (termed *EcoRI+O)* and a one-base selective nonlabeled *MseI* adaptor-specific reverse primer *(MseI+C).*

Sequential Digestion

1. Aliquot the appropriate volume of genomic DNA (approx 500 ng) into 0.5-mL reaction tubes, depending on the concentration of DNA. Add water to bring the final volume to 17 µL, if necessary.

2. In a separate tube, prepare an *MseI* enzyme master mix that contains the following reagents for each reaction: 2 μL of lox *MseI* reaction buffer, 0.2 μL of 100X BSA, 0.3 μL of 30 mg/mL Ribonuclease A, and 0.5 μL of 10 U/μL *MseI* enzyme.
3. Mix the reaction thoroughly.
4. Aliquot 3 μL of the master mix into each of the tubes with genomic DNAs.
5. Mix thoroughly and incubate at 37°C for 2 h.
6. In a separate tube, prepare an *EcoRI* enzyme master mix that contains the following reagents for each reaction: 0.5 μL of water, 1.7 μL of 0.5 M Tris-HCl (pH 7.2), 2.1 μL of 0.5 *M* NaCl, and 0.5 μL of 10 U/μL *EcoRI* enzyme.
7. Add 4.8 μL of the *EcoRI* master mix to each of the *MseI* digest.
8. Mix the reaction thoroughly and incubate at 37°C for a further 2 h.
9. Inactivate the endonucleases by heating at 65°C for 10 min.
10. Immediately cool the reactions on ice.

Annealing

1. The two individual strands (strand 1 and 2) of each of the adaptors are reconstituted to make a stock solution of 100 μM each.
2. To make up a 2-μM *EcoRI* adaptor solution, add 2 μL of *EcoRI* adaptor strand 1, 2 μL of *EcoRI* adaptor strand 2, and 96 μL of water.
3. Mix the reagents thoroughly.
4. Place the tubes in a block-based programmable thermocycler or a heating block set at 95°C for 10 min.
5. Slow cool to room temperature. This is achieved by leaving the tubes in a block-thermocycler or heating block and switching off the instrument until the tubes reach room temperature.
6. Aliquot the annealed adaptors and store at -20°C.
7. To make up a 20-μM *MseI* adaptor solution, add 20 μL of *MseI* adaptor strand 1, 20 μL of *MseI* adaptor strand 2, and 60 μL of water.
8. To anneal the two adaptor strands, follow steps 3-6 as for the *EcoRI* adaptor previously described.

Ligation

1. Following digestion with the two endonucleases, ligation is carried out using the previously annealed adaptors.
2. In a separate tube, prepare a ligation master mix that contains the following reagents for each reaction: 9.9 μL of water, 5.0 μL of 10X T4 ligase reaction buffer, 0.5 μL of 2 *μM EcoRI* adaptor, 0.5 μL of 20 *μM MseI* adaptor, 0.1 μL of 400 U/μL T4 DNA Ligase. Mix thoroughly.
3. Add 25 μL of the previously described master mix to the double-digested DNA in Sequential Digestion.

4. Mix the reaction thoroughly.
5. Incubate the reaction mixture at 12°C overnight.
6. Inactivate the reaction by heating at 65°C for 10 min.
7. Cool the reaction on ice. Once used for PCR, this double-digested ligated DNA can be stored at -20°C.

Amplification

1. PCR reactions are performed in 25 μL final volume.
2. Prepare a PCR-enzyme master mix that contains the following reagents for each reaction: 15.0 μL of sterile water, 2.5 μL of 10X *Taq* reaction buffer, 1.25 μL of 50 mM $MgCl_2$, 0.5 μL of 10 mM dNTPs mix, 2.0μL of 1 μM labeled *EcoRI* primer, 1.0 μL of 5 μM nonlabeled *MseI* selective primer, and 0.25 μL of 5 U/μL *Taq* DNA polymerase.
3. Mix the reagents thoroughly.
4. Aliquot 22.5 μL of the master-mix into thin-walled PCR tubes.
5. Add 2.5 μL of the double-digested/ligated reaction mix into individual tubes. Mix thoroughly.
6. Touchdown PCR is performed in a thermal cycler, and the cycling conditions used for amplification are as follows: initial denaturation at 94°C for 2 min (1 cycle) followed by 30 cycles of denaturation at 94°C for 20 s, primer annealing step for 30 s, and extension at 72°C for 2 min.

   The primer annealing temperature for the first cycle is 66°C, and for the next nine cycles, it is decreased by 1°C at each cycle. The annealing temperature for the remaining 20 cycles is 56°C. This is followed by a final extension at 60°C for 30 min.
7. Amplification products are separated by electrophoresis on a denaturing polyacrylamide gel. FAFLP reactions can be stored at -20°C for at least 6-9 mo.

Separation of Fragments

This can be achieved on various platforms available in the market. In our laboratory, we have primarily used the ABI Prism 377 automated gel-based DNA sequencer, but the Beckman Coulter capillary electrophoresis for fragment separation has also been used. FAFLP reaction mixtures for running on the ABI 377 sequencer contained I μL of the PCR sample, 1.25 μL of formamide, 0.25 μL of loading solution (dextran blue in 50 mM EDTA), and 0.5 μL of fluorescently labeled internal size standard (GeneScan-2500 or GeneScan-500 labeled with the red dye, ROX).

The sample mixture was heated at 95°C for 2 min, cooled on ice, and immediately loaded onto the gel. The running electrophoresis buffer was IX TBE, and the electrophoresis conditions were 2 kV at 51°C for 17 h, with the well-to-read distance of 48 cm.

FAFLP Fragment Analysis

Amplified fragments (AFs) separated on the ABI automated 377 sequencer are sized with GeneScan 3.1.0 software (Applied Biosystems), and the GenoTyper 2.5 software (Applied Biosystems) is used to generate a table in a binary format (with presence "1" or absence "0" of fragments) as a text file (tab-delimited) in MS Excel.

These data can be imported into suitable software packages for further analysis. For example, Bionumerics (Applied Maths) can be used for obtaining dendrograms to demonstrate relationships between the isolates.

POINTS TO REMEMBER

1. The appropriate combination of endonucleases is the primary determinant of the number and size range of amplified fragments, and the optimum distribution of these fragments on the polyacrylamide gel. In today's genomic era, published genome sequences can be used to model accurate *in silico* FAFLP analysis using online tools (e.g., http://193.129.245.227/aflp.html). This gives a rough indication of the number and sizes of amplified fragments that can be obtained using various endonuclease combinations.

   The combination *EcoRI* and *MseI* has been successfully used in our laboratory for *Staphylococcus spp., Streptococcus spp., Salmonella spp., Neisseria spp.*, and *Escherichia coli.* The other enzyme combination *HindIII* and *HhaI* has also been applied for FAFLP analysis of *Salmonella spp.* and *Campylobacter spp.*

2. Throughout the text, water refers to sterile water of molecular biology grade.

3. One of the primers, generally the one specific to the rare-cutting enzyme, *EcoRI,* is labeled at the 5'-end with a fluorescent dye, depending on the sequencer used for fragment separation. For the ABI instrumentation, either capillary or slab gel, 5-carboxyfluorescein (FAM, "blue" color), N,N,N',N'-tetramethyl-6-carboxyrhodamine (TAMRA, "yellow"), or 2',4'-dimethoxy-5',7'-dichloro-6-carboxyfluorescein (JOE, "green") can be used. In our laboratory, our primers were labeled with the blue dye, FAM.

   The internal size standard is labeled with 6-carboxy-x-rhodamine (ROX, "red") (Applied Biosystems). For the Beckman capillary sequencer, the fluorescent dyes Cy5 (color "blue") or Cy5.5 (color "green") can be used for labeling the primer and appropriate internal size standards labeled with fluorescent dye DI (color "red") are used for sizing.

4. The amplification of the restriction fragments can be made selective by the use of primers that have between one and three selective nucleotides on the 3'-end. Typically for smaller genomes such as prokaryotes, the total number of selective bases should be restricted to one or two.

   The number of selective nucleotides in the primers should be such that 50-100 fragments evenly distributed over a size range of 100 to 600 by are generated per sample. The selective primer is directed at the endonuclease restriction site, e.g., the *MseI* restriction site (5'-TTAA-3'), whereby it contains one or two additional nucleotides at its 3'-end. For example, a one-base selective primer, *MseI+C will* have a core sequence (complementary to the adaptor sequence) followed by 5'-TTAAC-3', where the first four bases are complementary to an *MseI* restriction site while the last nucleotide C at the 3'-end is selective.

   Hence, only fragments with the sequence 3'-AATTG-5' will be amplified. Assuming random base distribution, the addition of one selective nucleotide theoretically reduces the number of fragments amplified by a quarter, thus reducing the complexity of the resultant FAFLP pattern. A non-selective primer, e.g., *EcoRI+0 will* have a core sequence followed by the six bases (5'-GAATTC-3') complementary to the *EcoRI* restriction site.

5. The purity of genomic DNA is very important for successful FAFLP reactions. The first step of FAFLP reactions is the digestion of the DNA with two restriction endonucleases. The restriction endonucleases are sensitive to the presence of inhibitors like negatively charged polysaccharides, and organic and inorganic molecules in the DNA preparations.

   These impurities can result in partial digestion, resulting in nonspecific fragments detectable after amplification. Over the years, we have used several methods for obtaining pure bacterial DNA, ranging from manual extraction protocols like cetyl trimethyl ammonium chloride method, various commercial kits available on the market to using robotic extraction instruments such as the MagNa Pure LC instrument and the associated MagNa Pure DNA isolation Kit I or Kit III (Roche Diagnostics Ltd.).

6. The original method and subsequent researchers recommend double-digestion of genomic DNA with the two endonucleases. However, extensive experiments carried out in our laboratory have found that the best results are obtained by sequential digestion with the two endonucleases rather than double-digestion.

   The two endonucleases, *EcoRI* and *MseI,* require varying salt concentrations in their respective buffers for optimal activity. This is achieved by digesting the endonuclease requiring lower salt concentration first *(MseI)* followed by adjusting the salt concentration of the reaction (using small volumes of Tris-HCl and NaCl) for the second endonuclease *(EcoRI)* requiring the higher salt concentration.

   This results in significant decrease of partial digestion of genomic DNA that results in FAFLP profiles with very low (if any) background peaks. The endonucleases are also inactivated prior to digestion, to obtain clean FAFLP patterns.

7. Once a DNA extraction protocol has been standardized for a particular genus, the concentration of DNA for each bacterial preparation will be approximately the same. This will eliminate the need to add varied volumes of DNA for each of the bacterial strains.

   The necessary volume of water required to make up 500 ng of genomic DNA in a final volume of 17 μL can then be added to the master mix itself, thus reducing an additional pipetting step in the protocol.

8. While preparing master mix, add the reagents and mix thoroughly by briefly vortexing for a few seconds followed by a spin at 1000g for 15 s to collect the reaction mix to the bottom of the tube. All enzymatic reactions must be set up on ice.

9. It is very important to have equimolar concentrations of each of the two strands of the adaptor. Great care should be taken while pipetting precise volumes as excess of one of the strands (the "lower" strand) in the annealed adaptor can result in that strand acting as a primer in subsequent reactions.

   This can result in FAFLP profiles with fragments that are sized (n+1) bp, (n+2) bp, and so on, where "n" is the expected size of that fragment. The resulting electropherograms will display two or three additional peaks for each expected peak. These additional peaks were also observed when the concentration of the adaptors in the ligation reaction was increased 10-fold. We have also found that while annealing the adaptors, the slow cool step after the denaturation step is very important.

10. Following digestion, ligation is carried out in the same tube using the appropriate ligation buffer supplied with the ligase enzyme. Ligation is carried out at lower temperatures (12-16°C) overnight. However, it was found that 3-4 h was sufficient to achieve efficient ligation of adaptors to the digested sticky-ends generated by the two restriction endonucleases.

    The best results were obtained by using 400 U of T4 DNA Ligase per FAFLP reaction. The ligase is inactivated after ligation, and reaction mixes can be stored at -20°C. PCR from such ligations yielded reproducible FAFLP results even after 6 mo.

11. Touchdown PCR cycling conditions are used whereby higher than optimal annealing temperature is used for the first few cycles followed by 20 cycles at the optimum annealing temperature. This high-stringency PCR ensures amplification of only fragments that are complementary to the adaptor-restriction fragment sequence.

    For samples where the input DNA was very low, two rounds of PCR may be required. The first round includes a touchdown PCR using both nonselective primers i.e. EcoRI+O and *MseI+0*. For the second round PCR, use the amplification product from the first round PCR as template (1-2 µL in a 25- µL PCR reaction) and use the primers EcoRI+O and *MseI+C*.

12. On the ABI 377 automated DNA sequencer, amplification products are separated on a denaturing polyacrylamide gel using Long Ranger™ Singel™ packs (Cambrex BioSciences), containing all ingredients necessary for 5% Long Ranger acrylamide solution. Electrophoresis gels could be run for longer periods with better resolution of the fragments in the lower size range by using the easy-to-use Singel packs.

Index

A

B

C

E

I

K

L

N

T

U

V

W

X

Y